I0463301

Training Manual

APPLIED ENGINEERING PRINCIPLES MANUAL

NAVAL SEA SYSTEMS COMMAND   NAVY DEPARTMENT   **REV. 1, ACN-1, MAY 2003**

Date: May 2003

Document: Applied Engineering Principles Manual (AEP)

Revision: 1, ACN-1

1. **Description of Change:**

   Revision 1 to the AEP included video clips and animations that currently cannot be launched from a .pdf file. In response to a request for quick delivery of a CD ROM containing a .pdf version of the manual, this forwards ACN-1 to the AEP, a limited distribution issue of the AEP that does not include animations or video clips.

   It should be noted that the planned distribution date for a major revision to the AEP is early 2004. This new issue of the AEP will include sophisticated multimedia graphics, video clips, and animations that will illustrate the basic engineering concepts of the manual in a multi-dimensional fashion. The primary delivery medium for the next issue of the manual will be .pdf on CD ROM. Hard copy versions will be available upon request.

2. **Effective Date of Change:** Upon receipt.

   This endorsed form constitutes authority for recipients to use this CD ROM version of the AEP.

3. **Instructions for Entering Change:**

   a. Please use the AEP CD ROM in accordance with local security regulations. If applicable, destroy outdated copies of the AEP in accordance with local procedures.

4. **Approval:**

   a. Approval by Naval Reactors: N/A

   b. Approval of other Prime Contractor: Not Required

5. **Cancellation:**   None

6. **Authorized Signature:**

   R. E. Eshleman, Manager
   Prototype Training
   Operations Training
   Bettis Atomic Power Laboratory

Record of Revisions

Revision 1 (IETM issue only)	June 2001

Record of Advance Change Notices

| ACN-1 | May 2003 |

List of Effective Units

Text	Revision in Effect	Revision Date
Title Page	Revision 1, ACN-1	May 2003
i through xiv	Revision 1, ACN-1	May 2003
1-1 through 1-106	Revision 1, ACN-1	May 2003
2-1 through 2-94	Revision 1, ACN-1	May 2003
3-1 through 3-100	Revision 1, ACN-1	May 2003
4-1 through 4-58	Revision 1, ACN-1	May 2003
5-1 through 5-30	Revision 1, ACN-1	May 2003
6-1 through 6-34	Revision 1, ACN-1	May 2003
7-1 through 7-14	Revision 1, ACN-1	May 2003

Table of Contents

Table of Contents

Text   **Page Number**

Table of Contents

Text   **Page Number**

Table of Contents

Table of Contents

Table of Contents

Table of Contents

Table of Contents

Table of Contents

Table of Contents

Text   **Page Number**

List of References

Reference Number	Document Number	Title
1	NAVSEA S8921–TA-TXT-000/(R)	T-1, Reactor Core Materials
2	NAVSEA S8921–TJ-TXT-000/(R)	T-8, Plant Materials
3	NAVSEA S8921–TH-TXT-000(R)	T-9, Reactor Principles
4	NAVSEA S8921–TK-TXT-010/(N)	T-10, Water Chemistry Fundamentals, Volume I
5	NAVSEA S8921–TK-TXT-020/(R)	T-10, Water Chemistry Fundamentals, Volume II
6	NAVSEA S8921–TP-TXT-000/(R)	T-14, Applied Mechanical Principles
7	NAVSEA S8921–TS-TXT-000/(N)	T-17, Radiological Fundamentals
8	NAVSEA S8921–GA-TXT-000/(N)	T-20, Enlisted Mathematics
9	NAVSEA S8921–GD-TXT-000/(R)	T-23, Reactor Plant Technology
10	NAVSEA S8921–LA-TXT-000/(N)	T-30, Officer Mathematics

Chapter 1 — ELECTRICAL REVIEW

Section 1.1 — Fundamentals Of Electricity

1.1.1 — Definitions

<u>Current</u> (I). The flow of charge past a point (in reality, only electrons (negative charges) are free to move in a conductor, but because of convention, positive charges are used to describe current flow). Measured in **amperes** (amp).

1 amp = 1 coulomb per second

Throughout this text, the direction of current flow is defined as the direction in which positive charges flow; that is, from positive to negative, through a device which is receiving electrical energy such as a radio or a motor, and from negative to positive through a source of electrical energy such as a battery or generator. Exhibit 1-1 illustrates this sign convention. The short line of the battery symbol always represents the negative terminal regardless of current direction.

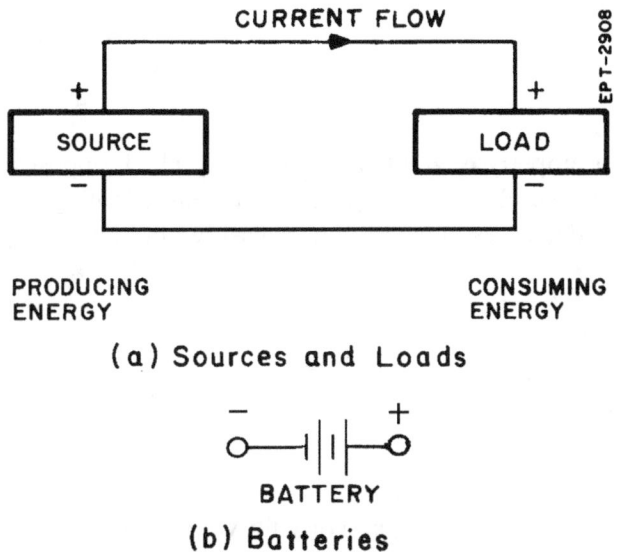

Exhibit 1-1 — Sign Conventions For (a) Sources And Loads And (b) Batteries

<u>Voltage</u> (E or V). Voltage is also called potential, potential difference, or electromotive force (emf). Defined as the work per unit charge necessary to move a charge from one terminal (or point) to the other.

Measured in **volts** (symbol V).

1 volt = 1 joule per coulomb

The symbol for a **voltage rise** (or source), as across a battery or generator, is e or E. A **voltage drop** is indicated by v or V.

Resistance (R). Defined as the property of a material which opposes the **flow of current** and thus makes it necessary to expend energy to move charge through it. Measured in **ohms** (symbol Ω). Schematic symbol

Ohm's Law. Valid for d-c or purely resistive a-c circuits.

Equation 1-1

$$V = IR$$

Capacitance (C). Defined as that property of an element or circuit which opposes any **change in voltage**. The capacitor acts as a load whenever the voltage across it increases in magnitude and acts as a source whenever the voltage across it decreases in magnitude. As long as the voltage remains constant, the capacitor stores

charge in its electric field. Measured in **farads** (symbol f). Schematic symbol

microfarads (μf) 1 μf = 10^{-6}f

picofarads (pf) 1 pf = 10^{-12}f.

Inductance (L). Defined as that property of an element or circuit which opposes any **change in current**. The inductor acts as a load whenever the current through it increases in magnitude and acts as a source whenever the current through it decreases in magnitude. Measured in **henries** (symbol h). Schematic symbol

millihenries (mh) 1 mh = 10^{-3} h

microhenries (μh) 1 μh = 10^{-6} h.

1.1.1.1 — Some Notation

Throughout this manual, a lower case letter such as e or v for voltage, or i for current, signifies that the quantity varies with time. When discussing values which remain constant, capital letters are used.

1.1.2 — Simple Circuits

Series Resistances

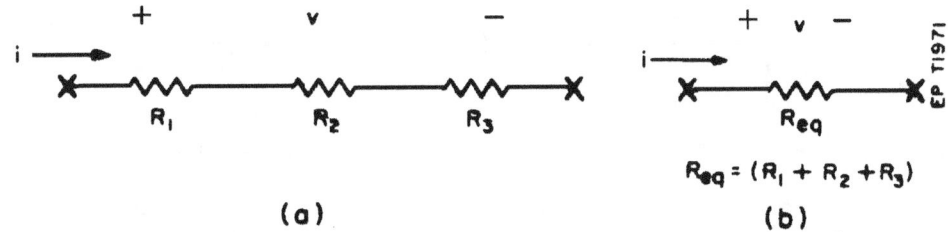

Exhibit 1-2 — Series Resistance Circuit

Parallel Resistances

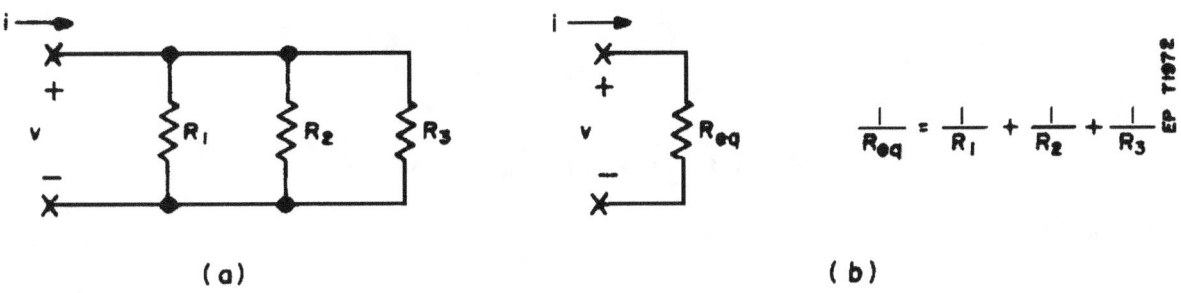

Exhibit 1-3 — Parallel Resistance Circuit

Special Case: two resistors in parallel

Equation 1-2

$$R_t = \frac{R_1 R_2}{R_1 + R_2}$$

Series Capacitors

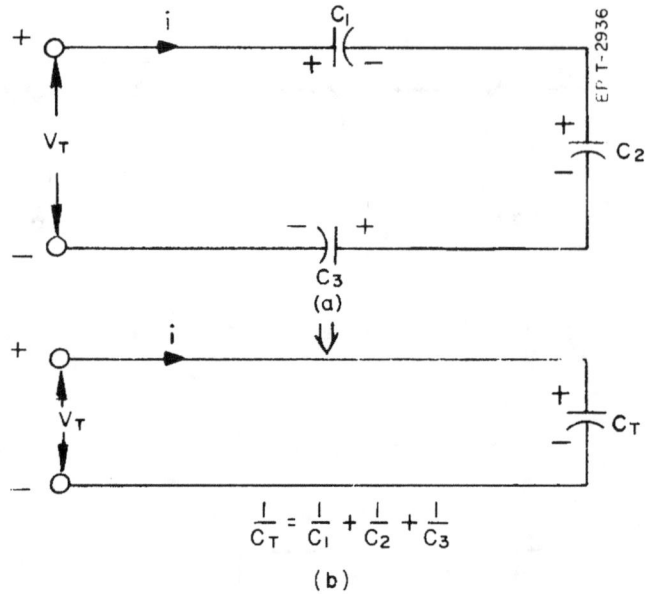

Exhibit 1-4 — Series Capacitor Circuit

Parallel Capacitors

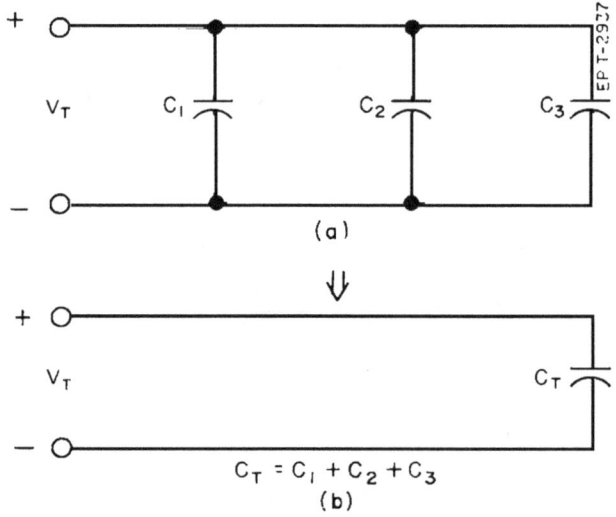

Exhibit 1-5 — Parallel Capacitor Circuit

Series Inductors

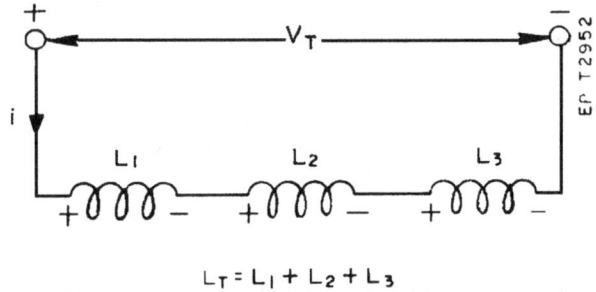

$$L_T = L_1 + L_2 + L_3$$

Exhibit 1-6 — Series Inductor Circuit

Parallel Inductors

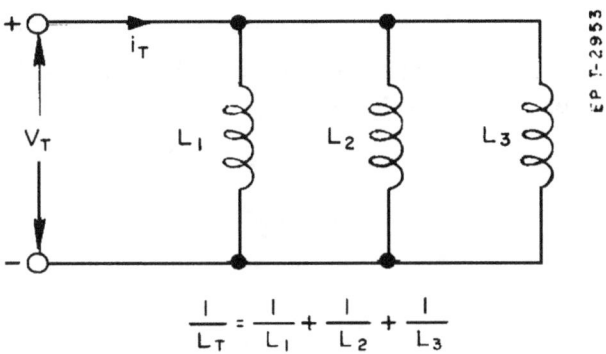

$$\frac{1}{L_T} = \frac{1}{L_1} + \frac{1}{L_2} + \frac{1}{L_3}$$

Exhibit 1-7 — Parallel Inductor Circuit

1.1.3 — Circuit Laws

<u>Kirchhoff's Current Law</u> (Exhibit 1-8). The algebraic sum of the currents at any node is zero. The sum of the currents entering a node must be equal to the sum of the currents leaving a node. Convention is current leaving the node is defined to be positive.

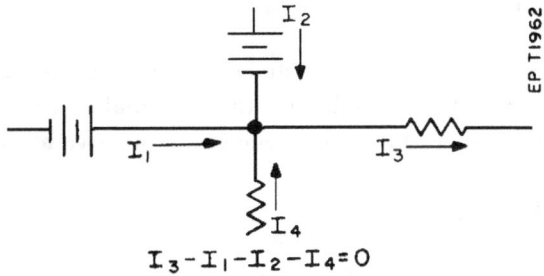

$$I_3 - I_1 - I_2 - I_4 = 0$$

Exhibit 1-8 — Kirchhoff's Current Law

Kirchhoff's Voltage Law (Exhibit 1-9). The algebraic sum of the potential differences around any closed loop is zero. The sum of the voltage rises must be equal to the sum of the voltage drops.

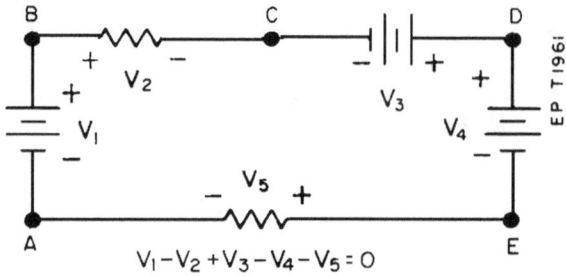

Exhibit 1-9 — Kirchhoff's Voltage Law

1.1.4 — D-C Power

Power is the rate at which work is done or the rate at which energy is expended. The unit for power is the **watt** which is equal to one joule per second. In a d-c circuit

Equation 1-3

$$P = IE$$

Alternate forms:

Equation 1-4

$$P = I^2 R$$

Equation 1-5

$$P = \frac{E^2}{R}$$

where P is in watts, I in amperes, E in volts, and R in ohms.

The definitions of current, voltage, and power should make the following statements clear:
1. Power is produced if current leaves the positive terminal of an element.
2. Power is expended (consumed) if current enters the positive terminal.

1.1.5 — Basic Magnetic Principles

The effects of a simple bar magnet are well known. At each end of the bar magnet there is a concentration of magnetic force. These areas of concentration are known as magnetic poles. Magnetic poles always exist in pairs. Like poles repel and unlike poles attract.

The area surrounding the magnet which exerts a magnetic force on magnetic materials is known as the **magnetic field**. For convenience purposes, a magnetic field is considered as being made up of magnetic lines of force or lines of flux. Flux lines always emerge from the north pole of a magnet and return to the south pole. The total number of magnetic flux lines making up a magnetic field is measured in **webers** and denoted by the symbol φ.

All magnetic fields are created by charge in motion, that is, by a flow of current. The magnetic flux set up by a current in a straight conductor is indicated in Exhibit 1-10. The direction of the magnetic field can be found by application of the Right-hand rule.

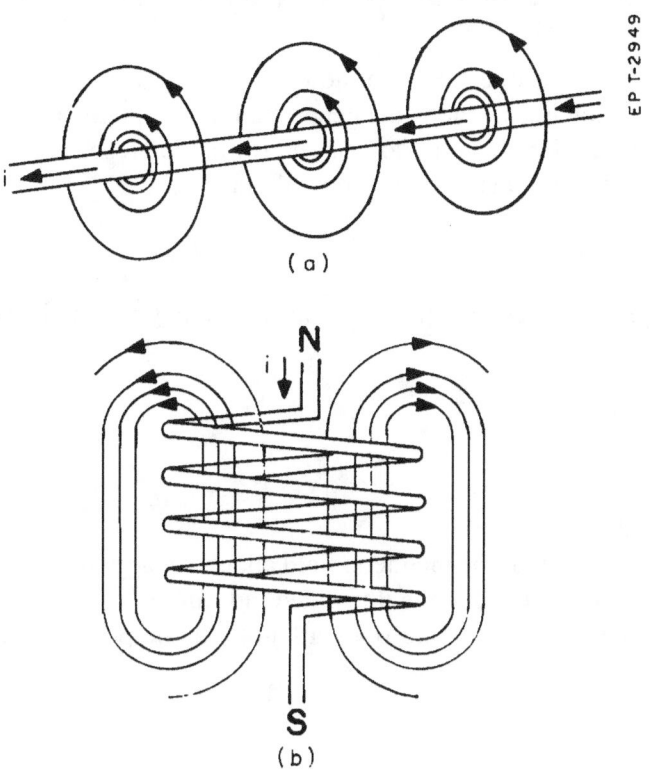

Exhibit 1-10 — Magnetic Fields Produced By Current Flow In A Straight Conductor And A Coil

1.1.5.1 — Right-Hand Rule For Conductors

The Right-hand rule for conductors states: **If a conductor is held in the right hand with the thumb indicating the direction of current flow, the fingers will curl in the direction of the magnetic field.**

1.1.5.2 — Right-Hand Rule For Coils

An inductor is formed by coiling a conductor as shown in Exhibit 1-10. The field inside the coil is uniform and much stronger than the field outside the coil. Hence, by coiling the conductor the magnetic effects have been greatly increased. The field inside the coil may be made even greater by winding the coil on an iron core. The direction of the field may be found by applying the Right-hand rule for coils.

The <u>Right-hand rule for coils</u> states: **If a coil is held with the fingers curled in the direction of current flow, the thumb points in the direction of the North pole of the magnetic field, that is, in the direction of the magnetic field within the solenoid.**

1.1.6 — Electrical Machinery

The electric machine is an electro-mechanical energy conversion device. An electric machine performs the conversion of electrical to mechanical or mechanical to electrical energy through the medium of a magnetic field. Although the conversion is reversible, when the operation is from mechanical to electrical the machine is called a <u>generator</u> and when the operation is from electrical to mechanical it is called a <u>motor</u>.

Regardless of whether the machine is operated as a motor or a generator, there are two inseparable electro-mechanical phenomena involved. One is the generation of a voltage due to a changing magnetic flux, and the other is the generation of a force by the flow of current in the presence of a magnetic field. These two fundamentals are essential to an understanding of electrical machinery.

1.1.6.1 — Induced Voltage

Recall from the Physics subject that a charge in motion through a magnetic field experiences a force given by the relation

Equation 1-6

$$\vec{F} = q\vec{V} \times \vec{B}$$

This force is a result of the interaction of the magnetic field about a charge in motion, as shown in Exhibit 1-11, and the applied magnetic field. The interaction is illustrated in Exhibit 1-11 where it is assumed that a positive charge is moving out of the page. When F, V and B are all mutually perpendicular, Equation 1-6 reduces to:

$$F = qVB$$

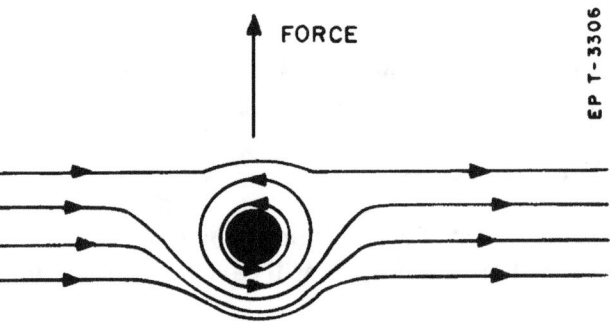

FORCE

EP T-3306

Exhibit 1-11 — Interaction Of A Magnetic Field And A Charge In Motion

Now consider a single conductor being moved at a constant velocity V, perpendicular to a constant magnetic field B, directed out of the page as shown in Exhibit 1-12.

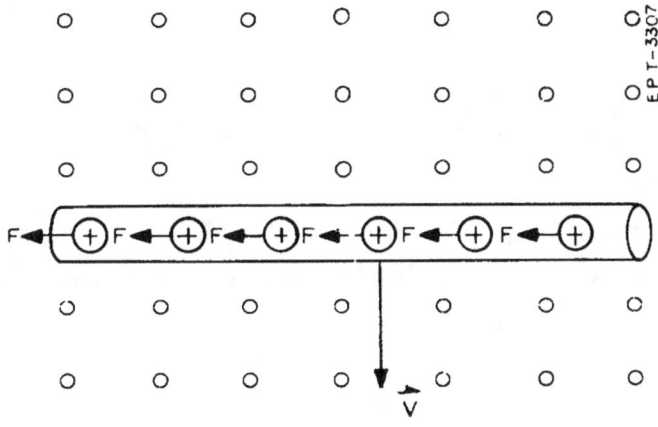

O DENOTES $\vec{B}$ FIELD DIRECTED OUT OF PAGE

Exhibit 1-12 — Faraday Induced Voltage

Positive charges in the conductor experience a force with a magnitude given by Equation 1-6. This force moves the charges toward one end of the conductor, creating a positive potential at that end. (In reality, only electrons (negative charges) are free to move in a conductor, but because of convention, positive charges are used to describe current flow.) A corresponding negative potential is created at the opposite end. Clearly, moving the conductor through the field has induced a voltage (potential difference) across the ends of the conductor. In general, an induced voltage appears whenever relative motion exists between a conductor and a magnetic field. Alternately stated, a voltage is induced whenever the conductor is subjected to a changing magnetic field. The relative motion may be obtained by moving a conductor through a uniform magnetic field or by varying the magnetic field around a stationary conductor. Both of these schemes are applied in practical electrical machines.

If an external load is connected to the ends of the conductor in Exhibit 1-12, the induced voltage will create a current flow through the completed path. The direction of current flow may be determined by applying the right-hand rule for generator action as illustrated in Exhibit 1-13.

In summary, an emf will be induced in a conductor moved through a magnetic field. The direction of the emf is given by the **right-hand rule for generator action** (Exhibit 1-13) which states: **If the thumb and the first two fingers of the right hand are oriented at right angles to one another, with the thumb placed in the direction of conductor motion and the first finger pointed in the direction of the magnetic field, then the second finger points in the direction of current flow.**

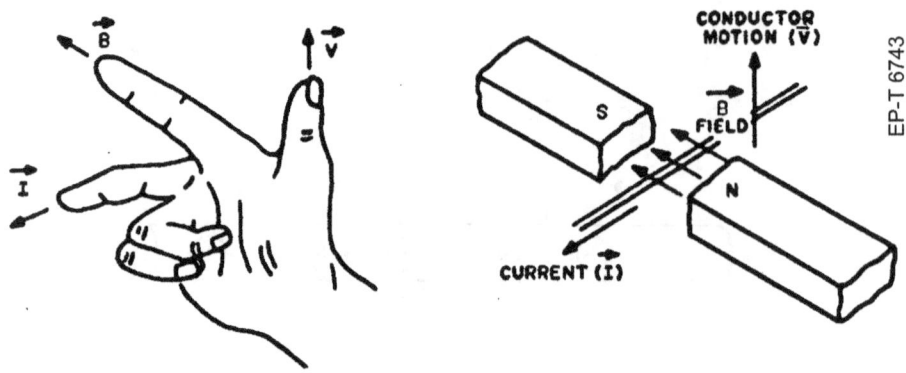

**Exhibit 1-13 — EMF Induced In Conductor Moving Through Magnetic Field:
Right-Hand Rule For Generator Action**

1.1.6.2 — Induced Force

If current i is passed through a stationary conductor in the presence of a magnetic field, a force will be induced on the conductor. This situation is depicted in Exhibit 1-14 where a constant magnetic field $\vec{B}$ is directed out of the page. Again, Equation 1-6 may be applied to determine the force on the charges. In this case, the velocity of the charges V, is created by the current i. The sum of the forces on the individual charges is an induced force acting on the conductor. This force must be balanced mechanically to hold the current-carrying conductor stationary.

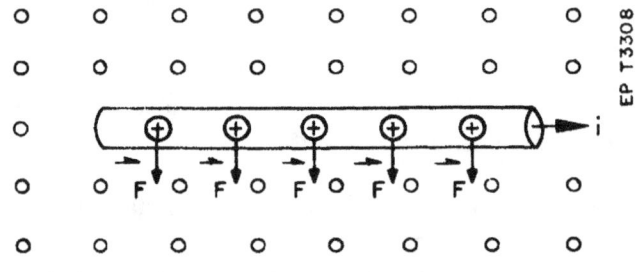

O DENOTES $\vec{B}$ FIELD DIRECTED
OUT OF THE PAGE

Exhibit 1-14 — Induced Force On A Current-Carrying Conductor

In general, when a conductor carries a current across a magnetic field, there is an induced force on the conductor. The magnitude of this induced force depends upon the strength of the magnetic field, $\vec{B}$, the length of the conductor, ℓ , and the amount of current, i, flowing in the conductor.

Equation 1-7

$$\vec{F} = i\,\vec{\ell} \times \vec{B}$$

To determine the direction of the induced force, the right-hand rule for motor action is used. In summary, a current carrying conductor will experience a force in a magnetic field. The direction of the force is given by the **right-hand rule for motor action** (Exhibit 1-15) which states: **If the thumb and first two fingers of the right hand are oriented at right angles to one another with the thumb pointed in the direction of current flow in the conductor and the first finger pointed in the direction of the magnetic field, then the second finger points in the direction of the induced force on the conductor.**

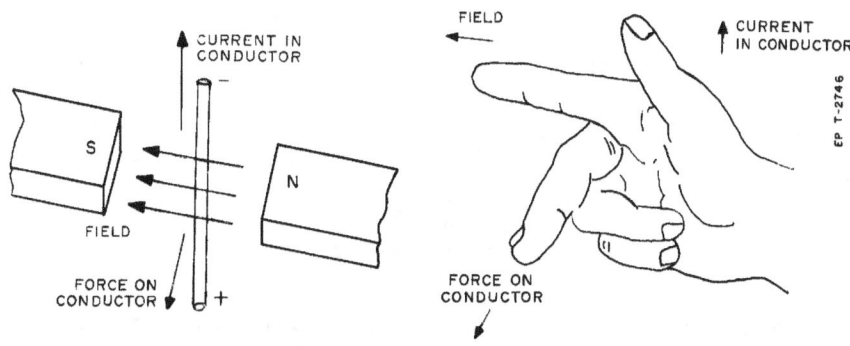

Exhibit 1-15 — Force On Current-Carrying Conductor Moving Through Magnetic Field: Right-Hand Rule For Motor Action

1.1.7 — The Linear Machine

In Paragraph 1.1.5 it was shown that mechanically moving a noncurrent-carrying conductor through a magnetic field created an induced voltage in the conductor. In Paragraph 1.1.6 it was shown that electrically passing a current through a stationary conductor in a magnetic field created an induced force on the conductor. Since a practical machine must carry current and provide motion at the same time, it is evident that both an induced voltage and an induced force will exist. To illustrate the inseparability between induced voltage and induced force, a hypothetical linear machine will be considered. The linear machine consists of a conductor that is free to move through a uniform magnetic field along two conducting strips which electrically connect the conductor to the external electrical circuit. Construction is such that the field, conductor motion, and current flow are all mutually perpendicular.

To operate the linear machine as a generator, an external source of mechanical power must be supplied. This external source or **prime mover** may be a steam turbine, diesel engine, gasoline engine, or an electric motor. In Exhibit 1-16, it is assumed that the conductor is being moved down the page by the mechanical power source P_{mech}. As discussed in Paragraph 1.1.5, the motion will create an induced voltage or electromotive force (emf) across the ends of the conductor. Since a complete electrical path exists, a current will flow through the load. The direction of this current may be determined by applying the right-hand rule of Exhibit 1-13. However, from Paragraph 1.1.6 it is known that there is an induced force on a current-carrying conductor. If the right-hand rule of Exhibit 1-15 is applied to Exhibit 1-16, it can be seen that the flow of current induces a force which opposes the applied mechanical force. The magnitude of the force depends upon the strength of the magnetic field and the amount of current flowing in the conductor as shown by Equation 1-7.

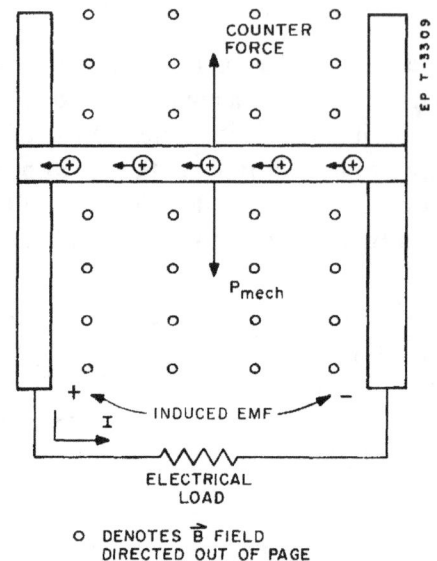

Exhibit 1-16 — The Linear Machine Operated As A Generator

This opposing or counter force is the means by which the electrical load regulates the amount of mechanical input power. In equilibrium, the generator operates at constant speed and the mechanical power input balances the machine losses and the power required to oppose the counter force. Thus, the counter force is equivalent to the amount of mechanical input power that is converted to electrical output power. Now if the electrical load is increased, a greater current will be demanded by the load. When the generator tries to meet this increased current demand, an increased counter force will result in accordance with Equation 1-7, which will cause a decrease in generator speed. The decrease in speed causes a corresponding decrease in induced emf, since the rate of change of flux is reduced. To maintain the original value of induced emf under the increased load, the mechanical input power must be increased until the generator again operates at the original constant speed.

In Exhibit 1-17, the input power is an electrical source P_{elec}, and the machine is being operated as a motor. The electrical source causes a current flow that creates an induced force in the direction given by the right-hand rule of Exhibit 1-15. The conductor motion produces an induced voltage of a polarity, given by the right-hand rule of Exhibit 1-13, which opposes the applied voltage. This opposing or **counter emf** is the means by which the mechanical load regulates the amount of electrical input. The magnitude of the counter emf depends on the strength of the magnetic field and the velocity of the conductor. The counter emf is also equivalent to the electrical input power that is converted to mechanical output power. The role of counter emf in a motor is analogous to the role of counter force in a generator.

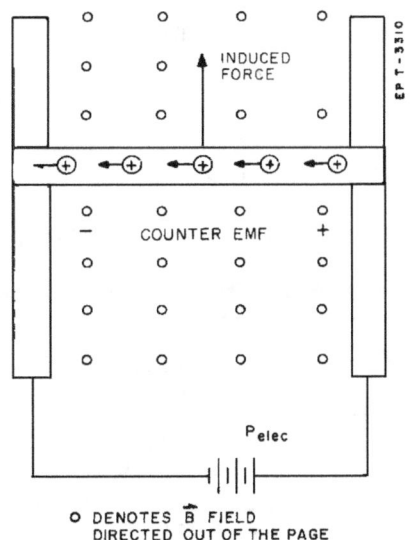

Exhibit 1-17 — The Linear Machine Operated As A Motor

Section 1.2 — Alternating Current Theory

1.2.1 — Alternating Current

An alternating current (a-c) is an electrical signal that is continuously changing in magnitude and reversing direction at regular intervals. This reversal of the signal illustrates the name, alternating current. A-C distribution circuits are used extensively in the ship's service electrical power system, providing advantages over d-c circuits, including:

1. the ability to easily change voltage and current levels, and provide electrical isolation by using transformers, and
2. no requirement for commutation, which results in smaller, more efficient rotating machines.

The most common form of a-c is the sinusoid. A sinusoidally varying current may be expressed mathematically as:

Equation 1-8

$$i(t) = I_p \sin \omega t$$

Similarly, a sinusoidal a-c voltage may be expressed mathematically as

Equation 1-9

$$V(t) = v_p \sin \omega t$$

Note that lower case symbols i(t) and v(t) are used in the equations above to denote time varying quantities. In addition, ω, is related to frequency by the following equation:

$$\omega = 2\pi f$$

where f is the frequency in hertz (discussed further below).

Exhibit 1-18 is a plot of one cycle of a sinusoidally varying a-c current.

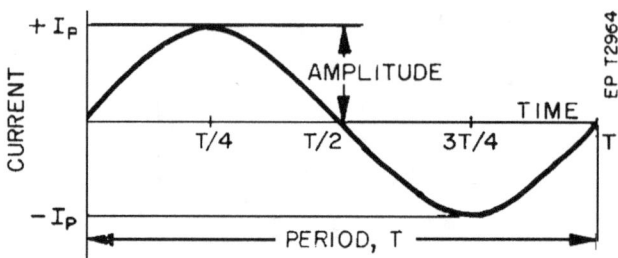

Exhibit 1-18 — One Cycle Of A Sinusoidally Varying A-C Current

1.2.2 — Measurements

The amplitude of the sine wave is I_p and it is equal to the peak or maximum excursion from zero. The instantaneous current passes through a complete set of positive and negative values in one cycle, after which the waveform repeats itself. The number of cycles completed in one second is the frequency, f, of the waveform. The time required to complete one cycle is the period, T.

Frequency is defined as:

Equation 1-10

$$f = \frac{1}{T}$$

where T is in **seconds** and f is in **hertz**.

The magnitude of the a-c quantity can be represented in several ways. The most obvious measure of the magnitude is probably the peak value, I_P. In some applications, particularly transistor amplifiers, the peak-to-peak value is convenient. The peak-to-peak value is twice the peak value. Two particularly useful values are the average value, and the root-mean-square (rms) or effective value.

The average value, useful in magnetic amplifier applications, is the mathematical average of the a-c quantity over one-half cycle. It is equal to the area under the curve divided by the length of the half-cycle. For sine waves, the value is conveniently related to the peak value by:

Equation 1-11

$$\text{Average value} = 0.637 \times \text{peak value}$$

The rms value is most useful since a-c power calculations can be performed directly using rms values. A one-ampere rms alternating current has the same heating effect in a resistor as a one ampere direct current would have in the same resistor. A-C meters are normally calibrated to indicate the rms values of voltages and currents. The rms value of a sine wave is related to the peak value by:

Equation 1-12

$$\text{rms value} = 0.707 \times \text{peak value}$$

The rms current or voltage can be used to find average power directly by means of Equation 1-4 or Equation 1-5.

1.2.3 — Aspects of A-C Quantities

In a-c circuits operating under steady-state conditions, all the voltages and currents vary with the same frequency. However, the voltages and currents may be out of phase. This is illustrated in Exhibit 1-19 where two alternating currents of different magnitude are drawn 90 degrees out of phase.

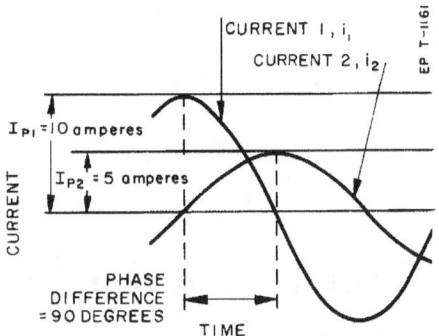

Exhibit 1-19 — Two A-C Sine Waves Shown 90 Degrees Out of Phase Representing Two Alternating Currents

To describe the phase relationship between two waves the terms lead and lag are used. The amount by which one wave leads or lags another is measured in degrees. Referring to Exhibit 1-19, current i_2 reaches its peak 90 degrees after current i_1 reaches its peak; thus, i_2 lags i_1 by 90 degrees. This can also be stated by saying that i_1 leads i_2 by 90 degrees.

1.2.3.1 — Impedance Calculations

A-C circuits in general contain capacitive, inductive, and resistive loads. The combined effect of these loads is called **impedance**. For a given a-c voltage, impedance is used to determine both the magnitude of the current and its phase relative to the voltage. A-C voltage and current are represented by rotating vectors. The magnitude of the vector is proportional to the peak value of voltage or current. The orientation of the vector represents the phase of the voltage or current. Each vector can be represented by a magnitude and angle (Example: E ∠θ) called **polar form** or by vertical and horizontal components (Example: $E_a + jE_b$) called **rectangular form** (Exhibit 1-20).

Conversion from polar to rectangular form

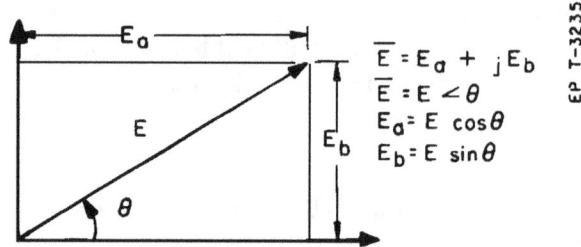

Exhibit 1-20 — Relationship Between Rectangular And Polar Coordinate Systems

Conversion from rectangular to polar form

Equation 1-13

$$\theta = \tan^{-1} \frac{E_b}{E_a}$$

Equation 1-14

$$E = \frac{E}{\sin \theta} = \sqrt{E_a^2 + E_b^2}$$

The **impedance** $\vec{Z}$ is the voltage $\vec{E}$ divided by the current vector, $\vec{I}$. For a given load,

Equation 1-15

$$\vec{Z} = \frac{\vec{E}}{\vec{I}}$$

the **impedance vector** is constant and does not rotate.

In general, for an a-c circuit, the voltage across a device and the current through it are out of phase. If the a-c voltage and current are represented by vectors, then the ratio of the vector voltage to the vector current is the impedance ($\vec{Z}$) of the device as shown in Exhibit 1-21. Unlike the voltage and current vectors, impedance is not a rotating vector. The ratio between voltage and current remains fixed with time. The following important characteristics of the impedance should become apparent:

1.   Impedance, $\vec{Z}$, always possesses both magnitude and direction.
2.   The magnitude of the impedance, denoted Z, is the ratio of the rms voltage to the rms current and is measured in ohms.

3. The direction or angle of the impedance, θ, is equal to the phase difference between the voltage and the current. If θ > 0, current lags voltage; if θ < 0, current leads voltage.

Thus, when the impedance of a circuit is known, the relationship between voltage and current is specified both in terms of magnitude and phase difference.

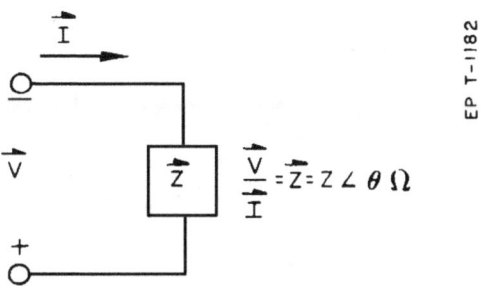

Exhibit 1-21 — Impedance Is The Ratio Of The Voltage Vector To The Current Vector

1.2.3.2 — Resistive Circuits

When an a-c voltage is applied across a resistance, the current that flows is directly proportional to the instantaneous value of the voltage. The current through the resistor will be zero when the voltage is zero and the current will reach its peak value when the voltage reaches its peak value. The sinusoidal a-c voltage is accompanied by a sinusoidal alternating current through the resistor. The voltage and current are in phase, as shown in Exhibit 1-22, and the ratio of the rms voltage to the rms current is the resistance. Since the voltage

and current are in phase, the voltage vector, $\vec{V}_r$, and the current vector, $\vec{I}_r$ point in the same direction.

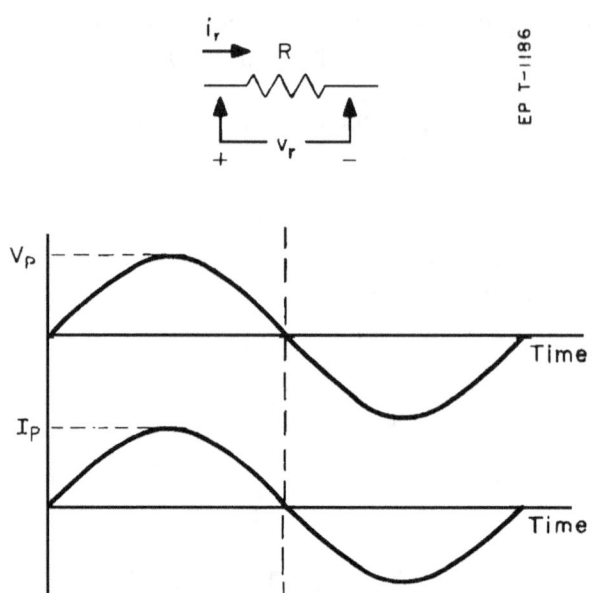

Exhibit 1-22 — The Voltage And Current In A Resistor Are In Phase

The impedance of a resistor is thus given by:

Equation 1-16

$$\vec{Z}_r = \frac{\vec{V}_r}{\vec{I}_r} = \frac{V_r}{I_r} \angle 0 = R\angle 0 \text{ ohms}$$

Note that the impedance angle must be zero since the voltage and current are in phase, that is, the phase difference equals 0 degrees. As a result of the current and voltage being **in phase** for a purely resistive circuit, all of the following laws applicable to d-c circuits apply in the same manner to a-c circuits:

• Ohm's Law E = IR
• Kirchhoff's circuit (current and voltage) laws
• Power law P = EI

A purely resistive load consumes power at the rate of P = EI by converting power to heat. A pure resistance cannot **store** energy.

1.2.3.3 — Inductance

When an alternating current flows in an inductor, the voltage across the inductor is an induced voltage produced by the changing magnetic field. This induced voltage is proportional to the rate of change of current:

$$v \propto \text{rate of change of current}$$

$$v = L \times (\text{rate of change of current})$$

Equation 1-17

$$v = L\frac{di}{dt}$$

where L is the inductance in henries.

The rate of change of current is the slope of the current curve; hence, the voltage across an inductor is proportional to the slope of the current curve as shown in Exhibit 1-23.

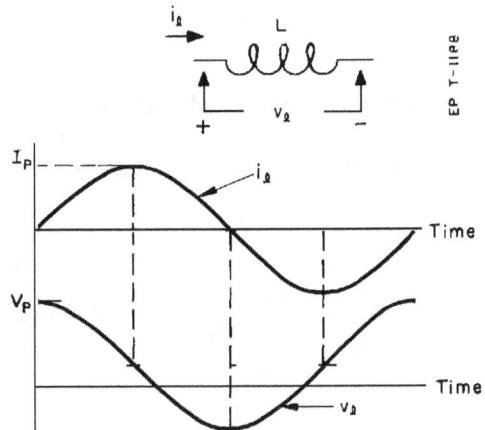

Exhibit 1-23 — Voltage And Current In An Inductor

Since the current lags the voltage by 90 degrees in an inductor, the vector representation of the voltage and current shows the current vector lagging the voltage vector by 90 degrees as in Exhibit 1-24.

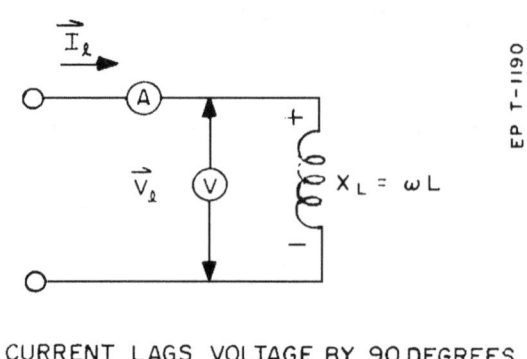

CURRENT LAGS VOLTAGE BY 90 DEGREES

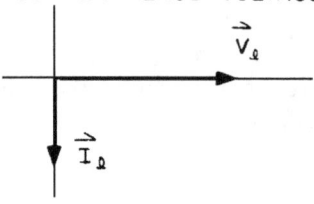

Exhibit 1-24 — The Vector Representation Of The Current And Voltage For An Inductor

The impedance of an inductor is the vector voltage divided by the vector current:

Equation 1-18

$$\vec{Z}_L = \frac{\vec{V}_i}{\vec{I}_i} = \frac{V_i \angle 0}{I_i \angle -90} = \frac{V_i}{I_i} \angle +90$$

$$= X_i \angle 90 \text{ ohms}$$

$$= \omega L \angle +90 \text{ ohms}$$

The magnitude of the impedance $X_L = \omega L$ is termed the inductive reactance. Note that the impedance angle for an inductor is + 90 degrees indicating that the current and voltage are out of phase by 90 degrees with the current lagging the voltage.

1.2.3.4 — Capacitance

When an a-c voltage is supplied to a capacitor, the current which flows is proportional to the rate of change of voltage:

$$i \propto \text{rate of change of voltage}$$

$$i = C \times (\text{rate of change of voltage})$$

Equation 1-19

$$i = C\frac{dv}{dt}$$

where C is the capacitance in farads.

The rate of change of voltage is the slope of the voltage curve; hence, the current in a capacitor is proportional to the slope of the voltage curve as shown in Exhibit 1-25.

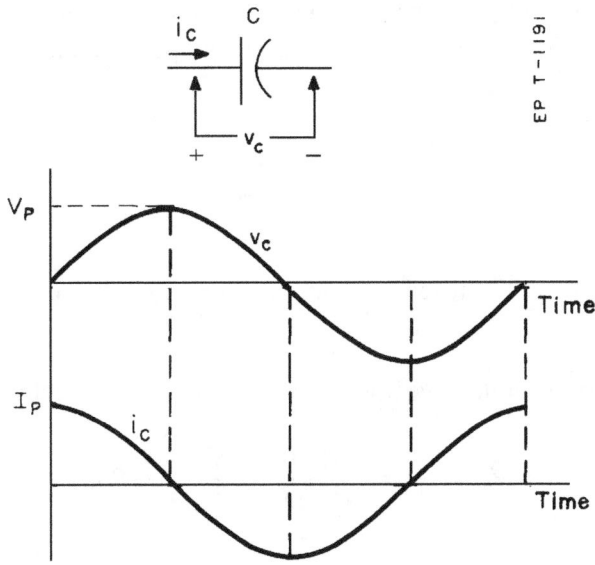

Exhibit 1-25 — Voltage And Current In A Capacitor

The current at every instant is equal to the capacitance, C, times the slope of the voltage curve. The current is a maximum at the instant the voltage is passing through its zero value. The zero values of current occur when the voltage is either at its negative peak value or at its positive peak value. The current waveform is a sine wave of the same frequency as the voltage but leads the voltage by 90 degrees. Therefore, for a capacitor, the alternating current leads the voltage by 90 degrees. (Exhibit 1-26).

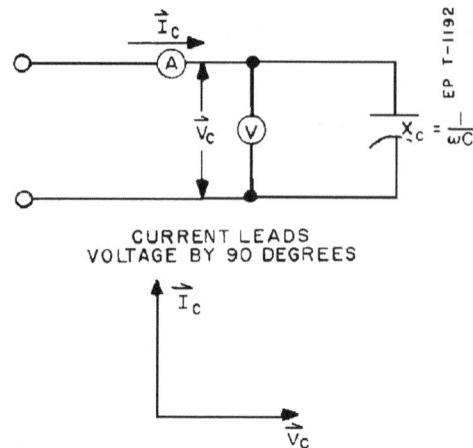

Exhibit 1-26 — The Vector Representation Of The Current And Voltage For A Capacitor

Capacitors oppose changes in voltage by acting as a temporary charge-storage device. When a changing voltage is applied to a capacitor, the capacitor draws a current proportional to the rate of change in voltage

Equation 1-20

$$i_c = C\frac{dv}{dt}$$

When a-c voltage is applied to a purely capacitive load, the current **leads** the voltage by 90 degrees (Exhibit 1-26).

The impedance of a capacitor is the voltage vector divided by the current vector:

$$\vec{Z}_c = \frac{\vec{V}_c}{\vec{I}_c} = \frac{V_c\angle 0}{I_c\angle 90} = \frac{V_c}{I_c}\angle -90$$

Equation 1-21

$$\vec{Z}_c = X_c\angle -90 = \frac{1}{\omega C}\angle -90 \text{ ohms}$$

Note that the impedance angle for a capacitor is a -90 degrees indicating that the current and voltage are out of phase by 90 degrees with the current leading the voltage.

1.2.4 — Power Factor

In purely capacitive or purely inductive circuits, loads absorb and supply energy to an a-c circuit in alternate quarter cycles. The average power (over an exact number of cycles) absorbed by these loads is zero since the amounts supplied and absorbed offset one another exactly. Only resistive loads consume real power continuously.

A typical a-c circuit consists of a combination of resistive, capacitive, and inductive loads. For these circuits, the phase difference is between zero and 90 degrees. The power consumed by such a circuit is somewhere between zero and the amount consumed if the same voltage and current were in phase.

pf, relates rms voltage and rms current to power as shown in Equation 1-22.

Equation 1-22

$$P = VI \cos \theta$$

The angle θ is the phase difference (leading or lagging) between voltage and current ($\vec{Z} = Z \angle \theta$). V and I are the rms voltage and current, respectively. (From now on voltage will be V). Example:

V	= 120 volts (rms)
I	= 40 amperes (rms) lagging V by 20 degrees
P	= 120 × 40 × cos 20 = 4.5 kilowatts.

pf is the ratio of power consumed to the product of V and I.

$$pf = \frac{P}{VI} = \frac{VI \cos \theta}{VI}.$$

Thus,

Equation 1-23

$$pf = \cos \theta$$

where θ is the impedance angle. The power factor is expressed as leading or lagging to indicate whether current leads or lags voltage. In a leading power factor circuit, the current leads the voltage; in a lagging power factor circuit, the current lags the voltage. In the preceding example, pf = cos 20º = 0.94 lagging.

Circuits that are inductive in nature are lagging power factor circuits since the current lags the voltage. This is the case most commonly encountered in a-c power systems since a-c induction motors, transformers, and coils contribute inductive components to the total impedance.

1.2.5 — Apparent Power

Apparent power (P_A) is the product of voltage and current without regard to the phase relationship.

Equation 1-24

$$P_A = VI$$

The units are volt-amperes (va) or kilovolt-amperes (kva). Apparent power equals real power, P, only if the power factor is unity, since

Equation 1-25

$$P = P_A \times (pf)$$

This equation is significant in that it shows that the maximum power output of an a-c generator depends on the phase difference between voltage and current (impedance angle of the load) as well as the current and voltage capabilities for the generator.

1.2.6 — Reactive Power (Q)

The power which is exchanged between reactive elements (inductors and capacitors) is called **reactive power**, Q, and is measured in vars (volt-ampere-reactive) or kilovars (kilovolt-ampere reactive). This power does no useful mechanical work, but must be alternately absorbed and supplied by the reactive elements.

Although there is often no need to calculate reactive power, it can be found by using Equation 1-26.

Equation 1-26

$$Q = VI \sin \theta = P_A \sin \theta$$

where::

Q	= reactive power, in vars
V	= rms voltage
I	= rms current
θ	= phase difference between voltage and current, and
P_A	= apparent power, in volt-amperes

1.2.7 — Summary of A-C Circuit Elements

Table 1-1 — Summary Of A-C Circuit Elements

	Resistor	Inductor	Capacitor
Symbol			
Measurement	ohms (Ω)	henries (h)	farads (f)
V-I relation	$v = iR$	$v = L\dfrac{di}{dt}$	$i = C\dfrac{dv}{dt}$

Table 1-1 — Summary Of A-C Circuit Elements (Cont)

	Resistor	Inductor	Capacitor
RMS V/I	$R = V/I$	$X_L = 2\pi f L$	$X_c = \dfrac{1}{2\pi f C}$
Impedance	$\vec{Z}_R = R\angle 0°$	$\vec{Z}_L = X_L \angle 90°$	$\vec{Z}_C = X_C \angle -90°$

Section 1.3 — Three-Phase Systems And Transformers

1.3.1 — Three-Phase Systems

A three-phase system consists of an arrangement of three single-phase systems in which the maximum values of the voltages occur 120 degrees out of phase with each other. The equipment used in three-phase systems has many advantages over that employed for single-phase service. These advantages include a reduction in the size of machinery for the same capacity, simpler machine construction, reduced cost, and increased reliability of power transmission. Also, three-phase power transmission lines are comparatively less expensive for the maximum number of lines needed than for three single-phase systems. This section discusses how loads are connected to adapt to three-phase voltage and the relation between currents and voltages of the system.

A three-phase distribution system is shown schematically in Exhibit 1-27. The three lines are denoted phase "a", phase "b", and phase "c", and the currents flowing in the three lines are I_a, I_b, and I_c respectively. These currents are also called line currents.

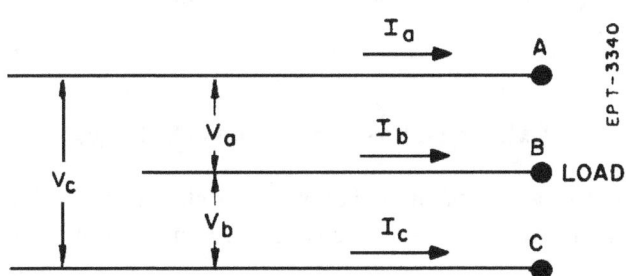

Exhibit 1-27 — Three-Phase Distribution System

The voltage differences between lines are labeled V_a, V_b, and V_c as shown on Exhibit 1-27, and are called line voltages. These voltages are identical in magnitude, but differ in phase by 120 degrees. (See Exhibit 1-28 for the relationship between these voltages and E_m)

Equation 1-27

$$V_a = E_m \ \sin \ \omega t$$

Equation 1-28

$$V_b = E_m \ \sin \ (\omega t - 120 \ \text{degrees})$$

Equation 1-29

$$V_c = E_m \ \sin \ (\omega t + 120 \ \text{degrees})$$

The rms magnitudes of the voltages are equal, and are designated V_{line}:

$$V_{line} = V_a = V_b = V_c = 0.707 \ E_m \ .$$

The phase relationships of the line voltages are shown in Exhibit 1-28.

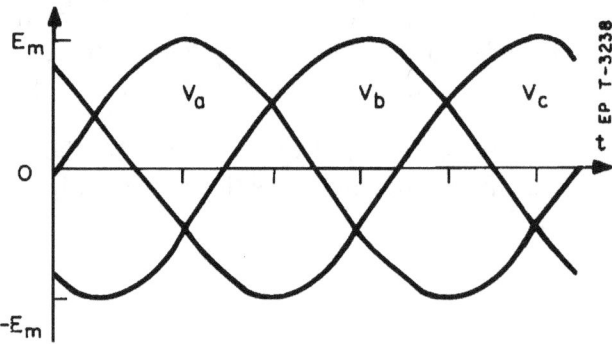

Exhibit 1-28 — Three-Phase Voltages

These phase relationships can also be represented vectorially as shown in Exhibit 1-29. In Exhibit 1-29, V_c leads V_a by 120 degrees and V_b lags V_a by 120 degrees, in agreement with the instantaneous expressions for the voltages.

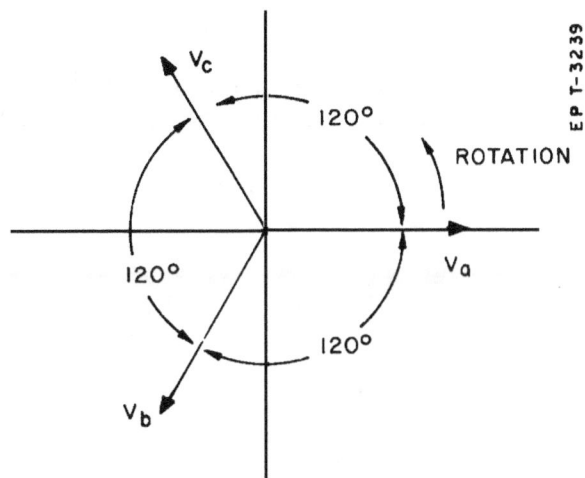

Exhibit 1-29 — Phase Diagram For Three-Phase Voltages

A three-phase load consists of three individual loads connected across A, B, and C of Exhibit 1-27. We shall only consider the case of the balanced load where the individual loads used in each leg of the three-phase connections will be identical.

Two methods of load connections are possible; these are called "delta", Δ, and "wye", Y, connections. The two connections result in different properties as described in Paragraph 1.3.2 and Paragraph 1.3.3.

1.3.2 — The Delta Connection

The delta connection is diagrammed in Exhibit 1-30.

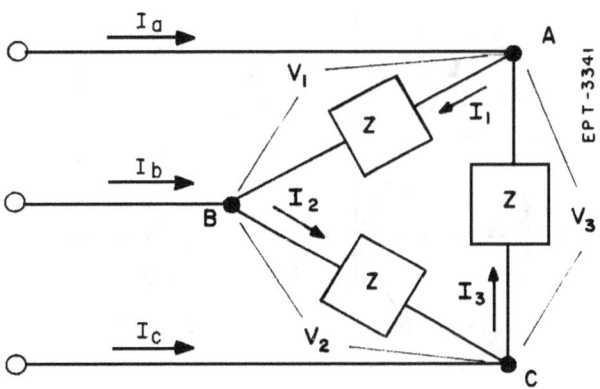

Exhibit 1-30 — Delta Connected Load

The connections to the three-phase distribution system are made at points A, B, and C, corresponding to the points on Exhibit 1-27. For the balanced load considered here, the impedance (Z) in each leg of the delta connection will be the same. We shall assume for simplicity that Z is purely resistive; however, the result will also be valid for reactive loads.

The voltages across the three loads are called phase voltages, and are designated V_1, V_2, and V_3 on Exhibit 1-30. The currents through the loads are called phase currents and designated I_1, I_2, and I_3. For balanced loads:

$$I_{phase} = I_1 = I_2 = I_3$$

$$V_{phase} = V_1 = V_2 = V_3$$

Comparison of Exhibit 1-27 with Exhibit 1-30 shows

$$V_{phase} = V_{line}$$

That is, $V_a = V_1$, $V_b = V_2$, and $V_c = V_3$. From Exhibit 1-30,

$$V_{phase} = I_{phase} Z$$

or

$$I_{phase} = \frac{V_{phase}}{Z}$$

For the resistive loads under consideration, Exhibit 1-29 can be redrawn to include the phase currents and voltages, as shown in Exhibit 1-31.

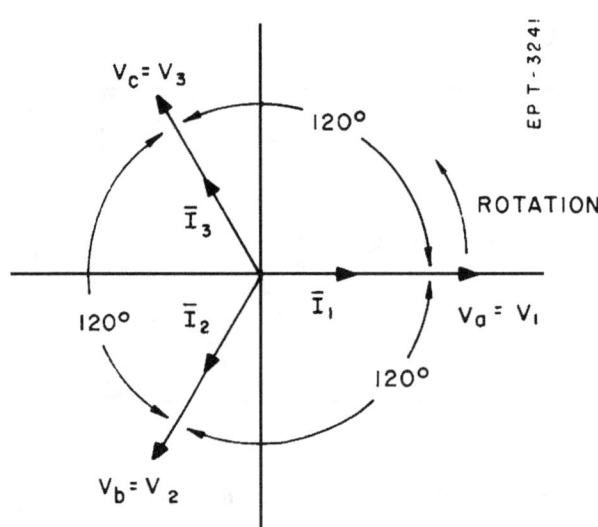

Exhibit 1-31 — Current-Voltage Relations For Delta Connection

Now consider the connection at point A. By inspection of Exhibit 1-27 and Exhibit 1-30 we can apply Kirchhoff's Current Law to point A and obtain the vector relation

$$\vec{I}_1 - \vec{I}_3 - \vec{I}_a = 0$$

Then I_a can be obtained as the vector difference

$$\vec{I}_a = \vec{I}_1 - \vec{I}_3 \ .$$

This operation is carried out as shown in Exhibit 1-32 using the $\vec{I}_1$, and $\vec{I}_3$ vectors from Exhibit 1-31.

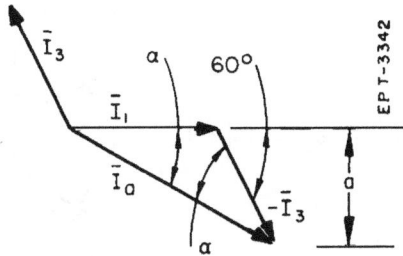

Exhibit 1-32 — Line Current For Delta Connection

The magnitude (length) of $\vec{I}_a$ can be calculated as follows. The interior angle between $\vec{I}_1$ and $\vec{I}_3$ can be seen to be 120 degrees. The triangle formed $\vec{I}_1$, $\vec{I}_3$, and $\vec{I}_a$ is isosceles because the magnitudes of $\vec{I}_1$ and $\vec{I}_3$ are equal. The remaining angles (α) each equal 30 degrees. By trigonometry it is seen that

$$I_a \ \sin \alpha = I_3 \ \sin 60^\circ = a$$

$$I_a = \frac{I_3 \ \sin 60^\circ}{\sin 30^\circ}$$

$$= \frac{I_3 \ \sqrt{3}/2}{1/2}$$

$$= I_3 \ \sqrt{3}$$

Thus for balanced delta loads,

Equation 1-30

$$I_{line} = I_{phase}\ \sqrt{3}$$

1.3.3 — The Wye Connection

The wye connection is diagrammed in Exhibit 1-33. The connections to the three-phase distribution system are made at points, A, B, and C, which correspond to the points on Exhibit 1-27. We shall again consider only the balanced load for which the impedance (Z) in each leg of the wye is equal. Also a purely resistive Z is assumed for simplicity although the result will be equally valid for reactive loads.

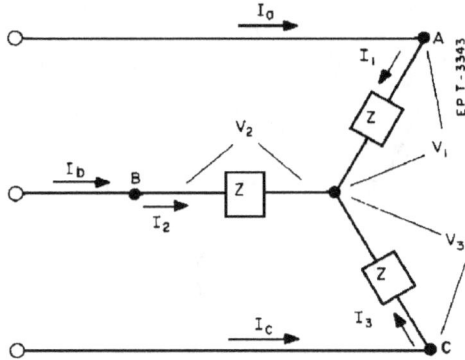

Exhibit 1-33 — Wye Connected Load

The currents I_1, I_2, and I_3 through the loads are again called phase currents, and the voltages V_1, V_2, and V_3 across the loads are called phase voltages. Since the line voltages applied to points A, B, and C are also equal, it is clear that

$$V_{phase} = V_1 = V_2 = V_3$$

Also by inspection of Exhibit 1-27 and Exhibit 1-33 it is clear that the line currents equal the phase currents.

$$I_{line} = I_{phase} = I_1 = I_2 = I_3$$

For the resistive loads under consideration, the phase voltages and phase currents will be related as shown in Exhibit 1-34, where each current is in phase with its corresponding voltage.

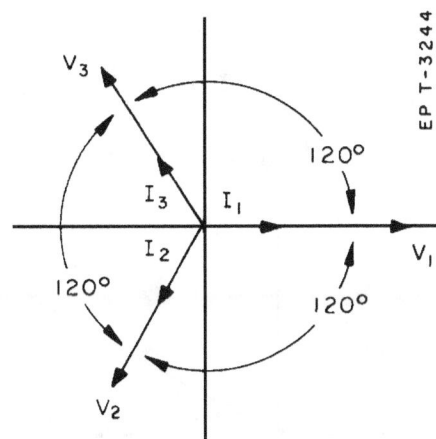

EP T-3244

Exhibit 1-34 — Current-Voltage Relations For Wye Connection

The only remaining problem is to relate phase voltage to line voltage. From Exhibit 1-27 and Exhibit 1-33 it can be seen that the voltage $\vec{V}_a$ across points A and B is equal to the vector difference

$$\vec{V}_1 - \vec{V}_2$$

that is

$$\vec{V}_a = \vec{V}_1 = \vec{V}_2 \,.$$

This subtraction is shown diagrammatically in Exhibit 1-35. The magnitude of $\mathbf{V_a}$ is obtained in the same manner as in the solution depicted by Exhibit 1-32. The angle α is seen to be 30 degrees; therefore,

$$\vec{V}_a \, \sin 30° = \vec{V}_2 \, \sin 60°$$

$$= \vec{V}_a \;\; = \frac{\vec{V}_2 \, \sin 60°}{\sin 30°}$$

$$= \frac{\vec{V}_2 \, \sqrt{3}/2}{1/2}$$

$$= \vec{V}_2 \, \sqrt{3}$$

Thus, for balanced loads,

Equation 1-31

$$V_{line} = V_{phase} \sqrt{3}$$

The same results could have been obtained for V_b or V_c.

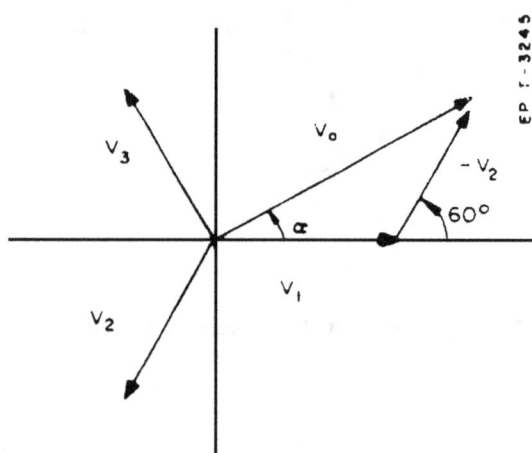

Exhibit 1-35 — Line Voltage For Wye Connection

1.3.4 — Power In Three-Phase Loads

Recall Equation 1-22 for power in an a-c circuit:

$$P = VI \cos \theta$$

Where we now include reactive loads ($\cos \theta \neq 1$) as well as the purely resistive loads considered in the previous two sections, for one leg of a wye or delta load, power is given by

Equation 1-32

$$P = V_{phase} I_{phase} \cos \theta$$

This expression gives the power consumed by one of the loads connected in a three-phase arrangement. Since the loads are balanced, the power consumed by all three loads, the **total power (P_t)**, will merely be three times this expression.

Equation 1-33

$$P_t = 3 V_{phase} I_{phase} \cos \theta \text{ (wye or delta connection)}$$

The results for wye and delta connected loads may be summarized as follows:

Table 1-2 — Voltage, Current, And Total Power Relationships For Wye And Delta Connected Loads

	Wye	Delta
Voltage	$V_{line} = V_{phase}\sqrt{3}$	$V_{line} = V_{phase}$
Current	$I_{line} = I_{phase}$	$I_{line} = I_{phase}\sqrt{3}$
Total Power (P_t)	$\sqrt{3}\,V_{line}I_{line}\cos\theta$	$\sqrt{3}\,V_{line}I_{line}\cos\theta$

1.3.5 — Transformers

A transformer is a device with no moving parts, which transfers electrical energy from one circuit to another through a magnetic coupling. It does not manufacture power, it merely transfers it. It may be used to step the voltage up and the current down, or vice-versa depending upon application.

Transformers depend on a changing magnetic field for their operation; therefore, they can function on alternating current or a current that is changing, but not steady-state direct current.

A transformer consists of two or more windings on a suitable core material. Power is normally applied to only one of the windings, and it is taken from the other winding(s). The windings to which power is applied are called the **primary** windings, and the windings from which power is taken are called the **secondary** windings. Exhibit 1-36 represents a simple transformer with two windings.

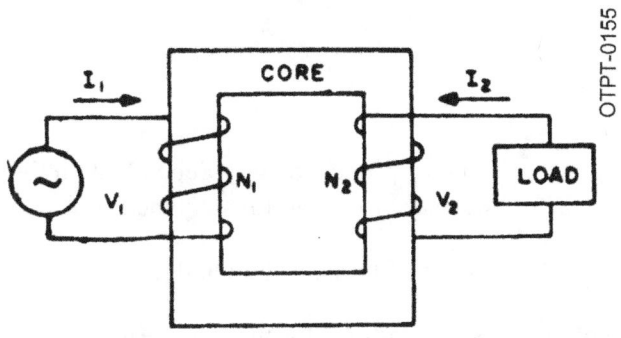

Exhibit 1-36 — Simple Transformer

In this review we will only consider the simple transformer (Exhibit 1-36) for which the following symbols apply:

V_1 = Voltage applied to the primary winding

V_2 = Voltage induced in the secondary winding by the changing magnetic flux in the core.

I_1 = Current supplied by the voltage source and flowing through the primary winding. This is the current that sets up the magnetic flux in the core.

I_2 = Current (due to the voltage V_2) induced in the secondary winding when a load is attached to it

N_1 = Number of turns in the primary winding

N_2 = Number of turns in the secondary winding.

If the secondary voltage (V_2) is larger than the primary voltage (V_1) the transformer is called a **step-up** transformer. If the secondary voltage is smaller than the primary voltage the transformer is then called a **step-down** transformer.

In the analysis of the relationship between the primary and secondary currents and voltages, it is assumed that the power factor of the secondary circuit equals that of the primary circuit and that no losses of power are incurred between the primary and secondary circuits. These assumptions do not introduce much error because most transformers approach 98 percent efficiency or better. The losses incurred in a transformer and the methods by which they are reduced will be discussed later.

Recall that the voltage induced in a conductor due to the change in flux lines linking the conductor is represented by

$$e = K d\phi/dt.$$

For the voltage induced in the winding of a transformer, this equation may be written as follows:

$$V = K'N \, d\phi/dt.$$

where:

K' = a constant of proportionality dependent on the properties of the core and the factor necessary to convert instantaneous to effective voltage

N = the number of turns of the winding.

By realizing that the core is common to both windings for the transformer of Exhibit 1-36, and that the flux linking both windings is the flux in the core, the equation may be rewritten for the voltages V_1 and V_2 as:

$$V = K'N \, d\phi/dt.$$

If a ratio is now taken of these two equations, the result is the relationship between the primary and secondary voltages of the transformer of Exhibit 1-36,

Equation 1-34

$$\frac{V_1}{V_2} = \frac{N_1}{N_2}$$

From the assumptions that the power factor of primary and secondary circuits is the same and that losses are negligible, the following relationship may be written:

$$V_1 I_1 = V_2 I_2$$

The relationship between the primary and secondary currents of the transformer is then:

$$N_1 I_1 = N_2 I_2$$

Three different sources of power losses are incurred in a transformer. One of them is due to the resistance in the windings of the transformer and is called **copper loss**. The other two are due to the core material and are called **hysteresis loss** and **eddy current loss**. The two losses due to the core are sometimes lumped together and given the common name **core losses**.

1.3.5.1 — Copper Losses

Copper losses are the power losses due to the resistance of the windings of the transformer. The copper loss of the primary winding equals $I_1^2 R_p$, where R_p is the resistance of the primary winding. The copper loss of the secondary winding equals $I_2^2 R_s$ where R_s is the resistance of the secondary winding.

Copper losses are reduced by choosing a material with a resistance as low as practicable, dependent on cost and reliability, and using a wire cross section as large as practicable, dependent on size and weight.

1.3.5.2 — Hysteresis Loss

Recall that, according to the magnetic domain theory, whenever a material is magnetized, its magnetic domains are brought into alignment. In a core material the magnetic field is changing, consequently the domains are constantly being turned. Since this constitutes work being done, a certain amount of power is consumed. The consumed power is called the **hysteresis loss**.

Hysteresis loss is kept to a minimum by using a core material in which the domains are aligned with relative ease and a small amount of power.

1.3.5.3 — Eddy Current Loss

A magnetic material that can be used as a core for a transformer is, by its nature, also a conductor and will be affected by the changing flux in the same manner that the windings on the core are affected. An emf will be induced, which will cause a current to flow in the core. This current is called an **eddy current** and is the cause of the eddy current losses. Eddy currents are circulating currents with a direction of motion that is at right angles to the direction of the magnetic flux.

Eddy current losses are reduced by using a core composed of laminated sheets of magnetic material that are bolted together instead of a solid core. The laminations offer high resistance to current flow; therefore, the eddy currents and the resulting losses are reduced.

Consider the transformer circuit of Exhibit 1-37 for which the applied voltage, secondary current, and load are known. Notice that the core is represented by the parallel lines between the windings. This is the manner in which transformers are normally represented schematically. The lines between the windings indicate that the two windings are linked magnetically by a ferromagnetic core.

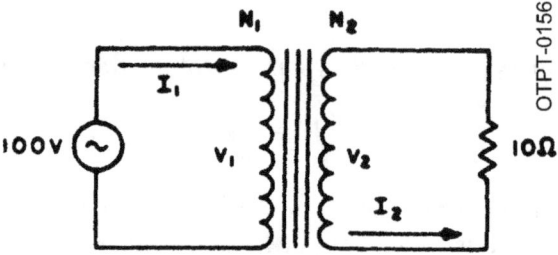

Exhibit 1-37 — Transformer Circuit

The secondary voltage may be found as follows (where, $I_2 = 2$ amps; $R_2 = 10$ ohms; and $V_1 = 100$ volts):

$$V_2 = I_2 R_2 = 2 \times 10 = 20 \text{ volts}.$$

The turns ratio of the transformer is found as follows:

$$\frac{N_1}{N_2} = \frac{V_1}{V_2} = \frac{100}{20} = 5.$$

The primary current of the transformer is found as follows:

$$N_1 I_1 = N_2 I_2$$

$$I_1 = \frac{N_2 I_2}{N_1}$$

$$I_1 = \frac{2}{5} = 0.4 \text{ amperes}.$$

The primary current could have also been found in the following manner:

$$V_1 I_1 = V_2 I_2$$

$$I_1 = \frac{V_2 I_2}{V_1}$$

$$I_1 = 20 \times \frac{2}{100} = 0.4 \text{ amperes}.$$

As can be seen, this particular transformer would be called a step-down transformer because the secondary voltage is less than the primary voltage. Power companies use transformers at both the generating and consuming ends of their power lines. At the generating ends they step the voltage up and the current down to transmit power through the lines at a low value of current and thereby reduce the I^2R loss of the lines. At the consuming end of the line a transformer is used to step the voltage down to a usable value such as 220 or 120 volts. Transformers are also used to obtain either a higher or lower value of a-c voltage than that which is available from generating equipment. Another use is simply to isolate one circuit from another.

Section 1.4 — Generators

1.4.1 — A-C Generators

A simple a-c generator would consist of a pair of fixed magnets with a loop of wire (armature) rotating in the air gap as in Exhibit 1-38.

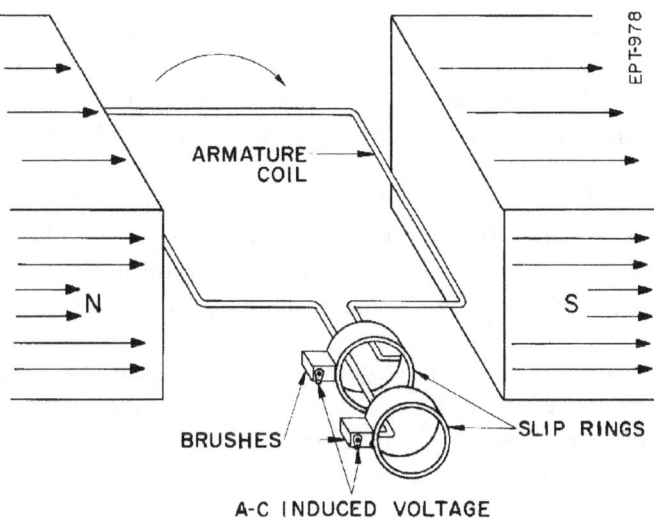

Exhibit 1-38 — Schematic Representation Of A Rotating-Armature A-C Generator

As the armature coil (mounted on a rotor) alternately intercepts the magnetic field lines, an a-c voltage is produced on slip rings, which rotate with the armature, and can be picked up on stationary brushes. The slip rings are simply a means of connecting the rotating coil to stationary terminals. The current-carrying capacity of the slip rings and brushes is a severe limitation in the construction of a generator. Since the field current is much less than the armature current, practical a-c generators place the field winding on the rotor and the armature winding on the stator. This overcomes the limitations of the slip rings and brushes (Exhibit 1-39) and avoids the need for high currents through the brushes. Note that d-c generators require that the armature be wound on the rotor in order to provide commutation.

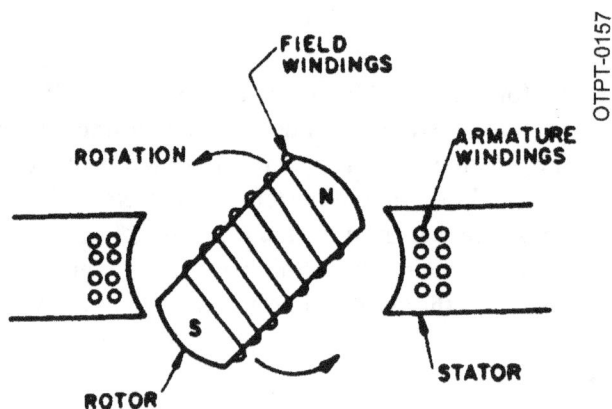

Exhibit 1-39 — Field And Armature Windings In A-C Generator

The rotating field winding is excited by direct current supplied through slip rings and brushes. The effect of the rotating field is the same as if a bar magnet was connected to the generator shaft. Rotating the field past the stationary armature winding causes a voltage to be induced in the armature. Recall that the requirement for a Faraday induced voltage is relative motion between the field and the conductor. A simple rotating field a-c generator is shown in Exhibit 1-39. If the pole pieces are properly shaped, the flux cutting the armature winding will vary sinusoidally from a maximum value to zero as the rotor is driven by a prime mover at constant speed.

One complete cycle is generated for each revolution of the two-pole rotor. Thus, the frequency of the induced voltage in cycles per second is **synchronized** with the mechanical speed and this is the reason for the designation **synchronous** generator. A two-pole synchronous generator must revolve at 60 revolutions per second × 60 seconds per minute = 3600 revolutions per minute (rpm) to produce a 60-hertz voltage.

If more poles are added to the rotor, the same frequency may be obtained at a lower speed. Exhibit 1-40 illustrates a four-pole synchronous generator. For each rotor revolution there will be two a-c cycles generated in the armature coil. Thus, a 60-hertz voltage will require only 30 revolutions per second or 1800 rpm. This relationship between frequency and speed for the synchronous generator may be generalized to:

Equation 1-35

$$f = \frac{PN}{120}$$

where:

   P   = number of poles

   N   = rotor speed, RPM

   f   = frequency of the induced voltage, hertz.

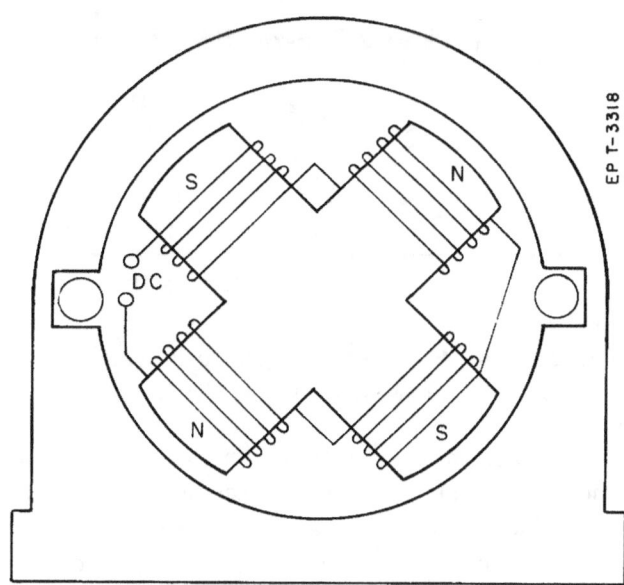

Exhibit 1-40 — Four-Pole, Three-Phase A-C Generator

As discussed previously, a considerable advantage is obtained by constructing three-phase a-c generators. Although a three-phase generator could consist of three separate a-c generators whose voltages are 120 degrees out of phase, these three generators can be easily constructed within a single generator housing as shown in Exhibit 1-41. Note that the armature consists of three sets of windings, called <u>phase groups</u>, which are equally spaced around the stator. The rotor shown is the same as that used in the single-phase a-c generator (Exhibit 1-39), containing one north pole and one south pole. Since the rotor contains two poles, the machine is designated as a two-pole generator. As the rotor is turned by the prime mover, it sets up a rotating magnetic field which cuts the armature windings at equal intervals. For every complete rotor revolution, a complete a-c cycle is induced in each armature winding. Thus, the speed of the rotor and the frequency of the a-c induced voltages are directly related by Equation 1-35 . Since there are three sets of armature windings, there are three separate sinusoidal a-c voltages. Due to the physical placement of the armature, windings will be 120 degrees out of phase. Thus, the generator shown in Exhibit 1-41 yields three phase voltages of equal frequency and magnitude and exactly 120 degrees apart in phase. The armature windings may be connected in either the wye or delta configuration.

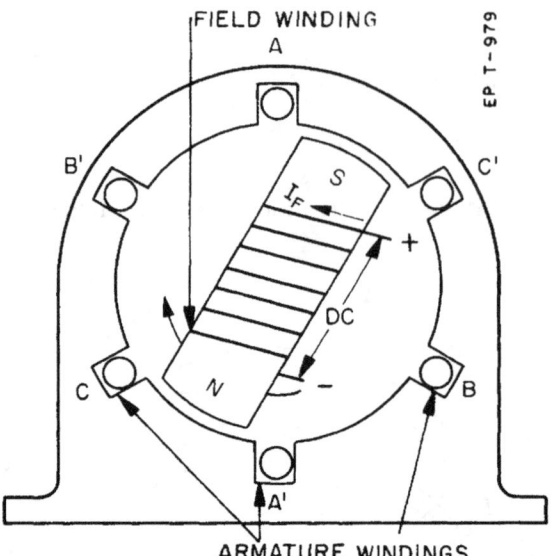

Exhibit 1-41 — Three-Phase, Rotating Field, Two-Pole, A-C Generator Showing Armature Phase Groups AA', BB', And CC'

A four-pole synchronous three-phase generator is illustrated in Exhibit 1-42. Note that the armature has two windings per phase. Since each winding will always see the same magnetic flux, the voltage induced in each coil will be the same. The windings of each phase are connected in series to increase the voltage output. The phases may be connected in either the wye or delta configuration. Connection diagrams for the generator of Exhibit 1-42 are illustrated in Exhibit 1-43.

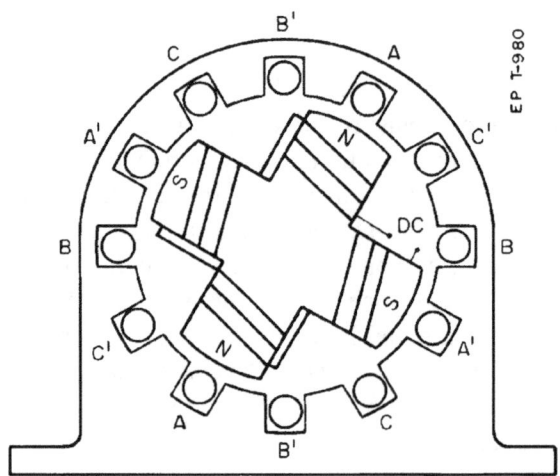

Exhibit 1-42 — A Three-Phase, Rotating-Field, Four-Pole A-C Generator Showing the Armature Phase Groups AA', BB', And CC' On The Stator

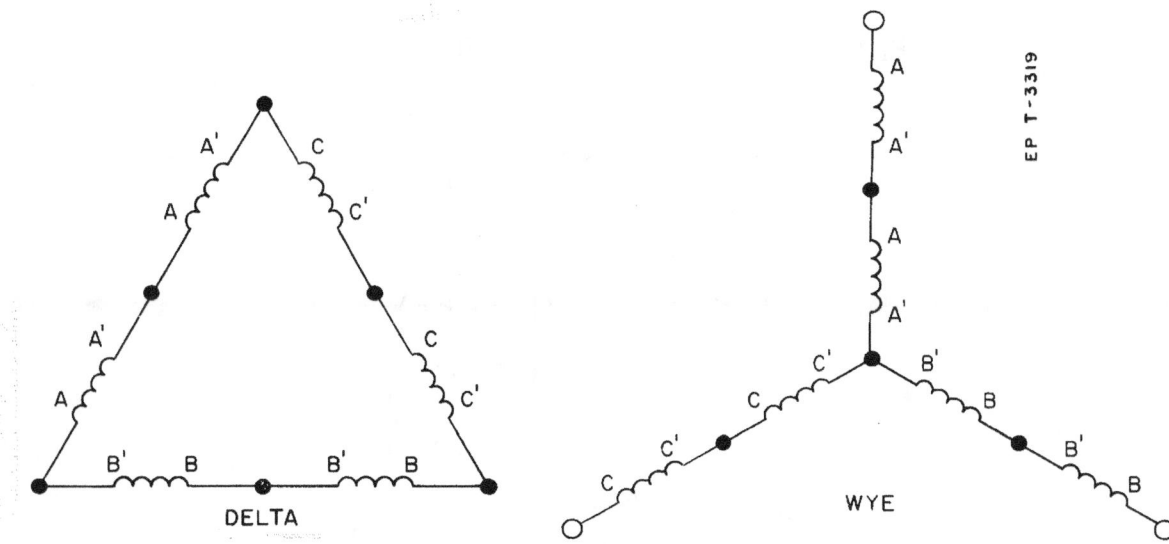

Exhibit 1-43 — Connection Diagrams For The Generator Of Exhibit 1-42

The advantage of an increased number of poles is that the rotor speed can be decreased for a given output frequency. Individual phases of a three-phase generator are generally connected in a delta configuration. This type of connection maintains phase-to-phase voltage relationships in case of failure of one phase.

1.4.2 — Armature Reaction

Armature reaction is the distortion of the field flux by the flux resulting from the flow of load current in the armature winding. Unlike d-c generators, the armature flux in a-c generators is not compensated for and its effect depends upon load phase angle as well as the magnitude of the load current.

The armature flux produced by a balanced three-phase current flowing in the armature windings may be represented by a single vector of constant magnitude rotating at synchronous speed. The resultant air gap flux, $\vec{\phi}_r$, will be the vector sum of the rotating armature flux, $\vec{\phi}_a$, and the rotating field flux, $\vec{\phi}_f$,

Equation 1-36

$$\vec{\theta}_r = \vec{\theta}_a + \vec{\theta}_f$$

The vector addition of Equation 1-36 is illustrated in Exhibit 1-44 where all vectors are revolving at synchronous speed N. Exhibit 1-44 is drawn for the case of a unity power factor load in which case the armature flux lags the field flux by exactly 90 degrees. For a fixed value of excitation, and consequently, field flux, the magnitude of the resultant air-gap flux will be proportional to the magnitude of the armature flux and hence the armature current. The magnitude of the resultant air-gap flux is also dependent upon the phase angle of the load, that is, the angle between the terminal voltage and the armature current. If the load phase angle is lagging, the angle between $\vec{\Phi}_f$ and $\vec{\Phi}_a$ will increase to 90 + θ degrees, where θ is the phase angle of the load. From Exhibit 1-45 it can be seen that as the load becomes more lagging the resultant air-gap flux is decreased.

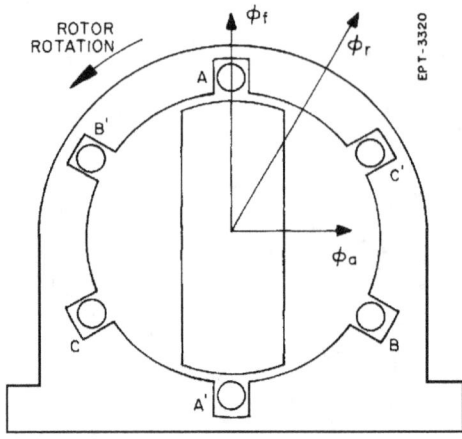

Exhibit 1-44 — Effect Of Armature Reaction For A Unity Power Factor Load

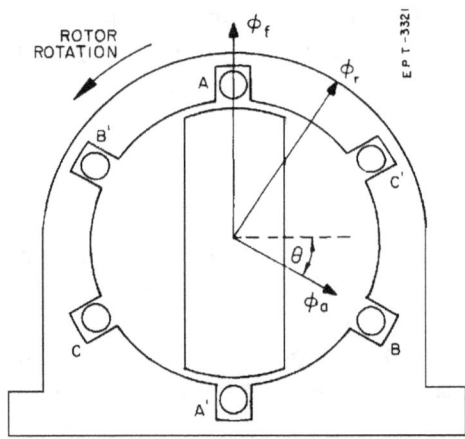

Exhibit 1-45 — Effect Of Armature Reaction For A Lagging Power Factor Load

Relative to the armature windings, the fluxes ($\vec{\Phi}_a$), ($\vec{\Phi}_r$), and ($\vec{\Phi}_f$) manifest themselves as induced voltages, lagging the flux that produced them by 90 degrees. Thus, Equation 1-36 may be rewritten in terms of induced voltages as:

Equation 1-37

$$\vec{E}_r = \vec{E}_a + \vec{E}_g$$

where:

$\vec{\mathbf{E}}_r$ = generated voltage

$\vec{\mathbf{E}}_a$ = armature reaction voltage

$\vec{\mathbf{E}}_g$ = no-load generated voltage.

In terms of generated voltage, the effect of armature reaction is to decrease the value of the generated voltage from its no-load value as the load becomes more inductive, that is, θ becomes greater and $\vec{\mathbf{E}}_a$ lags further behind $\vec{\mathbf{E}}_g$.

For a lagging load (generally the case in nuclear plants because of the large number of inductive loads), E_r decreases in magnitude with increasing load inductance. Since output voltage is proportional to E_r the net effect is a decrease in output voltage with increasing load inductance. In applications, this voltage "droop" (Exhibit 1-46) is corrected to a linear dependence by means of a voltage regulator which controls the field excitation.

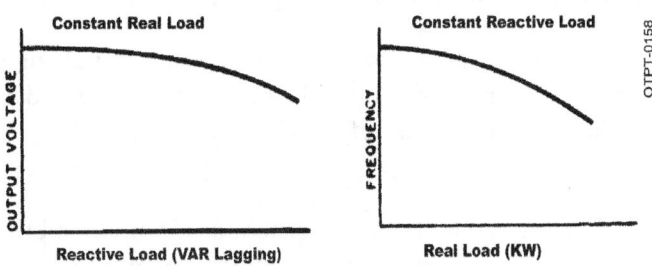

Exhibit 1-46 — A-C Generator Droop Characteristics

For constant reactive power, an a-c generator will reduce its speed as real load is increased. This dependence is a property of the prime mover (turbine or motor) rather than of the generator itself. The speed dependence is manifested to the operator as a reduction in a-c frequency with increasing real load. For a turbine-driven generator such as the SSTG, this droop is corrected to a straight line by a feedback controller which opens the turbine throttle to correct frequency.

1.4.3 — A-C Generator Operating Characteristics

The following quantities are used in evaluating generator operating characteristics. Note that this discussion is still on a per-phase basis and to obtain actual machine ratings the proper three-phase relations must be applied.

$$\text{Real power (per phase)} = P = V_t I_a \cos \theta$$

$$\text{Reactive power (per phase)} = Q = V_t I_a \sin \theta$$

$$\text{Apparent power (per phase)} = P_A = V_t I_a$$

In a system with several interconnected generators operating at a constant real power, the terminal voltage of an individual generator is a function of the reactive load supplied by that generator. A plot of the terminal voltage as a function of reactive load with constant real power and constant excitation is shown in Exhibit 1-47. This drooping characteristic is usually corrected to a linear droop by adding voltage regulators to the generator. Thus, in a regulated generator, adjustments in excitation are simply changes in voltage regulator settings.

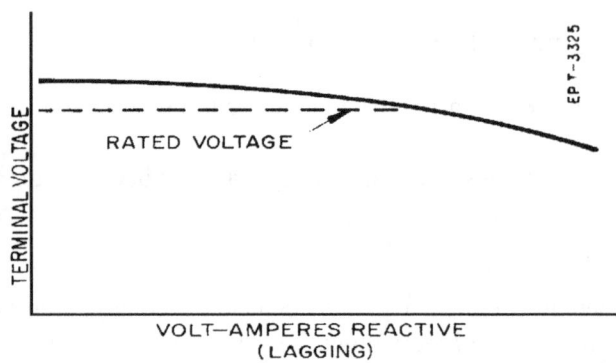

Exhibit 1-47 — Plot Of Terminal Voltage Versus Reactive Power For An A-C Generator With Constant Excitation And Supplying Constant Real Power

Now consider the case of constant reactive power to examine the effects of changes in true power load. It is a well known fact that when the load on most machines is increased the speed decreases. The same rule is true of generators, although the effect is a result of the characteristics of the prime mover, rather than of the generator itself.

When the electrical load on a generator is increased, the generator is compelled to produce more power. Since the output power of a generator is obtained from the prime mover, the prime mover is required to produce more shaft power. In the case of a steam turbine, an increase in the output power is accompanied by a speed decrease unless the steam throttle setting is changed.

For a prime mover operating at a constant throttle setting, the variation between speed and output power contains a slight droop as shown in Exhibit 1-48. The three different lines represent three different no-load turbine speeds that correspond to three different throttle settings. The higher curves represent throttle settings which admit more steam. Note that the lines for different throttle settings are parallel.

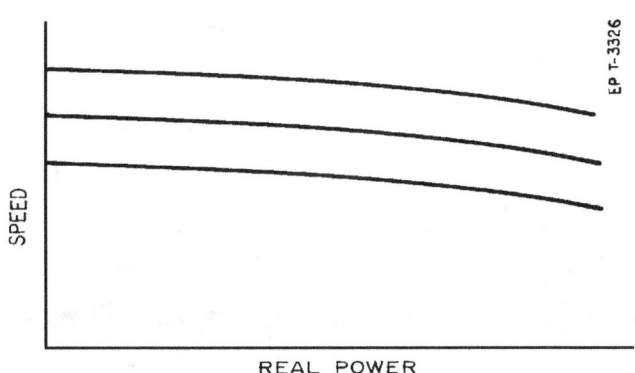

Exhibit 1-48 — Speed Versus Output Power For An Unregulated Steam Turbine At Three Different No-Load Speeds

The curves of Exhibit 1-48 would also apply to the generator connected to the turbine since their speeds are the same and the generator output differs from the turbine output only by the small percentage which represents the losses. However, as was seen previously, the speed of an a-c generator is directly linked to the frequency of the induced voltage. From an electrical operator's viewpoint, the frequency is of greater interest than speed so a plot of frequency versus output power for the generator is of greater value. Curves for different prime mover throttle settings are again shown in Exhibit 1-49.

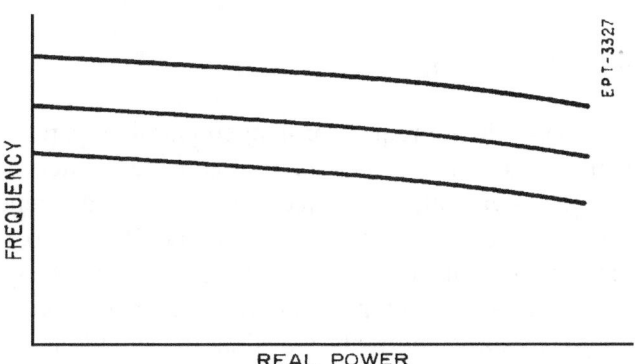

Exhibit 1-49 — Frequency Versus Output Power For A Typical A-C Generator With An Unregulated Prime Mover

By adding automatic speed regulators which adjust the steam throttle to change the turbine speed, a linear droop may be obtained. To fully understand a typical frequency versus output power curve for a speed-regulated generator, consider the solid line curve of Exhibit 1-50 which shows the relationship for a single throttle setting. The point where the curve crosses the vertical axis corresponds to zero output power and is called the <u>no-load frequency</u>. This is the frequency of the a-c voltage when the generator is disconnected from the load while the prime mover throttle is held fixed. When the generator supplies a 100-kilowatt load, the system frequency will be 60.2 hertz; for a 200-kilowatt load, the frequency will be 60.0 hertz; and for a 300-kilowatt load, the system frequency will drop to 59.8 hertz. If it is desired to supply 300 kilowatts at a system frequency of 60.0 hertz, the throttle must be opened until the dotted curve is reached. At this new throttle setting, the generator will have a no-load frequency of 60.6 hertz. Note that the line corresponding to the new throttle setting is parallel to the old one.

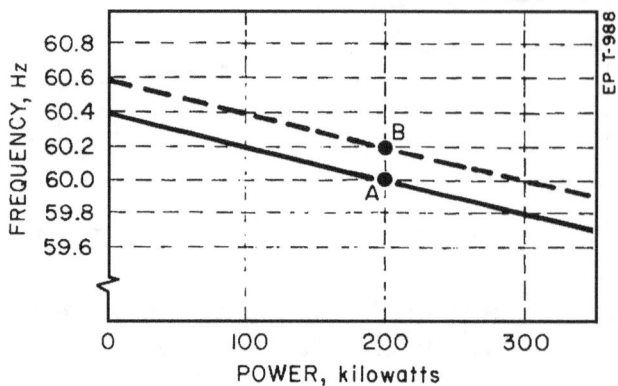

Exhibit 1-50 — Frequency Versus Output Power For A Typical A-C Generator Equipped With A Speed Regulator

If the load is constant, say at 200 kilowatts, the effect of increasing the throttle setting is to increase the system frequency as can be seen by comparing points A and B which represent a 200-kilowatt output at the two different throttle settings.

1.4.4 — Parallel Operation Of A-C Generators

Although normal operating procedures do not require prolonged parallel operation, procedures for placing two a-c generators in parallel and the certain system adjustments made to alter the load distribution, current distribution, system frequency, and system voltage are necessary to place the generators on line or to remove them without seriously disturbing the electrical system. For example, if the two generators being paralleled are not matched as nearly as possible in voltage and frequency, breakers may trip out on underfrequency, undervoltage, reverse power, or no-voltage. In addition, improper synchronization can cause damage to generators and breakers from high circulating currents. Consider the situation of, Exhibit 1-51 where generator G1 is initially supplying the load at rated voltage and frequency while generator G2 is disconnected.

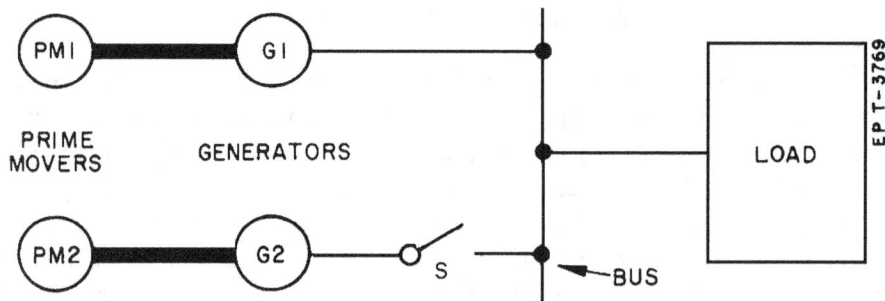

Exhibit 1-51 — Paralleling Synchronous Generators

Generator G2 may be paralleled with G1 by driving it at synchronous speed and adjusting its excitation so that the terminal voltages of the two generators are identical in all respects. This **synchronism** requirement entails four distinct conditions:
1.   Terminal voltages must be equal in magnitude.
2.   Terminal voltages must be in phase with each other.
3.   Terminal voltages must have the same frequency.
4.   Both generators must have the same phase sequence.

Switch S should be closed when the two voltages are momentarily in phase and the voltage across the switch is zero. The appropriate moment can be indicated by a synchroscope, such as that shown in Exhibit 1-52.

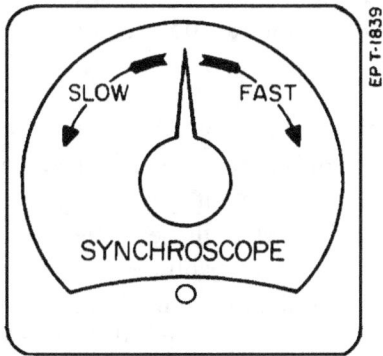

Exhibit 1-52 — Synchroscope

If switch S is closed when the instantaneous voltage values of the connected phases are not equal in magnitude, the resultant voltage difference causes extremely high currents to circulate between the machines, and may trip the circuit breaker and cause severe damage.

Equal voltages may be checked by using a voltmeter. A small three-phase induction motor may be used to check for proper phase sequence since the direction of rotation of the motor depends upon the phase sequence of the applied voltage. Opposite phase sequences may be corrected by interchanging any two power supply lines of the machine. The remaining conditions, equal frequency and phase relations are checked by using a synchroscope, as shown in Exhibit 1-52.

The synchroscope indicates the instantaneous angle of phase displacement between two voltage waves. If the two voltage waves have different frequencies, the angle between the two waves will always be changing causing the synchroscope pointer to revolve. When properly connected, the synchroscope indicates the frequency of the incoming machine with respect to the bus. If the synchroscope revolves in the direction marked SLOW, the incoming machine has a lower frequency than that of the bus. Correspondingly, revolution in the direction marked FAST indicates that the incoming machine has a higher frequency than that of the bus. If the frequency of the incoming machine is equal to the bus frequency, the pointer will not revolve and its fixed position will indicate the phase angle between the incoming machine and the bus in electrical degrees.

The following procedure is used for paralleling one a-c generator with another a-c generator or shore power:
1.   Place the synchronizing selector switch in the desired position.

2.   Set the incoming voltage equal to the running voltage.

3.   Set the incoming frequency equal to the running frequency, then fine adjust to make the synchroscope pointer rotate slowly (one turn every 6 seconds or more) in the FAST direction. Having the incoming frequency slightly higher ensures that the incoming generator will pick up real load.

4.   Close the output breaker of the oncoming generator when the synchroscope pointer is approximately at the "ten o'clock" position. This criterion allows time for operator response and time delay in the breaker mechanism.

5.   Disconnect the synchroscope to avoid overheating its coils.

With two generators operating in parallel, adjustments must be made to balance the real and reactive loads carried by the machines. The real load carried by each generator can be determined by examining the frequency versus real power curves discussed in the previous section. The real load carried by a generator is changed by adjusting its speed regulator setting.

With two generators operating in parallel, there is no guarantee that the speed regulator settings of both are the same. This means that they may operate at different no-load frequencies as shown in Exhibit 1-53. Note that values measured along the horizontal power axis in <u>both directions</u> from the origin are <u>positive</u>. Those to left of the frequency axis are the power outputs of G_2 while those to the right are the power outputs of G_1. A typical operating condition is represented by the load line for 60 hertz. The total length of this line, 300 kilowatts, represents the total load power. The portion of the line to the left of the vertical axis represents the output power of G_2, 200 kilowatts, and the portion of the line to the right of the vertical axis represents the output power of G_1, 100 kilowatts. The operating frequency of the generators must be the same since they are connected in parallel. By studying Exhibit 1-53, it should be apparent that if the load is increased to 500 kilowatts without a change in the speed regulator settings, the system frequency will drop to 59.8 hertz and G_1 will carry 200 kilowatts, while G_2 carries 300 kilowatts. On the other hand, if the total load remains at 300 kilowatts and G_1 suddenly trips off the line, G_2 will carry the full 300-kilowatt load, but at a system frequency of only 59.8 hertz.

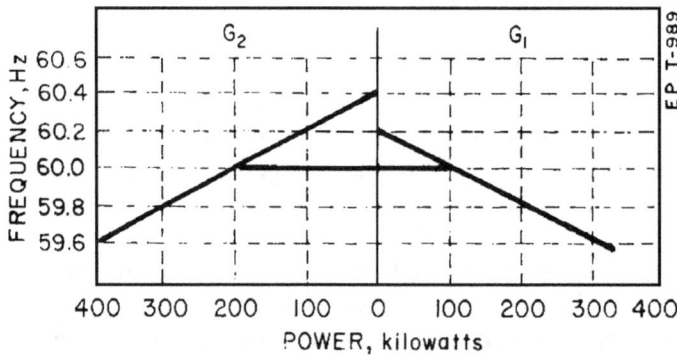

Exhibit 1-53 — Two Generators Operating In Parallel At Different No-Load Frequencies

Returning to the original condition described by Exhibit 1-53, that is, G_1 carries 100 kilowatts, G_2 carries 200 kilowatts, and the frequency is 60 hertz; it can be seen that if the kilowatt load is to be evenly distributed between the two generators at 60 hertz, that is, G_1 carries 150 kilowatts, G_2 carries 150 kilowatts, and the frequency is 60 hertz; the speed regulator setting of G_1 must be increased while the speed regulator setting of G_2 must be decreased until the no-load frequency for both generators is 60.3 hertz as shown in Exhibit 1-54.

This will cause a redistribution of the kilowatt load without varying either the system frequency or the speed of either generator. If both speed regulators had been raised or lowered together, the result would have been (Exhibit 1-55) a change in system frequency without altering the kilowatt load distribution between G_1 and G_2.

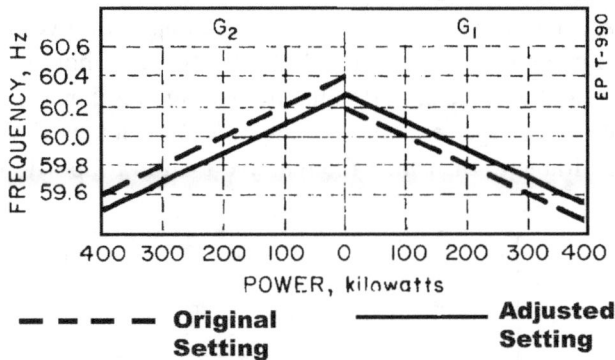

Exhibit 1-54 — Kilowatt Load Balancing

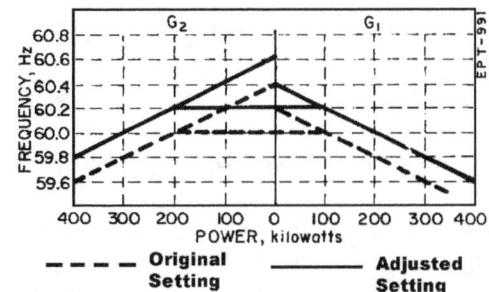

Exhibit 1-55 — System Frequency Adjusting Without Changing Load Distribution

When similar a-c generators are operating in parallel for any appreciable length of time, it is good operating procedure to distribute the kilowatt load in direct proportion to the generator ratings. This will ease the operating stresses on the generators and prime movers and minimize the possibility of severe system disturbances if one of the generators unexpectedly trips off the line. For example, if G1 was rated at 500 kilowatts and G2 at 1500 kilowatts, a total load of 1000 kilowatts would be distributed such that G1 carries 250 kilowatts while G2 carries 750 kilowatts.

When removing a generator from service, the load on that generator is first minimized by lowering its speed regulator setting while increasing the speed regulator setting on the other generator to maintain a constant system frequency.

A similar procedure is used to balance the reactive load. With two generators operating in parallel and supplying a constant real power, the reactive load carried by each generator can be determined by examining the terminal voltage versus reactive power characteristics. Different curves on a terminal voltage-reactive power characteristic represent different values of field excitation. For a fixed value of field excitation, the terminal voltage exhibits a drooping characteristic. Practical generators incorporate automatic voltage regulators which act on the generators' field circuitry to correct the terminal voltage-reactive power characteristic to a linear droop. Exhibit 1-56 illustrates the redistribution of reactive load. As with the frequency-power characteristic, values measured along the horizontal power axes in both directions from the origin are positive. Since paralleled generators must operate at the same voltage, the reactive load carried by each generator is found by the intersection of the generator characteristic for the particular voltage regulator setting in use and the operating voltage. For example, for the settings shown in Exhibit 1-56 at an operating voltage of 450 volts, generator G1 initially carries 600 KVAR and generator G2 carries 200 KVAR (curves shown dotted). To balance the reactive load while maintaining the terminal voltage at 450 volts, the voltage regulator setting on G1 is decreased so that G1 carries 400 KVAR and the voltage regulator setting on G2 is increased so that G2 carries 400 KVAR. As in the real load case, it is desirable to distribute the reactive load in proportion to the generator ratings.

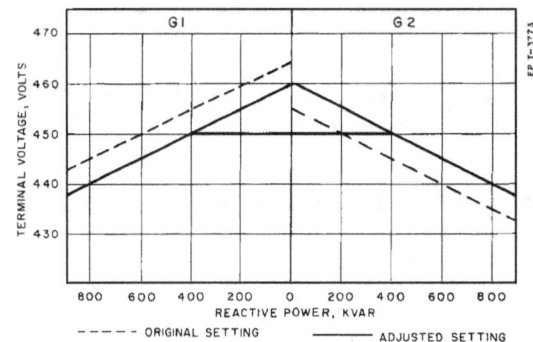

Exhibit 1-56 — Reactive Load Is Balanced By Adjusting Voltage Regulator Settings

To ensure system stability, the no-load settings on paralleled generators should be identical. No-load settings are the result of calibrating the voltage control rheostat of each generator for different generator terminal voltages between 440 and 470 volts in 5-volt increments with the generator breaker open. If the no-load settings of paralleled generators are different, it is possible that a reduction in reactive load can cause the output voltage of one generator to exceed the no-load voltage of the other generator. This causes the generator whose no-load voltage is exceeded to become a reactive load itself, supplied by the other generator. Becoming a reactive load means the generator is operating at a leading power factor. For a generator operating at a leading power factor, in order for the regulator to maintain an essentially constant voltage, the generator excitation must be reduced. If the generator's power factor becomes too leading, the field excitation will be reduced to a point where the generator can no longer carry the real load and the generator will lose synchronism, causing excessive current and voltage. When both paralleled generators are set at identical no-load settings, it is impossible for either generator to go leading unless the entire system reactive load goes negative, which is an impossible situation due to the inductive nature of typical plant loads.

Adjustment of the reactive load on a generator supplying a constant real load results in a change in the current output of the generator. Although the real load between two generators may be evenly split in proportion with the generator ratings, there are many possible values for the generator currents, depending upon the individual power factors of the generators. When the reactive power is properly balanced in proportion with the generator ratings, all generators will operate at the same power factor which is equal to the power factor of the load. Such a situation is illustrated in Exhibit 1-57, where the real load is balanced in both cases (the generator ratings are assumed to be equal) and the reactive load is only balanced in case (B). The importance of balancing the reactive load can be understood by realizing that the heating effect in the generator windings and the distribution lines is proportional to the output current squared. Since the total real power supplied to the load remains constant, it can be shown that the minimum combined heating effect on both generators occurs when the reactive load is balanced. Note for the example in Exhibit 1-57, $(100)^2 + (188)^2 > (136.5)^2 + (136.5)^2$.

When a single generator is paralleled with a large power system, such as the SSTG being paralleled with shore power, the power output of the oncoming generator is minimal compared with the power output of the entire system. Hence, the system may be represented as a constant-frequency, constant-voltage power source. Exhibit 1-58 illustrates the distribution of real and reactive load of a single generator paralleled with a large power system. Clearly, the load on the single generator can be set by adjusting the appropriate regulators. The remainder of the load required will be supplied by the power system. It may be important to properly adjust the reactive load on the single generator so that the current-carrying capacity of the shore power cables is not exceeded in the case of the SSTG in parallel with shore power. It is possible to have the single generator supply negative reactive power, thereby decreasing the total reactive load, improving the load power factor, and requiring less current for the same total real power.

	G1	G2	LOAD
V_T	450	450	450
KW	42.3	42.3	84.6
KVAR	15.3	73.35	88.65
KVA	45	84.6	122.85
pf	0.94 LAG	0.50 LAG	0.69 LAG
I	100	188	273

	G1	G2	LOAD
V_T	450	450	450
KW	42.3	42.3	84.6
KVAR	44.325	44.325	88.65
KVA	61.425	61.425	122.85
pf	0.69 LAG	0.69 LAG	0.69 LAG
I	136.5	136.5	273

EPT-3774

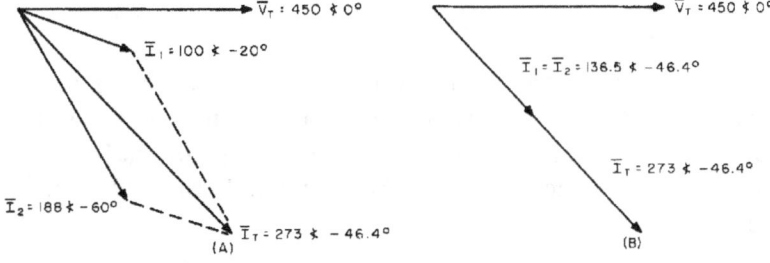

Exhibit 1-57 — Relationship Between Current Distribution And Reactive Power Distribution For Two Parallel-Connected Generators

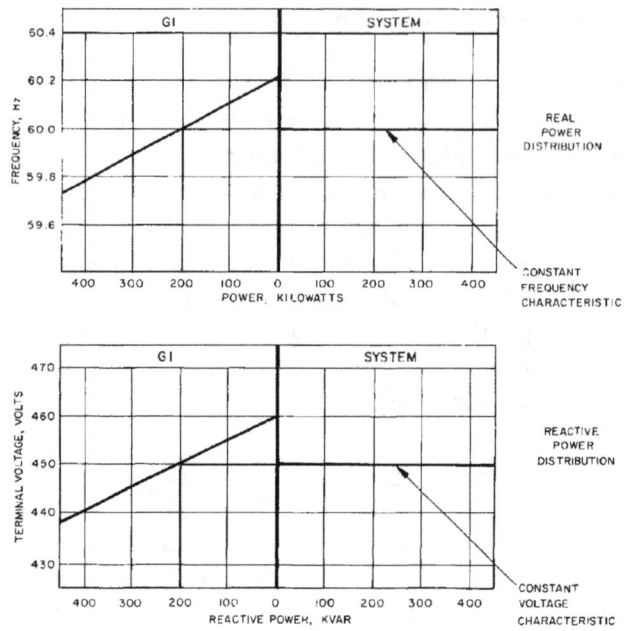

Exhibit 1-58 — Parallel Operation Of A Single Generator With A Large Power System

1.4.5 — D-C Generators

The field windings of a d-c generator, located on the stator, generate a fixed and nearly uniform magnetic field through which the rotor rotates. As the armature windings, located on the rotor, pass through the magnetic field an a-c voltage is induced in each armature winding. The resulting current is effectively converted to direct current and carried to the load by means of the commutator and brushes (Exhibit 1-59). The resulting (idealized) waveforms are shown in Exhibit 1-60.

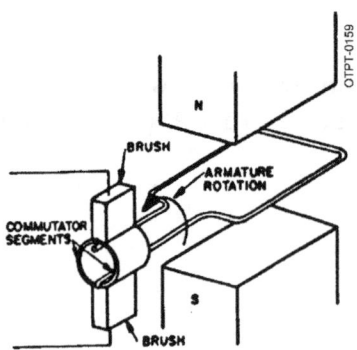

Exhibit 1-59 — D-C Generator Operation

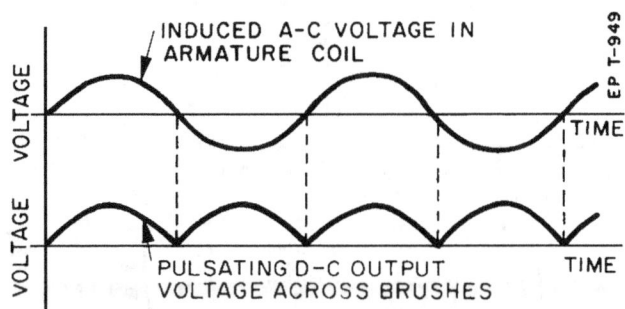

Exhibit 1-60 — D-C Generator Output

Various schemes are employed in a practical d-c generator to improve the quality of the d-c output. With properly shaped pole pieces, the flux can be made to remain uniform under each pole piece and then rapidly drop off to zero between poles. The resulting induced voltage in the armature is a flat-topped wave rather than the sine wave of Exhibit 1-60. The use of several sets of armature windings, equally spaced around the rotor, further smoothes the output by summing the voltages induced in each set of windings.

1.4.5.1 — Armature Reaction

When the generator is connected to a load, the current induced in the armature windings produces a magnetic flux $\vec{\phi}_A$ that is perpendicular to the field flux $\vec{\phi}_F$. These fluxes add vectorially to produce a distorted flux across the pole gap:

$$\vec{\phi}_T = \vec{\phi}_A + \vec{\phi}_F$$

This flux distortion has two significant effects on generator performance. First, the orientation of the "neutral plane," the plane in which an armature winding produces zero output voltage, is shifted from the no-load position. This results in sparking across the commutator-brush connection each time the armature connection is reversed. Second, the distorted flux across the pole gap causes a gradual reduction in output voltage with increasing generator load.

The commutation problem is usually solved by the addition of two small poles called "interpoles" in the no-load neutral plane. The armature current is passed through the interpole windings to produce a flux exactly equal to the armature field but opposite in direction. This technique assures that the neutral plane remains fixed and that commutation always occurs in the neutral plane regardless of load.

The distorted flux near the air gap between pole tips and rotor may be corrected, if desired, by the addition of "compensating windings" connected in series with the armature windings. The compensating windings develop a flux which opposes the armature flux in the air gap. This compensating flux varies with load in the same manner as the armature reaction; thus, the field flux in the air gap remains uniform regardless of load. However, compensating windings add to the cost of a generator and are not usually employed.

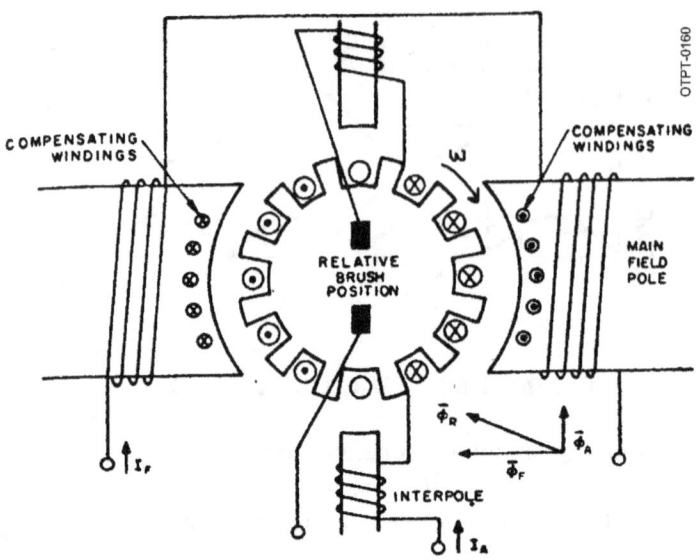

Exhibit 1-61 — Corrections For Armature Reaction

1.4.5.2 — Compounding Of D-C Generators

Compound generators employ a series field winding in addition to the shunt field winding. The series field coils are made of a relatively small number of turns of large copper conductor, either circular or rectangular in cross section, and are connected in series with the armature circuit. These coils contribute a magnetic field that influences the main field flux of the generator.

If the ampere-turns of the series field act in the same direction as those of the shunt field, the combined magnetic field is equal to the sum of the series and shunt field components. If load is added to a compound generator, armature current and thus, series field circuit current increases. The effect of the additive series field is to increase field flux with increasing load. The extent of the increased field flux depends on the degree of saturation of the field iron as determined by the shunt field current. Thus, the terminal voltage of the generator may increase with load or it may decrease depending upon the influence of the series field coils. This influence is referred to as the degree of compounding. A **flat-compounded** generator is one in which the no-load and full-load voltages have the same value. An **under-compounded** generator is one in which the full-load voltage is less than the no-load value. An **over-compounded** generator is one in which the full-load voltage is higher than the no-load value. The way the terminal voltage changes with increasing load depends upon the degree of compounding.

A variable shunt is connected across the series field coils to permit adjustment of the degree of compounding. This shunt is called a **diverter**. Decreasing the diverter resistance increases the amount of armature circuit current that is bypassed around the series field coils, thereby reducing the degree of compounding.

A field rheostat in the shunt field coil circuit permits adjustment of the no-load voltage of the compound generator. The diverter across the series field coils permits adjustment of the full-load voltage.

The variation of terminal voltage with load is indicated by the external characteristic curves of Exhibit 1-62. The no-load and full-load voltages for a flat-compounded generator are the same. Neither the diverter nor the field rheostat is altered in this test. The speed is maintained constant at the rated value. The hump in the curve is caused by the increased influence of the series ampere-turns on the field iron at half-load when the degree of saturation is reduced and the armature reaction and armature IR drop are approximately half their normal values.

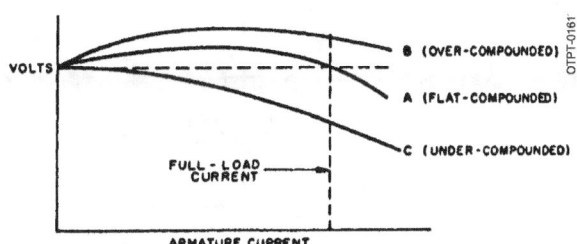

Exhibit 1-62 — External Characteristics Of Compounded Generators

As load is added to an over-compounded generator, its terminal voltage increases so that the full-load voltage is higher than the no-load value. Such a characteristic might be desirable where the generator is located some distance from the load and the rise in voltage compensates for the voltage loss in the feeder lines. This action holds the load voltage approximately constant from no-load to full-load by increasing the generator terminal voltage an amount that is just equal to the voltage drop in the feeder lines at full load.

The lowest curve represents the external characteristic for an undercompounded generator. The Navy uses a so-called stabilized shunt generator that has a small series field winding of a few turns in the coils to partially compensate for the voltage loss due to armature voltage drop and armature reaction. Thus, the external characteristic is almost the same as that of a shunt generator except that the terminal voltage does not fall off quite as rapidly with increased load.

Section 1.5 — Motors

Compound motors, like compound generators, have both a shunt and a series field. In most cases the series winding is connected as shown in Figure (a) of Exhibit 1-63. Motors of this type are called cumulatively compounded motors. If the series winding is connected so that its field opposes that of the shunt winding, as shown in Figure (b) of Exhibit 1-63, the motor is called a differentially compounded motor. Under full load, the ampere-turns of the shunt coil are greater than the ampere-turns of the series coil.

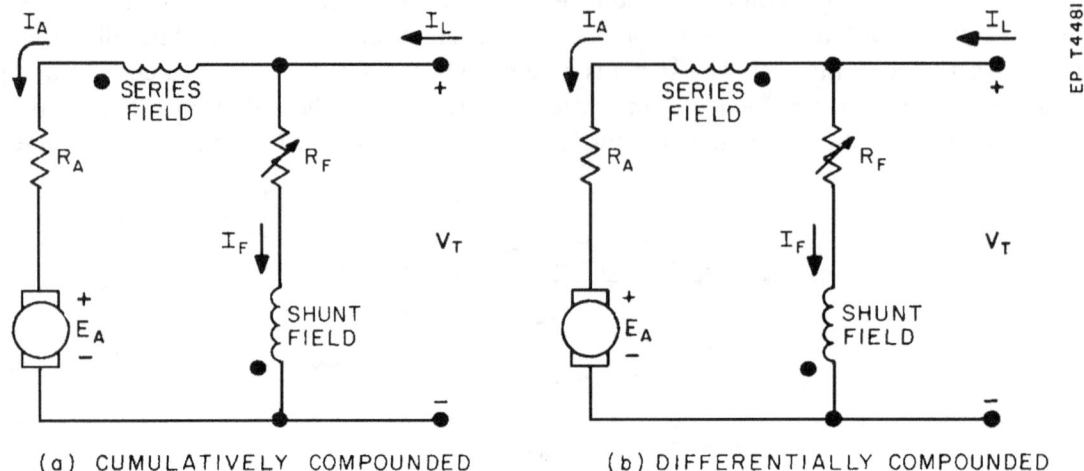

(a) CUMULATIVELY COMPOUNDED   (b) DIFFERENTIALLY COMPOUNDED

Exhibit 1-63 — Compounding Of D-C Motors

When a load is added to a cumulatively compounded motor, the speed decreases more rapidly than it does in a shunt motor, but less rapidly than in a series motor. Series field strength increases as in a series motor.

Equation 1-38

$$I_a = \frac{E - E_c}{R_a + R_s}$$

where I_a is armature current, E is applied voltage, E_c is counter emf, and R_a and R_s are armature and series field resistances, respectively.

To increase I_a, E_c must decrease. As in a series motor, the decrease in speed is necessary to allow the counter emf to decrease at the same time the field increases. Because the torque varies directly as the product of the armature current and the field flux, it is evident that the cumulatively compounded motor has a greater starting torque than does the shunt motor for equal values of armature current and shunt field strength. The performance of this type of motor approaches that of a shunt motor as the ratio of the ampere-turns of the shunt winding to the ampere-turns of the series winding becomes greater. The performance approaches that of a series motor as this ratio becomes smaller.

If the load is removed from this type of motor, it tends to speed up and the counter emf increases. The current in the series field is reduced and the greater portion of the field flux is produced by the shunt field coils. The compound motor then has characteristics similar to a shunt motor. Unlike the series motor, there is a definite no-load speed.

Because the total flux increases when there is an increase in load, there is a greater proportionate increase in torque than in armature current. Therefore, for a given torque increase, this type of motor requires less increase in armature current than the shunt motor, but more than the series motor.

In some operations it is desirable to use the cumulative series winding to obtain good starting torque; and when the motor comes up to speed, the series winding is shorted out. The motor then has the improved speed regulation of a shunt motor.

In a differentially compounded motor, because of the opposition of the series field to the shunt field, the flux decreases as the load and armature current increase. Because the speed is proportional to E_c/ϕ, if both factors vary in the same proportion, the speed will remain constant. This action may occur in the differentially compounded motor. If more turns are added to the series coil, it is possible to cause the motor to run faster as the load is increased.

More armature current is required of the differentially compounded motor than of the shunt motor for the same increase in torque. This results from the fact that an increase in armature and series-field current reduces the field flux. Since the torque acting on the armature is proportional to the product of armature current and field strength, a decrease in field strength must be accompanied by a disproportionate increase in armature current for the product to increase.

Under heavy load the speed of the differentially compounded motor is unstable; and if the overload current is very heavy, the direction of rotation may be reversed. Thereafter, the motor will run as a series motor with the danger of overspeed on no-load that is the inherent characteristic of all series motors.

The characteristics of the differentially compounded motor are somewhat similar to those of the shunt motor, only exaggerated in scope. Thus, the starting torque is very low, and the speed regulation is very good if the load is not excessive. However, because of some of the undesirable features, this type of motor does not have wide use.

Cumulatively compounded motors are used under conditions where large starting torque is necessary, where a relatively large change in speed can be tolerated, and where the load may be removed from the motor with safety. These motors are therefore used for such applications as driving hoists, punches, and shears.

1.5.1 — Induction Motors

A three-phase induction motor has a three-phase a-c current supplied to the stator windings. These stator windings are similar to those for a corresponding a-c generator. The supplied current results in a rotating magnetic field of constant magnitude in the air gap between the stator and the rotor. The stator field rotates at synchronous speed determined by the frequency and phase of the a-c current, and the number of poles in the stator winding.

Equation 1-39

$$N_S = \frac{120\, f_S}{P}$$

where:

N_S = Synchronous speed in RPM

f_S = Stator frequency in hertz

P = number of poles per phase on the stator.

The rotor winding of an induction motor is electrically closed with no external connections. The rotor may be one of two types: a squirrel-cage (shown in Exhibit 1-64) or a wound rotor. Because of its simplicity and because it meets a wider range of applications than any other motor, the squirrel-cage induction motor is the most common of all motors in use.

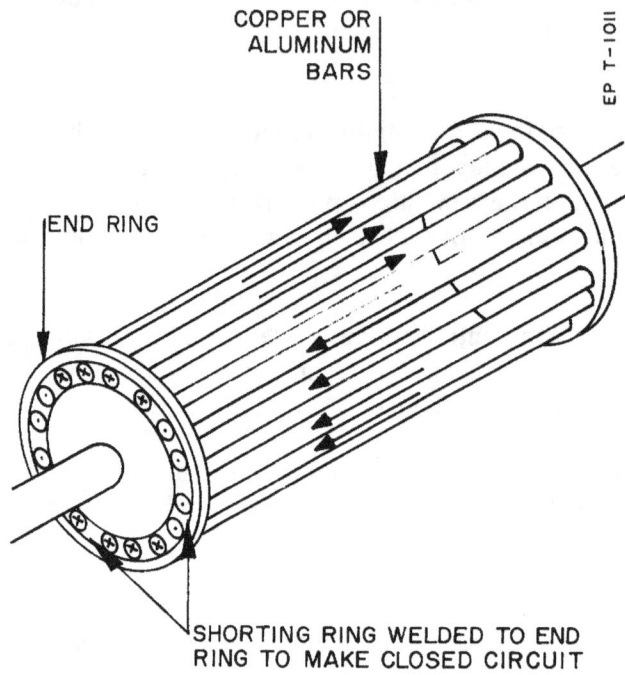

Exhibit 1-64 — Induction Motor Squirrel Cage Rotor

To understand the operation of the induction motor, consider the rotor at standstill. The function of the stator winding is to create a rotating magnetic field in the air gap as discussed above. The stator field rotates past the stationary rotor bars at synchronous speed N_S given by Equation 1-39 in the counterclockwise direction. This relative motion produces a Faraday induced voltage in the rotor bars whose polarity may be determined by applying the right-hand rule for generator action. Care must be taken in applying this rule because the motion of the magnetic field is equivalent to the motion of the rotor bars in the clockwise direction. Proper application of the hand-rule will yield the results shown in Exhibit 1-65, where the plus signs indicate the positive end of the induced voltage and the minus signs represent the negative end of the induced voltage. Since the rotor bars are all connected by the end resistance rings that complete their circuits, the rotationally induced voltages act in series addition to cause rotor currents to flow in the bars.

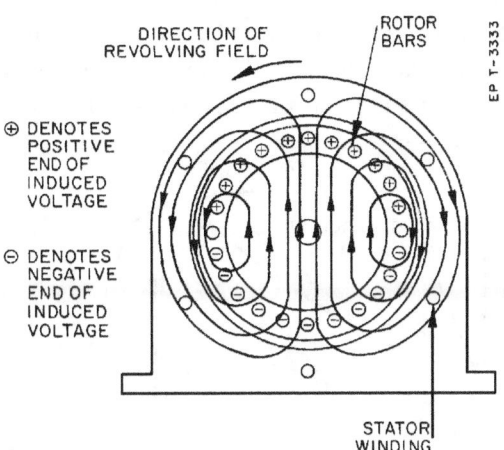

Exhibit 1-65 — Faraday Induced Voltage In The Induction Motor

The interaction between the magnetic stator flux and the current flowing in the rotor produces a force on the rotor bars. The right-hand rule for motor action may be applied to determine the direction of the forces on each rotor bar. If the starting torque developed exceeds the mechanical load torque the rotor will begin to turn.

When running, the rotor of the induction motor runs at a speed N_R, which is less than the speed of the rotating field N_S. This means that the rotating field will be moving past the rotor conductors and hence these conductors are cut by the rotating magnetic field. The relative speed between the rotating stator field and the rotor conductors is called the slip speed and is equal to the difference between the synchronous speed and the rotor speed.

Equation 1-40

$$\text{Slip speed} = N_S - N_R$$

The ratio of slip speed to synchronous speed is called the slip, s. When expressed as a percentage, it is called the percent slip, S.

Equation 1-41

$$s = \frac{\text{slip speed}}{\text{synchronous speed}} = \frac{N_S - N_R}{N_S}$$

$$S = \frac{N_S - N_R}{N_S} \times 100 \text{ percent}$$

The torque versus speed curve for an induction motor is given in Exhibit 1-66. When the motor is started, the rotor speed is near zero, the slip is near one, and in most induction motors, the inductive reactance is much greater than the resistance. As rotor speed increases, s will decrease and thus, the torque will increase because of decreasing rotor inductive reactance. Note that one rotor bar, located below the center line in Exhibit 1-67, has a current directed into the page while one rotor bar above the center line has a current directed out of the page. The torque produced by these bars cancels and; therefore, decreases the net torque on the rotor. Clearly as the rotor current is made more inductive the net rotor torque is reduced.

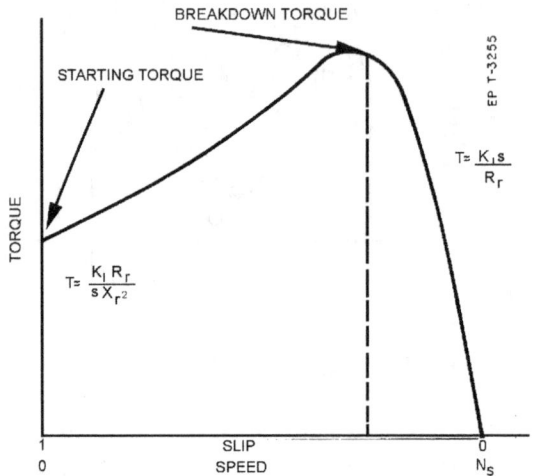

Exhibit 1-66 — Torque Versus Speed For Induction Motor

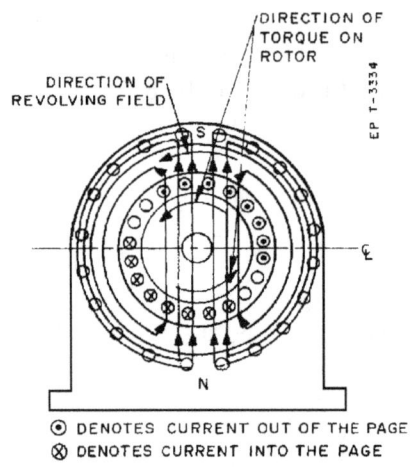

Exhibit 1-67 — Induced Currents In The Induction Motor At Standstill

The torque reaches its maximum value at what is called the **breakdown** or **pull-out torque** since it is the limiting torque which the motor can produce. This point and the value of the starting torque can be varied by changing the rotor resistance. Increased rotor resistance results in higher starting torque and a shift in pull-out torque to a lower rotor speed. The normal operating range of the induction motor is to the right of the pull-out torque point with slip values in the range of 3 to 10 percent. On this part of the curve, the motor is operating in a stable condition since any decrease in speed caused by an increase in load will increase the torque and cause the motor to pick up the new load at a lower speed. Once the pull-out torque is reached, however, the torque decreases with a speed decrease and the motor is unable to carry any increase in load.

Finally, the curve for induction motor torque is zero at synchronous speed, since the rotor "sees" a stationary stator field and there is no induced rotor voltage.

In the normal operating range of low percentage slip, the induction motor is essentially a constant speed motor. One method of speed control is to change stator coil connections so that the number of poles can be changed. The poles are generally changed in a ratio of 2 to 1 and the rotor is always a squirrel-cage type. Main coolant pumps on submarines employ pole changing speed control.

1.5.2 — Synchronous Motors

A synchronous motor is the same in construction as a three-phase generator. An a-c current supplied to the stator winding of a synchronous motor produces a rotating magnetic field the same as in an induction motor. A direct current is supplied to the rotor winding, thus producing a fixed polarity at each pole. If it could be assumed that the rotor had inertia and that no load of any kind were applied, then the rotor would revolve in step with the revolving field immediately upon application of power to both windings. This, however, is not the case; the rotor has inertia and, in addition, there is a load.

A synchronous motor has to be brought up to synchronous speed by special means. If the stator and rotor windings are energized, then as the poles of the rotating magnetic field approach rotor poles of opposite polarity, the attracting force tends to turn the rotor in the direction opposite to that of the rotating field. As the rotor starts in this direction, the rotating-field poles are leaving the rotor poles, and this tends to pull the rotor poles in the same direction as the rotating field. However, the rotor does not reach synchronous speed immediately because of its inertia: the rotating field passes it up and the rotor is again pulled in the reverse direction. Thus, the rotating field tends to pull the rotor poles first in one direction and then in the other, with the result that the average starting torque is zero.

The means usually employed for bringing a synchronous motor up to normal operating speed is to include a squirrel-cage or "damper" winding in the rotor pole heads. This allows the motor to develop starting torque by acting as an induction motor until it reaches nearly synchronous speed. The field windings are then excited (usually by a centrifugal switch) and a synchronous torque is developed. **A synchronous motor develops useful torque only at synchronous speed**. The power factor of a synchronous motor carrying a definite load may be unity or less than unity, either lagging or leading, depending on the d-c field strength.

The power factor for a synchronous motor is obtained as shown in Exhibit 1-68. The applied voltage ($\vec{E}_a$) is considered constant in magnitude because it is presumably obtained from a regulated source. The counter emf ($\vec{E}_c$) is nearly 180 degrees out of phase with ($\vec{E}_a$), but lags the 180 degree angle by α, where α is the torque angle of the rotor. The voltage drop ($\vec{E}_L$) across the rotor impedance must satisfy the relation

$$\vec{E}_a = \vec{E}_L - \vec{E}_c$$

where the minus sign accounts for the fact that the counter emf opposes the applied emf. Thus $\vec{E}_L$ is the vector sum of $\vec{E}_c$ and $\vec{E}_a$ as shown in the figure. The impedance of the armature is almost entirely inductive, so the armature current will lag $\vec{E}_L$ by 90 degrees, and lag $\vec{E}_a$ by the angle θ as shown. The power factor is equal to cos θ and in Exhibit 1-68 is lagging.

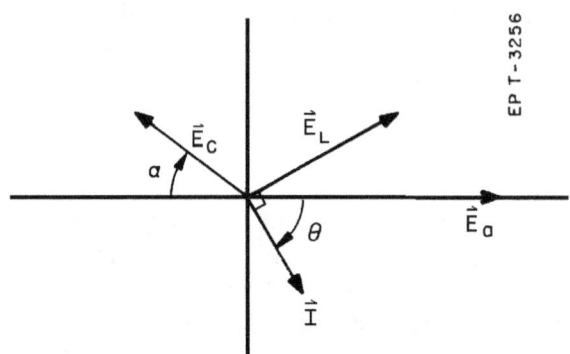

Exhibit 1-68 — Lagging Power Factor For Synchronous Motor

The power factor can be made less lagging, unity, or even leading by increasing the magnitude of $\vec{E}_a$ and/or the torque angle α. The magnitude of $\vec{E}_a$ is increased by increasing the field excitation current applied to the rotor. Exhibit 1-69 shows how unity or leading power factors are obtained. These figures are drawn for the same real load P = I E_a cosθ.

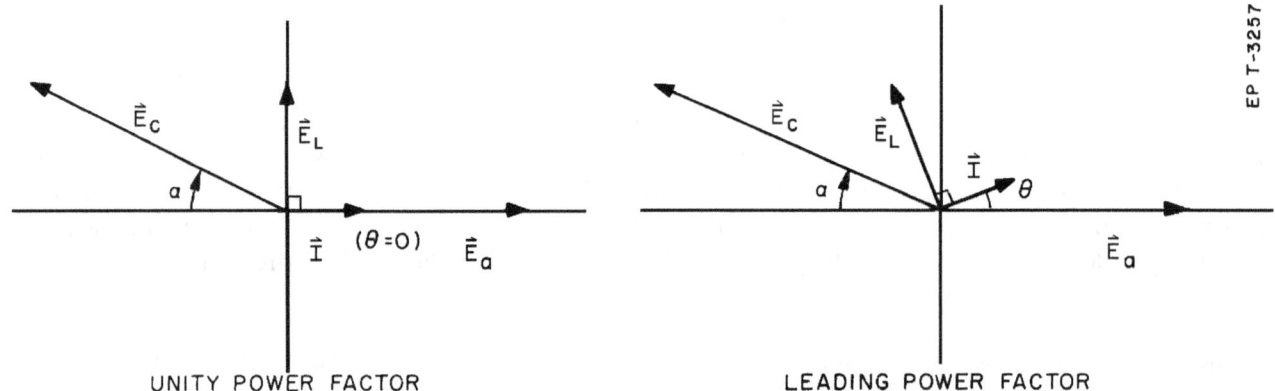

Exhibit 1-69 — Unity And Leading Power Factors For Synchronous Motor

Thus the power factor for a synchronous motor can be adjusted by means of the field excitation. A higher field excitation causes a more leading power factor. This occurs even for a constant load on the motor.

One important use of synchronous motors in nuclear submarines is the a-c end of the motor-generator sets. When a submarine is operating on shore power, the synchronous motor can be used to change the power factor of the system to which it is paralleled by adjusting the field excitation. If operated without load, its power factor may be adjusted to a value as low as 10 percent leading. When operated under this condition, the motor is generally referred to as a synchronous condenser because it draws a leading current in the same manner as capacitors. In this case the synchronous condenser requires only enough real power from the line to supply its losses. At the same time, it supplies a high leading reactive load, which offsets the lagging reactive power taken by the parallel inductive loads, and the system power factor is thereby improved.

1.5.3 — Reluctance Motors

The term **reluctance** refers to the ease with which a material passes magnetic flux. The reluctance of a particular object depends upon its composition and shape. The total reluctance of a magnetic path determines the amount of magnetic flux which is present for a given coil carrying a given current. Since a magnetic material has a much lower reluctance than a quantity of air having the same physical dimensions, the air gap between rotor and stator in electrical machinery is kept to a minimum so that the reluctance of the magnetic path will be minimal and the flux a maximum.

Magnetic lines of flux not only seek the path of lowest reluctance, but exert forces to minimize the reluctance whenever possible. This principle is the basis for the reluctance motor as illustrated in Exhibit 1-70. As the magnet is turned, the nail will turn with it to maintain a minimum air gap in the flux path. The rotating magnet sets up a rotating magnetic field which the nail tries to follow.

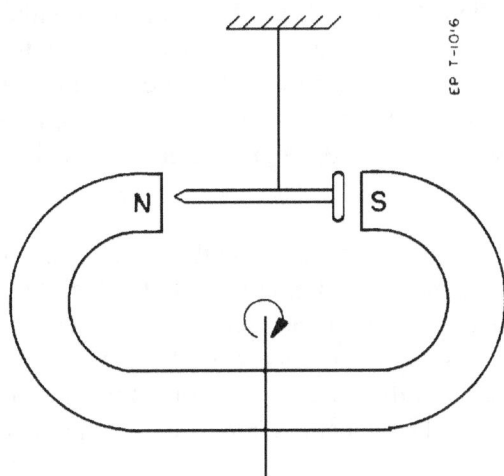

Exhibit 1-70 — Reluctance Motor Operating Principle

Exhibit 1-71 shows a reluctance motor under an operating condition. The flux flows through the rotor, but in so doing it traverses an air gap which is longer than the minimum. The result is a force on the rotor which tries to align the rotor with the magnetic field and thus minimize the reluctance of the magnetic path. The rotor axis lags the rotating field by a torque angle (δ). The magnitude of the torque depends on the size of this angle, increasing as the angle increases up to a maximum torque angle of 90 degrees. The rotor follows the field at synchronous speed.

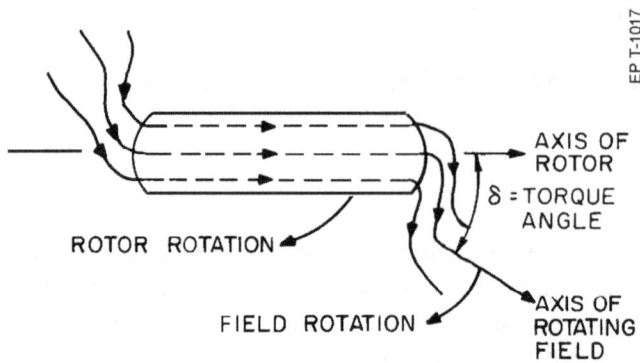

Exhibit 1-71 — Reluctance Motor Under An Operating Condition

The stator of a reluctance motor may be three-phase or single-phase. The rotor is usually similar to an induction-motor rotor. Anything which makes the reluctance of the air gap a function of angular position will produce reluctance torque when the rotor is revolving at synchronous speed. A common rotor for a reluctance motor is an ordinary squirrel-cage induction-motor rotor with some of the teeth removed. This motor will start as an induction motor and for light loads will speed up to a small value of slip. At a small slip, a reluctance torque is produced. This reluctance torque alternates slowly in direction; the rotor accelerates during positive half-cycles of torque variation and decelerates during negative half-cycles. If the rotor inertia and the mechanical load are small, the rotor will lock-in at synchronous speed during an acceleration half-cycle of reluctance torque. The rotor will then pull into step and continue to run at synchronous speed, producing a steady reluctance torque.

The reluctance motor produces less torque for its size than the synchronous motor. (Note that there is a small reluctance torque produced even in a synchronous motor due to the fact that practical rotors are not perfectly cylindrical.) The reluctance motor has a simpler construction than the synchronous motor since there are no electrical connections to the rotor. Reluctance motors are used extensively as small motors, such as electric clock motors, and they find application in the nuclear plant in control rod drive mechanisms, where the constant-speed property and no electrical rotor connections are essential features for the application.

1.5.4 — D-C Motor Operation

D-C motors are constructed in the same manner as d-c generators. They contain field windings on the stator and armature windings on the rotor. The armature windings are connected to the terminals through a commutator and brush combination. Since construction of d-c motors and d-c generators is the same, a d-c generator may function as a d-c motor if it is provided with an electrical power input. This reversibility of d-c machines is important in the operation of motor-generator sets.

Practical motors are rotating machines to provide for continuous motion. Torque is the name for the rotational effect produced when a force is exerted at some distance from the axis of rotation. The induced force on a current-carrying conductor supplies the torque for a motor. Thus, the torque developed in a motor is proportional to the product of the armature current and the magnetic flux:

$$T = K\phi I_A$$

where:

K = proportionality constant dependent upon machine construction.

ϕ = the field flux per pole pair

I_A = armature current

A simple d-c motor with one armature coil and a commutator with only two segments is shown in Exhibit 1-72. If current is sent through the coil as shown, the force of motor action will be directed upward on the left conductor and downward on the right conductor. These forces produce a torque (T) which will rotate the coil clockwise. When the coil reaches the vertical position and is perpendicular to the magnetic field, the force on the top conductor will be directed upward and the force on the bottom conductor will be directed downward, resulting in a net torque of zero. However, due to its inertia, the coil will continue to rotate beyond this position; but unless the direction of the current in the coil is reversed, the torque will then be counterclockwise and will force the coil back to the vertical position. Hence, a commutator is employed to change the direction of current through the armature coil each time the coil passes through the vertical position.

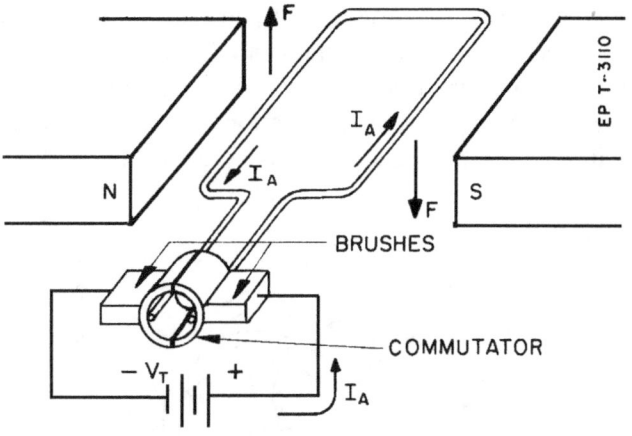

Exhibit 1-72 — Basic D-C Motor

Extra armature coils and multi-segmented commutators are used in practical d-c motors to produce both larger and steadier torques. As the rotor turns, the commutator-brush combination in the d-c motor switches connections to the armature coils in such a manner that all conductors on one side of the neutral plane carry current in the same direction. This is illustrated in Exhibit 1-73, where the directions of the induced forces are also illustrated. Since all conductors have forces exerted on them in the same direction, a unidirectional torque exists. The coils undergoing commutation are outside the main field and have no force exerted on them.

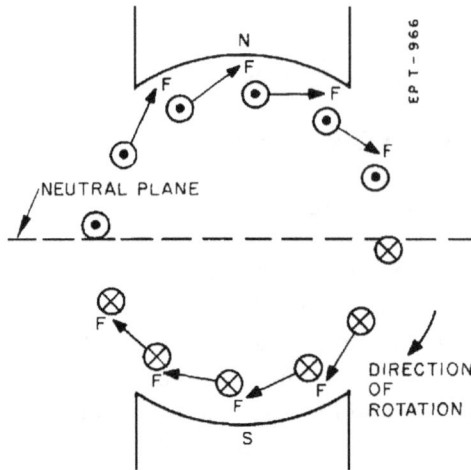

Exhibit 1-73 — D-C Motor

The action of the commutator can be understood by examining the more realistic armature winding shown end view in Exhibit 1-74. The commutator is represented by the ring of numbered segments in the center of the figure. The brushes are indicated by the black rectangles within the commutator. The connections of the coils to the commutator are shown by the circular arcs. The back end connections are shown dotted for the conductors in slots 1 and 7. The remaining end connections are not shown; however, they follow the same pattern. Each conductor in the bottom of a slot is connected to the conductor in the top of the slot diagrammatically opposite it.

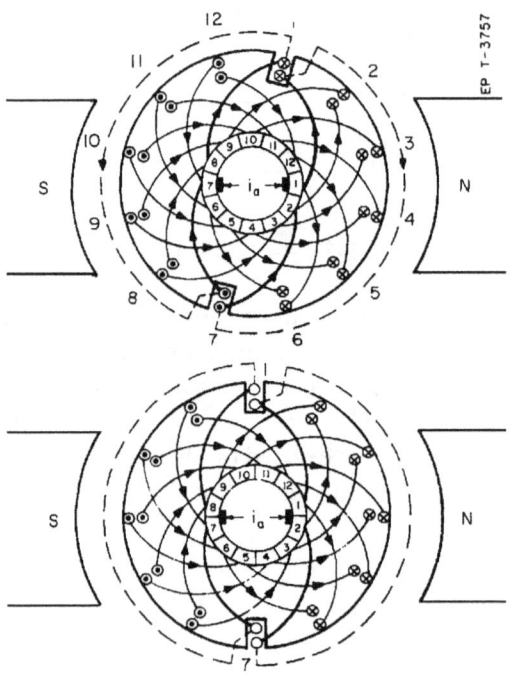

Exhibit 1-74 — Commutator Action In A D-C Motor

In the top diagram of Exhibit 1-74, the brushes are in contact with commutator segments 1 and 7. Current enters through the right-hand brush and divides equally between two parallel paths through the armature winding and out the left-hand brush. As the rotor revolves, the commutator shorts the coils in slots 1 and 7 at the instant the brushes span equal areas of the commutator segments, as shown in the bottom diagram of Exhibit 1-74. Further rotation of the rotor would remove the short. The brushes would now be in contact with commutator segments 2 and 8 and full armature current would be flowing in the reverse direction in the coils in slots 1 and 7.

The current flowing in the armature coils of a d-c motor produces a flux, ϕ_A, perpendicular to the flux, ϕ_F, produced by the field poles as shown in Exhibit 1-75. The resultant flux, ϕ_T, is shifted in direction, which causes a shift in direction of the neutral plane. This phenomenon, called **armature reaction**, is similar to the armature reaction in a d-c generator. Such a shift would require a repositioning of the brushes in response to load variations if steps were not taken to overcome the effects of the armature current.

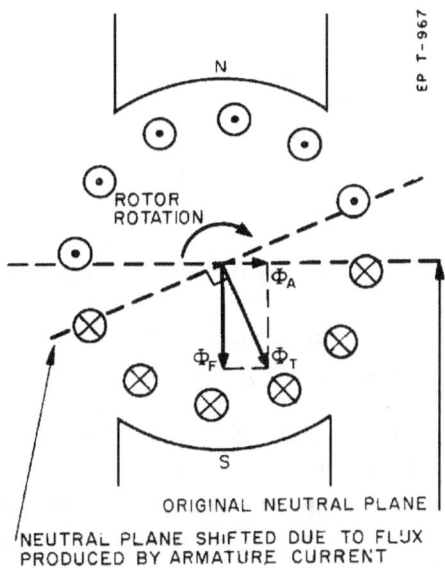

Exhibit 1-75 — Rotation Of Neutral Plane

Interpoles or compensating windings are used to ensure that commutation, and hence current reversal in the armature windings, takes place in the neutral plane for all values of armature current. These function in a manner similar to that in d-c generators by canceling the armature current flux and hence restoring the net field direction to be that of the field current flux, ϕ_F, only.

As the rotor turns under the influence of the induced forces, the armature conductors cut the magnetic flux lines, and a voltage (the counter emf discussed in Paragraph 1.1.7) is induced in the armature coils. By applying the right-hand rule to each of the armature conductors in Exhibit 1-75, it is seen that the polarity of the counter emf opposes the flow of armature current. The counter emf is the same induced voltage given by ($E_g = K\phi N$) which stated that its magnitude depends on the motor speed and the magnetic flux. The product of the armature current and the counter emf represents the mechanical power produced in the rotor by the electric current.

$$P_{MECH} = I_A E_g$$

The armature equivalent circuit for a d-c motor includes the resistance of the armature circuit, R_A, and the counter emf, E_g.

Applying Kirchhoff's voltage law around the armature equivalent circuit of Exhibit 1-76 yields

Equation 1-42

$$V_T = E_g + I_A R_A$$

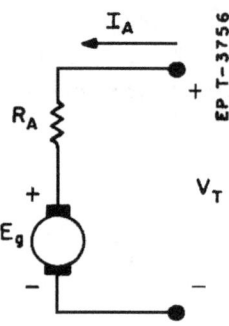

Exhibit 1-76 — Armature Equivalent Circuit For A D-C Motor

The terminal voltage, V_T, is greater than the counter emf, E_g, by an amount equal to the voltage drop across the armature resistance, R_A. Recall that in a generator, E_g was greater than V_T by an amount equal to the R_A drop.

At startup, when the speed is zero, the counter emf is zero ($E_g = K\phi N$) and the armature current is limited only by the armature resistance. Since this resistance is very low, a large armature current would result if full terminal voltage were applied directly to the armature. This large current could overheat the armature windings and cause damage to the motor. To prevent this excessive starting current in the armature, additional resistance (called **starting resistance**) is placed in series with the armature at startup and removed after the rotor comes up to speed and a counter emf is produced. Exhibit 1-77 illustrates a starting resistance (R_{START}) in the armature circuit and shows the shorting switch which is closed to remove the resistance when the motor is operating.

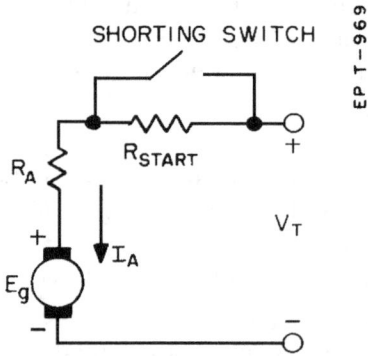

Exhibit 1-77 — Addition Of Starting Resistors In The Armature Circuit Limits Starting Current

1.5.4.1 — Shunt D-C Motors

D-C motors are classified according to the manner in which the field coils are connected to the armature. The same equivalent circuits which applied for d-c generators are used in conjunction with Kirchhoff's laws to determine the operating characteristics of d-c motors. However, the current flow will be **into** the motor and the output will be specified in terms of mechanical torque.

The **shunt d-c motor** has its field connected in parallel with the armature as shown in Exhibit 1-78.

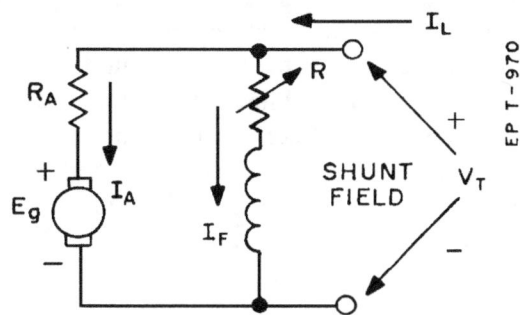

Exhibit 1-78 — Equivalent Circuit Of A Shunt D-C Motor

Since the field circuit is connected directly across the line, the field current depends only on the field resistance, R_F, and the terminal voltage, V_T. Since the terminal voltage, or the line voltage, is normally constant, the field flux may be varied by changing the setting of the field resistance (rheostat). If R_F is increased, the field current will decrease, and the magnetic flux will decrease. Conversely, if R_F is decreased, the field current will increase, and the magnetic flux will increase.

Under conditions of constant field resistance, the field flux is constant, and it can be seen from the following equation:

$$T = K\phi I_A$$

that the torque, T, becomes directly proportional to the armature current, I_A.

Equation 1-43

$$T \propto I_A$$

Equation 1-42 relating the counter emf and the terminal voltage can be rearranged as

$$I_A = \frac{V_T - E_g}{R_A}$$

Since the terminal voltage and the armature resistance are constant, the armature current will be controlled by the counter emf. In a practical motor, the armature resistance is very small, typically 0.0125 ohm, and thus a small change in counter emf will produce a large change in armature current.

Since the field resistance and hence the field flux is constant, any change in generated voltage must cause a corresponding change in motor speed, because

$$E_g = K\phi N.$$

The torque-speed characteristic for the shunt d-c motor is shown in Exhibit 1-79. At no-load, the torque (T_{NL}) is just sufficient to overcome rotational losses. When the motor is suddenly connected to its rated load, the no-load torque is not sufficient to carry rated load and hence the rotor slows down. This speed decrease causes a corresponding decrease in counter emf and hence a large increase in armature current. The speed will continue to decrease until the armature current has increased to a value sufficient to supply the necessary torque. The equations which follow show the relationship described above:

$$\downarrow E_g = \overleftrightarrow{K}\overleftrightarrow{\phi}N\downarrow$$

$$\uparrow I_A = \frac{\overleftrightarrow{V}_T - E_g\downarrow}{R_A}$$

$$\uparrow T = \overleftrightarrow{K}\overleftrightarrow{\phi}I_A\uparrow$$

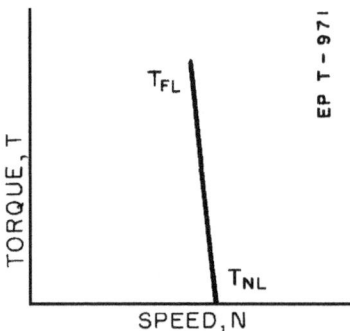

Exhibit 1-79 — Torque-Speed Curve For A Shunt D-C Motor

The effect of changing the excitation by changing the setting on the field resistance is to shift the entire torque-speed characteristic, as shown in Exhibit 1-80. For a decrease in R_F causing an increase in excitation, the torque-speed characteristic is shifted towards the origin. Adjusting the excitation provides a means of maintaining the motor speed for different torques.

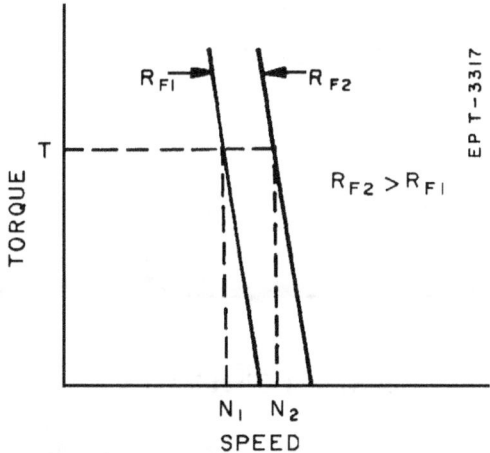

Exhibit 1-80 — Speed Control By Adjustment Of Field Rheostat

An important consideration in the use of the d-c shunt motor is the necessity to prevent the field from becoming open-circuited while the armature is connected to the line. Opening the field reduces the excitation to the low value supplied by residual magnetism. Initially, this causes a drastic reduction in counter emf and a corresponding inrush of armature current. This increased armature current causes the rotor speed to increase to a high value. This high rotor speed may cause excessive mechanical stresses on the armature windings and physically damage the rotor.

1.5.4.2 — Series D-C Motor

$$I_f = I_A$$

$$T = K\Phi I_A$$

In the series d-c motor the field is connected in series with the armature as shown in Exhibit 1-81. Note that the field current is the same as the armature current. Since the field conductors carry the full line current to the motor, they must be large, low resistance conductors. Thus, the resistance of the field winding is usually included in the armature resistance, R_A, as was done in the series generator.

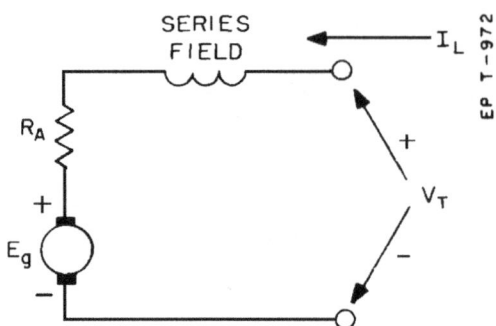

Exhibit 1-81 — Equivalent Circuit Of A Series D-C Motor

Since the field current and the armature current are the same, there is no way to vary the field current independently to control motor speed as was done with the shunt motor. Both field and armature current then depend upon the value of the counter emf.

When the load on a series motor is increased, the motor undergoes a greater drop in speed than would a comparable shunt motor. This is because as the armature current increases to yield a greater output torque, the field excitation is also increased. For the counter emf to remain at essentially the same value (remember, large changes in armature current require only small changes in counter emf), the motor speed must drop. The torque-speed characteristic for the series d-c motor is shown in Exhibit 1-82. Note the large torque available at startup when motor speed is low. The series motor is capable of starting much heavier loads than a comparable shunt motor. When no shaft load is applied, the torque required is low and the motor speed will be high.

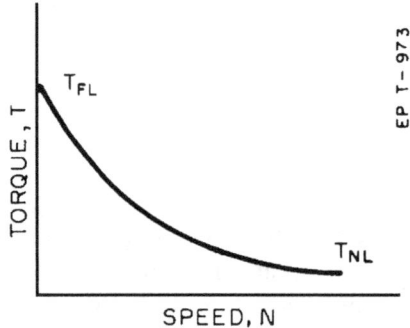

Exhibit 1-82 — Torque-Speed Curve For A Series D-C Motor

Because of the wide fluctuation of speed with load changes, a load should always be applied to the shaft of a series d-c motor to prevent overspeeding.

In summary, the series motor has the advantage over the shunt motor of producing a greater starting torque but has the disadvantages of a widely varying speed with varying load and the absence of independent speed control.

Section 1.6 — Motor Controllers

1.6.1 — LVP

In the Navy, three major types of alternating current across the line controllers are used. These are low voltage protection (LVP), low voltage release (LVR), and low voltage release effect (LVRE).

The purpose of an LVP controller is to disconnect the motor from the power supply upon a reduction or loss of voltage and keeping it disconnected upon a return of normal voltage (Exhibit 1-83) until the operator restarts the motor. For this reason, the LVP controller is used for motors which are not required to be restarted immediately upon restoration of power. LVP controllers are also used in applications which require that certain prerequisites must be satisfied before the motor is restarted.

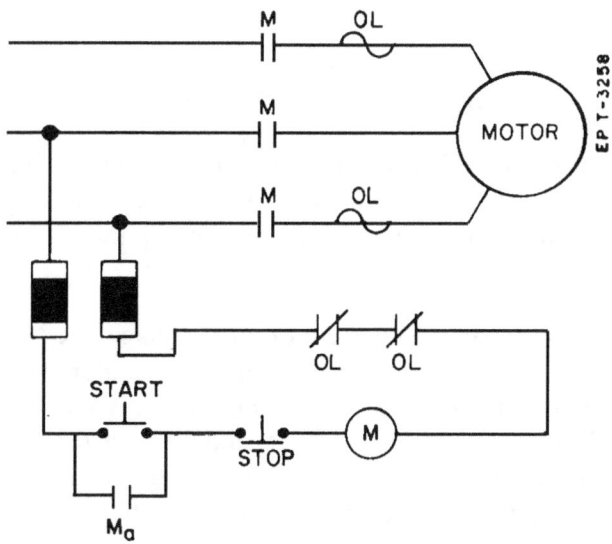

Exhibit 1-83 — LVP Controller

The operating principles can best be understood by referring to Exhibit 1-83. To start the motor, the START button is momentarily depressed, completing the circuit through the normally closed STOP button and the two normally closed overload (OL) contacts. This energizes the main (M) coil, closing the low voltage contacts (M_a) and the mail line contacts (M), thus energizing the motor. If at any time the coil voltage drops below a specified value, the main line and low voltage contacts open. Even if normal voltage returns, the main coil will not energize unless the START button is depressed.

1.6.2 — LVR

The purpose of an LVR controller is to disconnect the motor from the power supply upon a reduction or loss of voltage, keeping it disconnected until power returns, and then automatically reconnecting the motor to the power line to restart it upon restoration of voltage. This type controller (Exhibit 1-84) is found mostly on vital service motors which must be restarted automatically upon restoration of voltage.

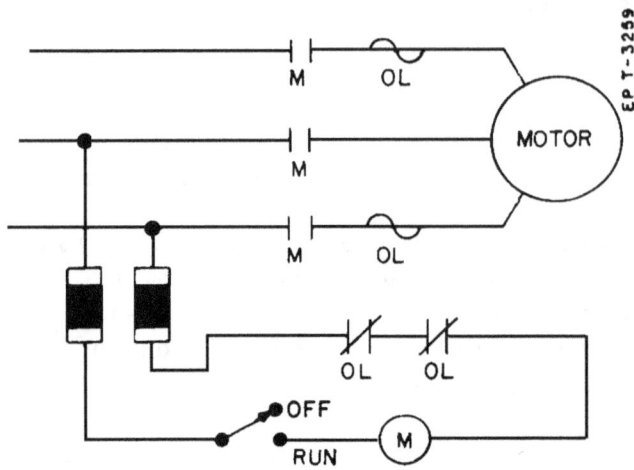

Exhibit 1-84 — LVR Controller

The operating principles of the LVR controller may best be understood by referring to Exhibit 1-84. Placing the master switch in the RUN position completes the path of current flow through the main (M) coil and the two normally closed overload (OL) contacts. Energizing the M coil allows it to close the main line (M) contacts, energizing the motor.

If the supply voltage drops to approximately 75 percent of rated voltage, the M coil no longer has sufficient strength to keep the M contacts shut. When these contacts open, the motor is deenergized. If the supply voltage returns to approximately 85 percent of rated value, the M coil will reclose the contacts. In case of an overload, the overload coils will open the overload contacts and deenergize the M coil. In order to restart the motor, the overload contacts must be reset. In some cases, an emergency run feature is provided whereby the controller may be closed by holding in an emergency run button that bypasses the overload contacts. **This is for emergencies only**.

1.6.3 — LVRE

The LVRE motor controller keeps the motor across the line at all times. This type controller is found mostly on auxiliaries, which must start automatically upon restoration of voltage (Exhibit 1-85). This applies to motors with small starting currents.

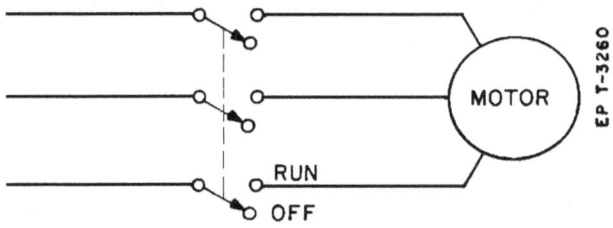

Exhibit 1-85 — LVRE Controller

The motor controller functions discussed here are basic. There are also many automatic control functions that the controller may perform: for example, start and stop a motor in response to pressure in a tank.

Section 1.7 — Electrical Safety

1.7.1 — First Aid For Electrical Shock

Everyone has at one time or another been **shocked** by electricity, either by a static charge built up by sliding across a car seat or from an electrical appliance. The amount of current that may pass through the body without danger depends on the current quantity, type, path, and length of contact time. The more important factors determining the amount of damage to the body from electricity depend upon several factors:

1. Type of Circuit and Its Frequency
    a. Alternating current will cause a greater muscular contraction than direct current; that is, the ability to **let go** is affected more with a-c current.
    b. 60-hertz alternating current is the most dangerous frequency to the body.
2. Voltage
    a. High voltage circuits are universally recognized as hazardous.
    b. Low voltage circuits are also dangerous. Fatalities have occurred with voltages as low as 30 volts d-c.
3. Resistance of the Body
    a. The outer layer of skin, the epidermis, offers a high resistance to the flow of electricity because it contains no blood vessels.
    b. The inner skin, or derma, is rich in blood capillaries and is a good conductor of electricity.
    c. Wet skin is a much better conductor than dry skin. The body resistance may be as low as 300Ω for wet skin and as high as $500K\Omega$ for dry skin. Besides the condition of the skin, the actual body resistance depends on the points of contact.
4. Pathway Through the Body
    a. A vital organ in the path of current flow presents the possibility of serious or fatal injury. Vital organ areas are the chest and the head, where the central nerve areas, blood pumping, and breathing mechanisms are located or controlled.
    b. If no vital organs are in the current path, resulting injury is negligible or confined to burns.
5. Duration

    The resistance of the victim decreases with time so the longer the contact, the less the chances of recovery
6. Current
    a. The amount of current flowing through the body determines the extent of resultant injury. The real measure of shock intensity is the amount of current that is forced through the body.
    b. A normal person can first feel a tingling at the point of contact when the current reaches a value of one milliampere, or 1/1000 of an ampere. This slight shock is not dangerous to the body, but an involuntary jerk away can cause secondary injuries, such as a workman falling from a ladder or other high places.
    c. Current as low as 5 milliamperes can be dangerous. If the palm of the hand makes contact with the conductor, a current of about 10 milliamperes will tend to cause the hand muscles to contract, freezing the body to the conductor. Such a shock may or may not cause serious

damage, depending on the contact time and the victim's physical condition, particularly the condition of the heart.

d. Currents between 100 and 200 milliamperes are lethal. In this current range, ventricular fibrillation of the heart occurs. Ventricular fibrillation is the uncoordinated action of the walls of the heart's ventricles, disrupting the regular pumping rhythm and causing loss of the pumping action of the heart. Fibrillation will continue until some force is used to restore the coordination of the heart's actions.

e. Above 200 milliamperes, the resulting muscular contractions are so severe that the heart is forcibly clamped during the shock, preventing ventricular fibrillation. Severe burns, unconsciousness, and stoppages of breathing are produced by currents above 200 milliamperes. These currents usually do not cause death if the victim is given immediate attention. The victim will usually respond if rendered resuscitation in the form of artificial respiration.

f. Current in the range of one to several amperes, **nerve block current**, blocks brain control of the heart and lungs. Again, death is the result.

Electric shock is a jarring, shaking sensation resulting from contact with electric circuits. The victim usually feels that he has received a sudden blow; if the resulting current is sufficiently high, the victim may become unconscious. Severe burns may appear on the skin at the place of contact. These burns extend deep into the body and are usually found in two places, the point of entrance and the point of exit. Look for both places for proper treatment. Muscular spasm may occur, causing the victim to clasp and be unable to turn loose the apparatus or wire from which the current is flowing. If an electric shock occurs, the power should be disconnected and the victim removed as quickly as possible. If the power switch is inaccessible, use a dry stick, rope, belt, coat, blanket, or any other nonconductor to drag or push the victim to safety. Do not attempt to cut an electrical wire. If one attempts to cut the wire, he will be much too close to the electrical charge and may receive severe burns from the flash or be wrapped into the ends of the wire when they whip free. After removing the victim from electrical contact, determine whether the victim is breathing. Keep him lying down in a comfortable position and loosen the clothing about his neck, chest, and abdomen so that he can breath freely. Protect him from exposure to cold, and watch him carefully. Keep him from moving about. In this condition, the heart is very weak, and any sudden muscular effort or activity on the part of the victim may result in heart failure. Do not give stimulants or opiates. Send for a medical officer at once and do not leave the victim until he has adequate medical care. If the victim is not breathing, it will be necessary to apply artificial respiration at once, even though he may appear to be lifeless. Do not stop resuscitation until a medical authority has pronounced the victim beyond help. It may take as long as 8 hours to revive the patient. There may be no pulse and a condition similar to rigor mortis may be present; these are a result of the shock and are not necessarily an indication that the victim has succumbed. The following is the American National Red Cross recommended procedure for performing mouth-to-mouth artificial respiration:

1. Wipe any obvious foreign matter from the mouth quickly.

2. Tilt the victim's head backward so that his chin is pointing upward.

a. In the unconscious patient in the supine position, the tongue may drop back and block the throat. To open the air passage, place one hand on the victim's forehead to tilt head back and lift on jaw to open airway.

b. Maintain the head in this position since it clears the airway by moving the tongue away from the back of the victim's throat. If additional airway opening is required, it can be achieved by thrusting the lower jaw forward into a jutting-out position

3. Pinch the victim's nostrils shut with the thumb and index finger of your hand that is pressing on the victim's forehead. This action prevents leakage of air when the lungs are inflated through the mouth.
4. Blow air into the victim's mouth.
   a. Open your mouth widely.
   b. Take a deep breath.
5. Watch the victim's chest to see when it rises. If air does not go in, reposition head and try again.
6. Stop blowing when the victim's chest is expanded; raise your mouth; turn your head to the side and listen for exhalation.
7. Watch the chest to see that it falls.
8. Repeat the blowing cycles.

If the victim has suffered an electric shock and has no heartbeat, he has had a cardiac arrest. There will be a complete absence of any pulse at the wrist or in the neck. Associated with this, the pupils of the eyes will be very dilated and respiration will be weak or stopped. The victim may appear to be dead. Under these circumstances, severe brain damage will occur in 4 minutes unless circulation is reestablished. Cardiopulmonary resuscitation (CPR) is the combination of artificial respiration and manual artificial circulation that is recommended for use in case of cardiac arrest. CPR involves the following steps:
1. Airway opening
2. Breathing restored
3. Circulation restored.

External cardiac compression consists of the application of rhythmic pressure over the lower half of the sternum. This pressure compresses the heart and produces a pulsatile artificial circulation. External cardiac compression must always be accompanied by artificial ventilation. CPR requires special supplemental training in the recognition of cardiac arrest and in the performance of CPR. Only qualified persons should perform CPR.

Burns caused by electrical shock will in general be third-degree burns. The usual signs are: (a) deep tissue destruction, (b) white or charred appearance, and (c) complete loss of all layers of the skin. Treatment of third-degree burns is outlined as follows:
1. Do not remove adhered particles of charred clothing.
2. Cover burns with thick, sterile dressings or a freshly ironed or laundered sheet or other linen.
3. If the hands are involved, keep them above the level of the victim's heart.
4. Keep burned feet or legs elevated. (The victim should not be allowed to walk.)
5. Have victims with face burns sit up or prop them up and keep them under continuous observation for breathing difficulty. If respiration problems develop, an open airway must be maintained.
6. Do not immerse an extensive burned area or apply ice water over it, because cold may intensify the shock reaction. However, a cold pack may be applied to the face or to the hands or feet.
7. If medical help will not be available within an hour or more and the victim is conscious and not vomiting, give him a weak solution of salt and soda: one level teaspoonful of salt and one-half level teaspoonful of baking soda to each quart of water, neither hot nor cold. Allow the victim to sip slowly. Give about 4 ounces over a period of 15 minutes. Discontinue fluid if vomiting occurs. Do not give alcohol.
8. Do not apply ointment, commercial preparations, greases, or other home remedies.

Reference for artificial respiration, burns, and CPR information: **Standard First Aid and Personal Safety**, American National Red Cross, Third Printing, Doubleday and Company, Inc., Garden City, New York, March 1975.

1.7.2 — Fuse Replacement And Electrical Troubleshooting

1.7.2.1 — Fuses and Fuse Replacement

A fuse is a protective device used to open an electric circuit when the current flow exceeds a safe value. Fuses are made in many styles and sizes for different voltages and currents, but they all operate on the same general principle. Each fuse contains a soft metal link that melts and opens the circuit when overheated by excessive current.

Most Navy electricians are familiar with only the cartridge-style fuse; therefore, we will limit our discussion solely to this type fuse. A cartridge fuse consists of a zinc-alloy link enclosed in a fiber, plastic, ceramic, or glass cylinder. Most fiber and plastic cylinders are filled with a nonconducting powder, normally a silicon compound. When the fuse blows, this powder melts and forms an insulating barrier between the two ferrules of the fuse.

Cartridge fuses are made in capacities of 1 through 1000 amperes for voltages of 125, 250, 600, and 1000 volts. Fuses intended for 600- and 1000-volt service are longer and will not fit the same fuse holders as fuses intended for 250-volt service. Fuses of different ampere capacity are also designed for different sizes of fuse holders. For example, fuses of 1 through 30 amperes fit one size of holder, and fuses with a capacity of 35 through 60 amperes fit a different size holder. Fuses above 60-ampere capacity have knife-blade contacts and are larger in diameter and length as the capacity increases.

Special type silver-sand fuses have been developed in physical sizes identical to the standard fuses and are rated at 500 volts, a-c or d-c. These fuses have silver-coated ferrules. They have the same current range as the standard fuses for the same size. However, they have a short-circuit current interrupting capability of 68,000 amperes. This characteristic permits their use in ship service power systems. **Silver-sand fuses are the only authorized fuses for use in power distribution systems**.

Certain safety precautions must be observed when dealing with fuses:
1. Replace blown fuses only with fuses of the same size, voltage, and current ratings, and having silver-plated ferrules (S designation). Always inspect the replacement fuse for damage such as a broken casing or a cracked ferrule.
2. **Never** short out a fuse.
3. All possible efforts should be made to deenergize the circuit prior to fuse replacement or removal. If circumstances dictate that the circuit must remain energized (tactical situation, procedure, ship's safety, etc.), only fuses of 10–ampere rated capacity or less should be removed from or replaced in energized circuits.
4. Fuses larger than 10–ampere rated capacity should be removed or replaced only when the circuit is **correctly** deenergized. After the fuse replacement, the circuit should be energized only when the cover over the fuses is closed. Neglect of this precaution has led to injuries caused by explosion of a fuse when the circuit was energized.
5. Only authorized fuse pullers made of insulating material should be used for the removal or replacement of fuses unless the fuse is mounted in a threaded, plug-type fuse holder. These fuses are removed by unscrewing the plugs.

6.   Always use rubber gloves and only one hand when removing or replacing fuses. Extreme caution must be used when dealing with the large type bus fuses. These fuses are very bulky and hard to handle with fuse pullers. Care should be taken not to short out the circuit line side with a fuse during replacement.

The following is a description of the designations found on cartridge-style fuses (Exhibit 1-86). Electricians should become familiar with the different designations and their meanings.

Capacity Designations

500V   = 500 volts

30A   = 30 amperes

S   = silver-plated ferrules

Note – The 30A after the 500V is the current rating of the fuse in all cases.

Size/function designation:

F60 denotes specific fuse style, construction, and dimensions. The letter that follows (A, B, or C) denotes the overload blowing characteristic. "A" indicates a normal overload blowing characteristic and blows at 135 percent of the fuse ampere rating in a specific time band, depending on the fuse style involved. "B" indicates a time delay overload blowing characteristic. The percent current at which the fuse blows and the time delay involved varies, depending on the particular fuse style involved. For example, an F15B fuse blows at 500 percent of the fuse ampere rating in 10 to 25 seconds. "C" also indicates a normal overload blowing characteristic and blows at 135 percent of the fuse ampere rating in a specific time band, depending on the fuse style involved. And, in addition, "C" fuses have a high interrupting rating of 68,000 amperes on circuits up to 500 volts, while "A" and "B" style fuses have various interrupting ratings not exceeding 10,000 amperes at various voltages up to 450 volts.

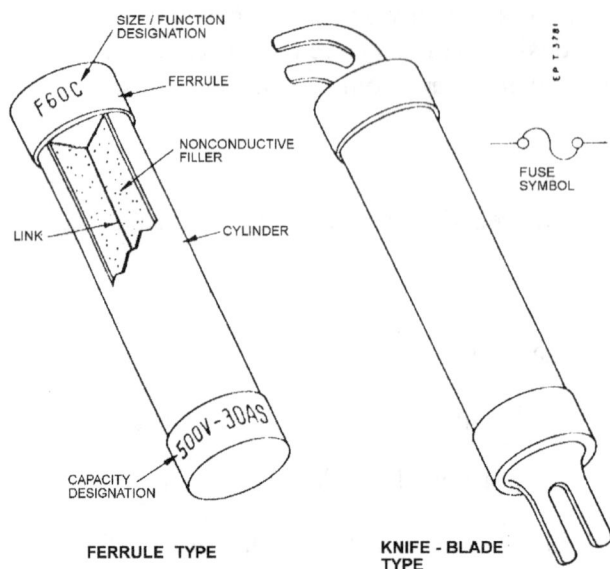

Exhibit 1-86 — Cartridge-Style Electrical Fuses

1.7.2.2 — Electrical Troubleshooting Procedure

The six-step troubleshooting procedure outlined here provides a logical guide to electrical troubleshooting and will save time wasted when a hit-or-miss method or relying on memory of past experiences is used for a troubleshooting guide.

Never rely on memory alone in troubleshooting any problem. Use equipment manuals which contain diagrams and information about the equipment. Recalling past equipment failures may not be helpful, since there are usually many possible problems which can cause similar symptoms.

1. Symptom Recognition

    Be thoroughly familiar with the operational characteristics of the equipment. Know how the equipment should behave under normal conditions and how well the equipment is performing. It is impossible to determine if the equipment is malfunctioning if the correct operational characteristics of the equipment are not known. A trouble symptom is an indication of some disorder or malfunction. Symptom recognition is the act of identifying a problem when it appears.

2. Symptom Elaboration

    This is the process of obtaining a more detailed description of the trouble symptom. The primary reason for performing this step is that any given symptoms can be caused by a large number of equipment faults. In order to proceed efficiently, it is necessary to decide which faults are the most probable for the specific symptom in question. System controls, indications, or built-in warning devices should be used to determine if there is power in the system, if the system's controls are set properly for the function to be tested, and if the trouble symptom(s) varies for different control settings. Sometimes, all that is needed is to flip the on-off switch to ON, or to plug in the power cord.

3. Fault Estimation

    This step uses the information obtained in Steps 1 and 2 and knowledge of the functional units of the equipment to make an educated estimate of where the trouble might be located.

List all possible faulty functional units that could produce the trouble symptom. For each functional unit ask yourself, "Would a fault in this unit cause the indicated symptom?" If the answer is yes, list that unit as a possible faulty functional unit. Use this procedure until all functional units have been considered. The selection of possible faulty functional units must be **logical** and must be technically substantiated by the information obtained during symptom elaboration, the relationship between signal paths and functional units, and operational knowledge of the equipment.

4.   <u>Localizing the Faulty Function</u>
    The fourth step involves determining which one of the faulty functional units obtained in Step 3 is actually malfunctioning. This is accomplished by systematically checking each probable faulty unit until the actual faulty unit is found. The proper test equipment should be used to check the input and output of each suspected faulty unit. You may isolate the fault with your first check. If not, the information gained from the test should be used to decide which possible faulty unit to check next.

5.   <u>Localizing Trouble to the Circuit</u>
    Use proper signal-tracing procedures and check the input and output of each circuit and stage within the faulty functional unit until the trouble is isolated to one particular circuit or stage. Always ensure that all safety precautions are observed when working on high-voltage equipment, that is, voltage in excess of 30 volts.

6.   <u>Failure Analysis</u>
    This final step includes (1) isolation of the bad and improperly adjusted component(s) and (2) verification of the troubleshooting findings. Once the fault has been isolated to the specific circuit, you will have to isolate the trouble to a specific component. Again, signal tracing and voltage and resistance readings are good methods to use. Prior to replacement of the suspected component, the entire sequence of indications and measurements should be analyzed to verify that the selected component could produce the observed symptoms and variations.

1.7.3 — Shipboard Ungrounded Electrical Systems

For the reasons given in the following paragraphs, all electrical systems can be potential killers. Shipboard "ungrounded" electrical systems are actually capacitively grounded to the extent that lethal currents can flow through a man's body if he touches a live conductor while in contact with ship's ground. The capacitance that causes this electrical ground leakage current to flow is inherent in the design of electrical equipment and cable, and cannot be eliminated by practical technical means. All personnel should be aware of the potential hazards, and those who work on electrical equipment or systems should be completely knowledgeable of the hazards, precautions, first-aid techniques, and theory of electric shock.

Ungrounded electrical distribution systems of both 450 and 120 volts a-c are provided on naval ships to achieve maximum system reliability and continuity of electrical power under combat conditions. If one line of the distribution system is grounded due to battle damage or deterioration of system insulation resistance, the circuit protective devices (circuit breakers, fuses, etc.) will not deenergize the circuit having the ground and electrical power will continue to be delivered to vital load equipment without further damage to the system. Frequent and proper use of system ground detectors provided on the ship service switchboards and certain power panels allow maintenance personnel to locate a ground and make repairs to remove the ground from the system as operating conditions permit. The primary advantage of an ungrounded system is that power can be maintained to a piece of vital load equipment, even when a ground occurs on one line of the electrical circuit supplying power to the equipment. If the system were designed as a grounded system, a ground on one power line would result in immediate tripping of the circuit protective devices, possibly deenergizing a piece of vital equipment when it is most needed.

1.7.3.1 — Shock Hazard For Shipboard Ungrounded Electrical Systems

Many personnel believe that since the electrical system is supposed to be ungrounded, it is possible to touch one bare conductor without danger since there is no electrical path for current to flow and, therefore, no electrical shock hazard. This possibly deadly belief is fostered by a misconception of what an ungrounded system is. The perfectly ungrounded single-phase, two-wire distribution system as shown in Exhibit 1-87 consists of a generator, distribution cable, and load equipment.

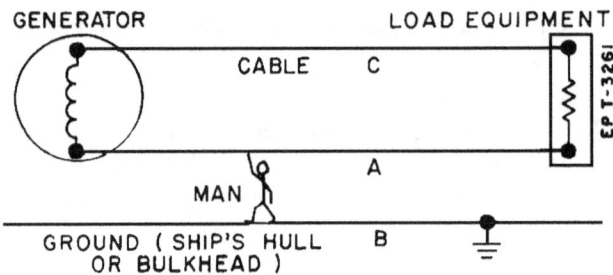

Exhibit 1-87 — Perfectly Ungrounded Electrical System

A perfectly ungrounded system is one in which the insulation is perfect on all cables, switchboards, circuit breakers, generators, and load equipment; one in which no radio frequency interference (RFI) filter capacitors are connected from ground to any of the conductors, and one in which none of the system equipment or cables have any inherent capacitance-to-ground. If all these conditions are met, there is no path for electrical current to flow to ground from any of the system conductors. Exhibit 1-87 shows that if a man touches a live conductor at point A while standing on the deck or ground at point B, there is no completed path for current to flow from conductor A to conductor C through the man's body, and thus he will not sustain an electrical shock. However, shipboard electrical power distribution systems do not and cannot meet the criteria for a perfectly ungrounded system. If we examine a typical shipboard "real" ungrounded system (Exhibit 1-88) there are additional factors that must be considered, some of which are not visible.

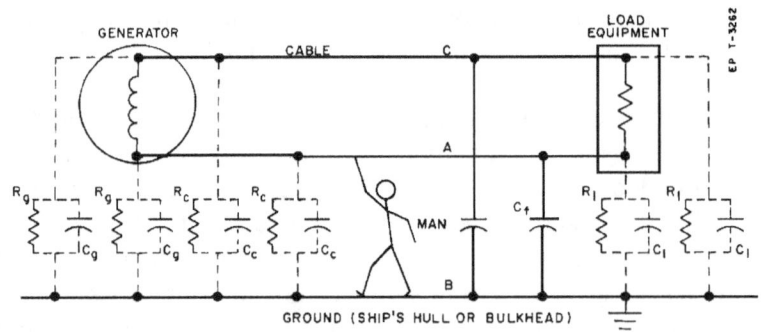

Exhibit 1-88 — Shipboard Ungrounded Electrical System

The additional components in a shipboard electrical system can be grouped into two categories. In the first category: R_g is the generator insulation resistance, R_c is the electric cable insulation resistance, and R_1 is the load insulation resistance. These resistances, when combined in parallel, form the insulation resistance of the system. Insulation resistance is periodically measured by the crew using a 500-volt d-c megger. The reading obtained is an indication of the integrity or quality of the insulation. These resistors, (R_g , R_c , R_1) cannot be seen as physical components, but are representative of small current paths through equipment and cable electrical insulation. The values of these resistances are measured in ohms; the higher the resistance, the better the system insulation and the less current will flow between conductor and ground. Typical values would be:

R_g = 500,000 ohms*

R_c = 50,000 ohms* for large system

R_1 = 1,000,000 ohms* or greater

* These values are typical of a large operating system, but can vary widely depending on the size of the ship and the number of electrical circuits connected together

The second category of components (Exhibit 1-88) are capacitors; C_g is the capacitance of the generator to ground, C_c is the capacitance of the distribution cable to ground, and C_1 is the capacitance of the load equipment to ground. As before, these capacitances cannot be seen, since they are inherent in the design of electrical equipment and cable. As an example, if we consider an electrical conductor surrounded by insulation and mounted on a metal bulkhead, we have two pieces of metal separated by an insulating material. Then, since on shipboard systems a potential difference (voltage) exists between the conductor and metal bulkhead or ground, we have established, in effect, a capacitor as shown in Exhibit 1-89.

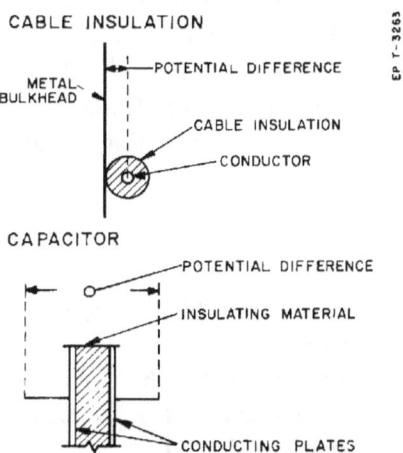

Exhibit 1-89 — Capacitance Developed Between Shipboard Cable And Bulkhead

The value of the capacitance thus generated is determined by the radius of the conductor, the distance between the conductor and the bulkhead, the dielectric constant of the material between the two, and the length of the cable. Similar capacitance exists between various pieces of load equipment and ground. Since capacitors ideally have an infinite impedance to d-c current, their presence cannot be detected by a megger or insulation resistance test.

In addition to the nonvisible system capacitance, typical shipboard electrical systems contain radio frequency interference (RFI) filters that contain capacitors (C_1) connected from the conductors to ground. These filters may be a part of the load equipment or mounted separately, and are used to reduce interference to communications equipment. The impedance of this capacitance to electrical current flow, also measured in ohms, is determined by the relation

$$X_c = \frac{\text{capacitive reactance in ohms}}{1} = \frac{1}{2\pi\,(\text{frequency})(\text{capacitance in farads})}$$

Typical values of X are:

X_g = 26,000 ohms (C_g = 0.1 μf/phase)

X_c = 1,060 ohms (C_c = 2.5 μf/phase)

X_f = 53 ohms (C_f = 50 μf/phase)

X_1 Very high unless equipment contains filter capacitors.

If we reexamine Exhibit 1-88, we notice that the capacitances (C_g, C_c, C_f, and C_1) are in parallel with the system insulation resistances (R_g, R_c, and R_1) and form an additional path for electrical current flow from the conductors to ground (ship's hull or bulkhead). If a man should accidentally touch a conductor at point A, current would flow through his body to ground at point B and back through the system resistances and capacitances to the other conductor at point C, thus completing the electrical circuit, and presenting a serious shock hazard.

1.7.3.2 — Resistance Versus Capacitance

If we were to megger the system of Exhibit 1-88 and obtain a system value of insulation resistance of approximately 50,000 ohms, we would conclude rightly that no low-resistance grounds exist on the system and wrongly that the system is a "perfect" ungrounded system. We have forgotten the system capacitance that exists in parallel with the resistance. If we look at the typical values of system capacitances given in Paragraph 1.7.3.1, we see that if the system contains many RFI filters, the capacitive reactance of the system to ground may be as low as 50 ohms; while, even without these filters, the inherent capacitance of generators and distribution cable would result in a capacitive reactance of approximately 1000 ohms. What this means to the person who is careless and touches a live electrical conductor is shown in the following example. Exhibit 1-90 shows a simplified circuit assuming no RFI filters connected to the system.

Example 1-1 —

Assume the man's body resistance of 600 ohms (possible under work conditions when hands are wet from sweat); then, combine the total system insulation resistance (R) in parallel with the system capacitive reactance (X_c) to obtain the system impedance (Z) to ground:

$$Z = \frac{(R)(X_c)}{\sqrt{R^2 + X_c^2}} = \frac{(50{,}000)(1{,}000)}{\sqrt{(50{,}000)^2 + (1{,}000)^2}} = \text{Approximately } 1{,}000 \text{ ohms}.$$

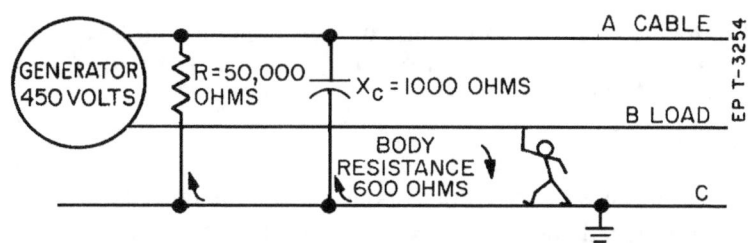

Exhibit 1-90 — Ground (Ships Hull Or Bulkhead)

If the man touches conductor B, the current will flow through his body (600 ohms) to ground, then through the system impedance (Z = 1000 ohms), and back to conductor A completing the electrical circuit. The total opposition to this current flow is:

$$\sqrt{1000^2 + 600^2} \text{ or } 1166 \text{ ohms.}$$

The amount of shock current that would flow is given by Ohm's Law:

$$I_s = \frac{450 \text{ volts}}{Z_T} = \frac{450 \text{ volts}}{1166 \text{ ohms}} = 0.386 \text{ ampere}.$$

According to the following general guidelines from NAVSHIPS 250-660-42, **Electric Shock - Its Causes and Its Prevention**:
- At 1.0-ma shock is felt.
- At 10-ma, man may be unable to let go.
- A 100-ma shock is fatal if it lasts for 1 second or more.

Thus, in the example even if no RFI filters were connected to the system, even if a megger test shows the system to be ungrounded, a current of almost **four times that required to kill a man** would flow through his body and death would be the result. The capacitance used above (C_g, C_c, and C_1) cannot be eliminated since they result from inherent laws of electrical theory associated with the practical design of electrical equipment and cable. This condition has existed on all previous ships constructed with a-c distribution systems, it exists on all present ships, and will exist on all future ships. However, unlike the system insulation resistance, the system capacitance will not change with time (unless equipment containing capacitors connected to ground is added) since it is a function of equipment and cable design. Therefore, no need exists to measure capacitance periodically as we measure the insulation resistance.

The addition or elimination of RFI filter capacitors connected to the system makes no difference from a safety standpoint. It only takes so much current to kill you, and there is more than enough without RFI filters.

REMEMBER:
1. Never touch a conductor of an electrical circuit, "ungrounded" or grounded. High system insulation readings from a megger test do not make the system safe to touch – NOTHING DOES.
2. Insulation resistance tests are made to ensure the system will operate properly, not to make the system safe.
3. Know and follow the electrical safety instructions contained in **NAVSHIPS Technical Manual (NAVSHIPS 0901–960–000),** Chapter 9600. Also, **NRF 1630,** Section 4103.

1.7.3.3 — Isolated Receptacle Circuits

To reduce the inherent hazard of these leakage currents on receptacle circuits, where portable tools or appliances are plugged and unplugged and personnel are more likely to receive an electrical shock, isolated receptacle circuits are installed on all new-construction ships. These circuits are individually isolated from the main power distribution system by transformers, and each circuit is limited to 1500 feet in length to reduce the capacitance to an acceptable level. This design is intended to limit ground leakage currents to 10 ma which would produce a nonlethal shock. To maintain this low level of leakage current and provide personnel safety, it is extremely important that the isolated receptacle circuits be maintained free of all resistance grounds. Ships already in the fleet were provided information for installation of either fixed or portable isolation transformers in the receptacle circuits in 1960.

The use of isolated receptacle circuits and equipment design improvements, have materially reduced the hazard of using portable tools and appliances. However, the best safety device is a respect for the deadly hazard of all electrical systems, grounded or ungrounded, low or high voltage, alternating current or direct current for they are all potential killers of the careless or inexperienced. Each crewman should be familiar with the electrical safety precautions.

Section 1.8 — Storage Batteries

1.8.1 — Lead-Acid Batteries

A storage battery is a device for storing electrical energy in the form of chemical energy. When the electrical energy is used, a chemical reaction occurs inside the battery; when the flow of energy is reversed, supplying electrical energy to the battery, the chemical reaction proceeds in the opposite direction.

The most common storage battery is the lead-acid storage battery. The battery is made of a series of cells. Each cell is constructed of alternating lead oxide (PbO_2) and spongy lead (Pb) plates immersed in an electrolyte solution of dilute sulfuric acid (H_2SO_4). The basic chemical reaction occurring at the negative plate when the cell discharges is the loss of electrons by lead (oxidation).

$$Pb + SO_4{}^{2-} \rightarrow PbSO_4 + 2e^-$$

At the positive plate, lead oxide gains electrons and passes into solution as Pb^{2+} ions (reduction). The Pb^{2+} ions combine with $SO_4{}^{2-}$ ions again forming $PbSO_4$.

$$PbO_2 + 4H_3O^+ + SO_4{}^{2-} + 2e^- \rightarrow PbSO_4 + 6H_2O$$

These equations show that an excess of electrons is produced at the negative plate and that electrons are consumed at the positive plate. Thus, a flow of electrons (current) occurs when an external path is provided between the negative and positive plates. The equations also show that as the cell discharges, a coating of insoluble lead sulfate ($PbSO_4$) builds up on both the positive and negative plates. The $PbSO_4$ causes an expansion of the materials into the voids or pores of the plates and results in a gradual clogging. If discharges are prolonged, excessive expansion may take place with the result that the plate may swell, creating mechanical stresses which reduce battery life. To minimize this condition, care must be taken to avoid discharging any cell beyond the low voltage limit.

Another effect during discharge is the consumption of $SO_4{}^{2-}$, causing the acid concentration of the electrolyte to decrease. This results in a decrease in the specific gravity of the electrolyte. When so much of the active material has been converted into lead sulfate that the cell can no longer produce sufficient current, the cell is discharged.

If the discharged cell is properly connected to a d-c charging source of slightly higher than battery voltage, a current will flow through the cell in the opposite direction to that of discharge, and the cell is charging. On charge, lead sulfate goes to lead at the negative plate and to lead oxide at the positive plate, resulting in a reduction in weight and a gain in porosity of the plates. At the same time, the sulfate is restored to the electrolyte with the result that the specific gravity of the electrolyte increases. Exhibit 1-91 summarizes the chemical action of a lead-acid cell.

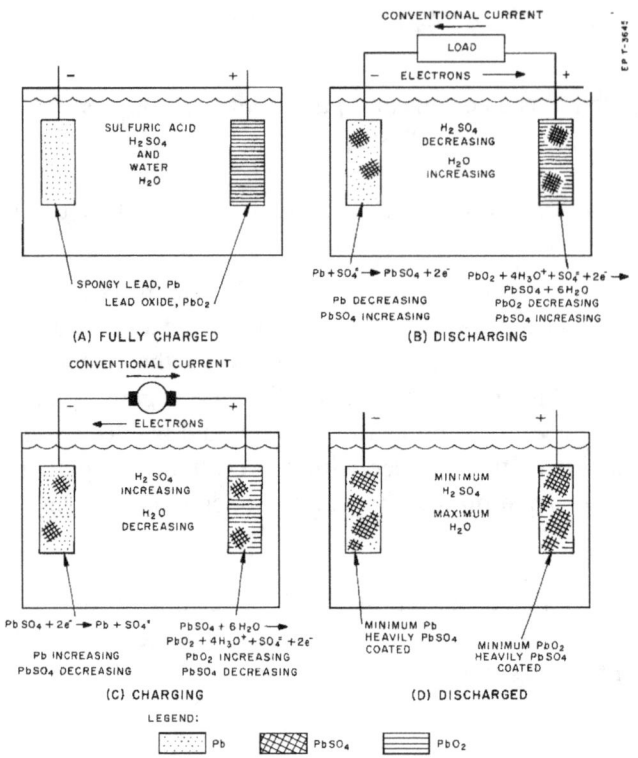

Exhibit 1-91 — Basic Chemical Action Of A Lead-Acid Storage Battery

Since the amount of sulfuric acid combining with the plates at any time during discharge is in direct proportion to the ampere-hours of discharge, the specific gravity of the electrolyte is a guide in determining the state of discharge of the lead-acid cell. The ratio of the weight of a certain volume of liquid to the weight of the same volume of water is called the specific gravity of the liquid. The specific gravity of pure water is 1.000. Sulfuric acid has a specific gravity of 1.830; thus, sulfuric acid is 1.830 times as heavy as water. The specific gravity of a mixture of sulfuric acid and water varies with the strength of the solution from 1.000 to 1.830. The specific gravity of the electrolyte in sample pilot cells is routinely measured to determine the state of charge of the battery. The specific gravity of the electrolyte is affected by temperature. The electrolyte expands and becomes less dense when heated and its specific gravity reading is lowered. The electrolyte contracts and becomes denser when cooled and its specific gravity reading is raised. Hydrometers used to measure specific gravity are calibrated at a specific temperature and are provided with a temperature correction factor, which must be applied to obtain correct readings when used at other than the calibrated temperature.

The addition of sulfuric acid to a discharged lead-acid cell does not recharge the cell. Adding acid only increases the specific gravity of the electrolyte and does not convert the lead sulfate on the plates back into active material and consequently does not bring the cell back to a charged condition. A charging current must be passed through the cell to recharge it.

As a cell charging operation nears completion, hydrogen and oxygen gases evolve. During discharge and idle periods, only hydrogen is produced in significant quantities. Toxic chlorine can evolve if a battery becomes contaminated with salt water, irrespective of the state of battery charge. The maximum safe concentration of hydrogen is 3 percent, a concentration of 4 to 8 percent is combustible, and a mixture of more than 8 percent is explosive. To limit the effects of gas evolution, the battery must be charged in accordance with temperature voltage-gassing (TVG) conditions. Exhibit 1-92 illustrates a typical TVG curve which gives the maximum charging cell voltage at a given cell temperature for which excessive gases will not develop. As long as the cell voltage is below the TVG voltage, charging may be accomplished at the highest current within the capacity of the charging generator. Once the TVG voltage is reached, the charging current must be kept at or below the approved finishing rate to avoid excessive gas development.

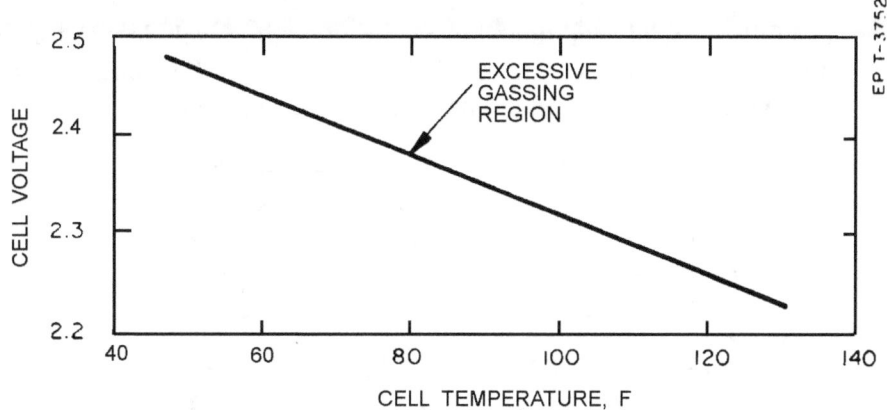

Exhibit 1-92 — A Temperature-Voltage-Gassing (TVG) Curve

1.8.2 — Typical Submarine Storage Battery

A typical lead-acid storage battery used for backup power on submarines is composed of approximately 100 to 150 individual cells. Each cell develops a maximum voltage of about 2.05 volts. The battery voltage is increased by connecting individual cells in series. The current capacity of the battery is increased by connecting individual cells in parallel. In the submarine battery, the individual cells are connected together in series by copper bars bolted from the positive terminal of one cell to the negative terminal of an adjacent cell. The individual cells are arranged in several rows within the battery compartment. Cell connections regularly cross over between rows to minimize magnetic field development and preclude magnetizing the ship's steel hull. Since the cell groups are staggered, battery current creates opposing magnetic fields which tend to neutralize each other. High temperatures can have harmful effects on battery operation, particularly in that hydrogen evolution during discharge increases very rapidly as the temperature rises. The principal sources of heat in the battery are:

1. Heat due to the electrical resistance of the electrolyte, plates, and cell connectors
2. Heat due to the heat of formation of molecular hydrogen and oxygen during the gassing phase of charging
3. Heat due to the chemical reactions occurring in the cell. Because of the harmful effects of high temperature, the electrolyte temperature should not be allowed to exceed the stated maximum value ($\sim 130°F$, typically).

1.8.3 — Battery Charging

Proper charging is the most important factor in determining the life and performance of a storage battery. A battery is considered on charge whenever electrical energy is being put into it. At periodic intervals, a charge is applied to maintain the battery voltage at full or near full capacity.

Charging is started at the highest current within the capacity of the generator and cables. This current is held constant until the battery voltage reaches the limit set by the TVG curve. As charging continues, the temperature rises and the voltage is reduced so as to ride the TVG curve. The charging current decreases as the battery becomes closer to a fully charged condition, thus requiring progressively less charging current. When the current decreases to the approved finishing rate, it is held constant at the finishing rate until the charge is completed. Exhibit 1-93 shows the time variation of charging current and voltage.

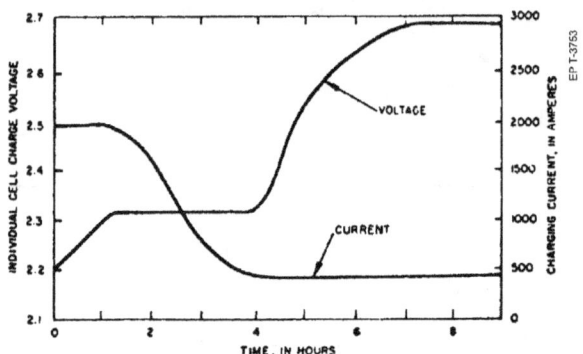

Exhibit 1-93 — Time Variation Of Voltage And Current During Battery Charging

There are three major types of charges that can be used on a battery:
1. normal
2. equalizing
3. partial

A **normal** charge is a routine charge given at regular intervals (typically not greater than 7 days) to restore a partly or fully discharged battery to a substantially fully charged condition.

An **equalizing** charge is a normal charge extended at the finishing rate, given periodically (typically at intervals not greater than 30 days), to ensure that all the sulfate is driven from the plates and that all cells are restored to a uniform, maximum capacity.

A **partial** charge is a charge which is insufficient to satisfy the requirements of a normal charge. It may be given to recharge a battery sufficiently to meet some special demand when time is lacking for a normal charge, or it may result from an interruption of a normal charge caused by a breakdown, or by reaching the established temperature limits.

A practice of minimal discharges followed by daily full normal charges is undesirable. The overcharge accumulated by such a practice is out of proportion to the amount of work the battery does and can lead to a reduced lifetime. Battery charges are performed when specific discharge conditions exist. The conditions that must be satisfied for the various battery charges depend upon the type of cell, and the approved manufacturer's instructions should be consulted to determine these conditions for a particular battery.

Battery support systems include a ventilation system, a hydrogen detection system, a battery airflow indicator system, an electrolyte agitation system, and a battery freshwater system. Instruments are provided for remote monitoring of battery operation, and individual cell voltages are indicated at a local voltage and ground resistance panel.

The number of ampere-hours delivered by a battery on discharge is equal to the time of discharge, in hours, times the average discharge current, in amperes. The ampere-hour capacity of a battery is the number of ampere-hours it is capable of delivering on discharge before the discharge must be terminated because a low voltage limit has been reached. Ampere-hour capacity depends upon a number of factors in addition to the type of battery and its construction. Among these are:

1. The age of the battery and its past history
2. The specific gravity of the electrolyte when fully charged
3. The rate of discharge
4. The temperature of the electrolyte at the start of a discharge.

Test discharges are conducted at the initial installation and at six-month intervals thereafter to determine the capacity of the battery. The test discharges made at the time of installation are made to determine the initial capacity and to help build up capacity. The battery does not develop rated capacity until it has gone through a number of charging and discharging cycles.

A battery ground is a leakage path for current which extends from a conducting path of the ship's structure to a conductor in the battery. Current flow through these paths is indicated on the ground detector. Leakage paths may also exist which extend between cells without any connection to ground. Current flow through these paths is not indicated on the ground detector. Battery cleanliness is necessary to minimize danger from this source.

Leakage paths are undesirable for the following reasons:

1. A break in the leakage path, which might occur due to the rolling of the ship or to evaporation of the electrolyte film caused by air currents, may furnish the spark necessary to ignite an explosive hydrogen-air mixture, if present.
2. A battery ground may furnish a return path for the current from another ground or any of the ship's power or lighting circuits. Under adverse conditions, this might cause a disastrous fire.
3. Battery grounds or leakage paths between cells may cause partial discharge of some cells or of the battery as a whole.
4. Grounds create the danger that a man may receive an electric shock if he makes contact with ground and a live conductor.

Section 1.9 — Electrical Measuring Instruments

Electrical measuring instruments are vital to the maintenance and operation of the electric plant. A good understanding of the functional design and operation of electrical instruments is important to ensure their proper application. The instruments most used to measure and monitor parameters such as voltage, current, and power are electric meters.

The function of an electric meter is to convert an electrical input into a mechanical output, usually the angular displacement of a pointer as shown in Exhibit 1-94. The pointer moves in front of a calibrated scale to indicate the magnitude of the measured signal.

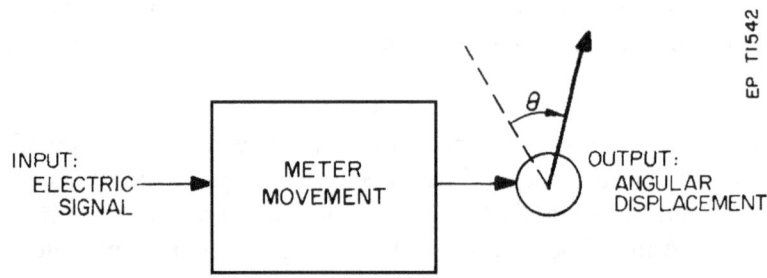

Exhibit 1-94 — Electric Meter Block Diagram

1.9.1 — Meter Movements

A meter movement is the electromechanical device which provides the mechanical motion of the pointer in response to an applied electrical signal. The motion is produced by the interaction of magnetic fields produced by the voltage or current to be measured.

The force acting on a current-carrying conductor in a magnetic field is directly proportional to the product of the field strength (B), the length of the conductor (L), and the magnitude of the current flowing in the conductor (I).

$$F \propto BIL$$

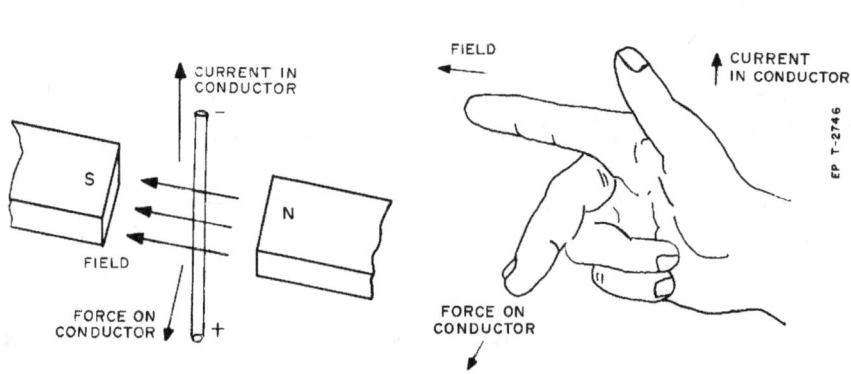

Exhibit 1-95 — The Right-Hand Rule Is Used To Determine The Direction Of An Induced Force

If a single loop of wire having a rectangular shape is placed into the field and a direct current is circulated through the wire, a force is exerted on each of the segments (L) with one of the segments being forced upward and the other downward (Exhibit 1-96). The end segments (W) are parallel to the flux field and no force is exerted on them. The torque imparted to the coil is directly proportional to the product of the moment arm (W/2) and the force (F) on each of the two segments.

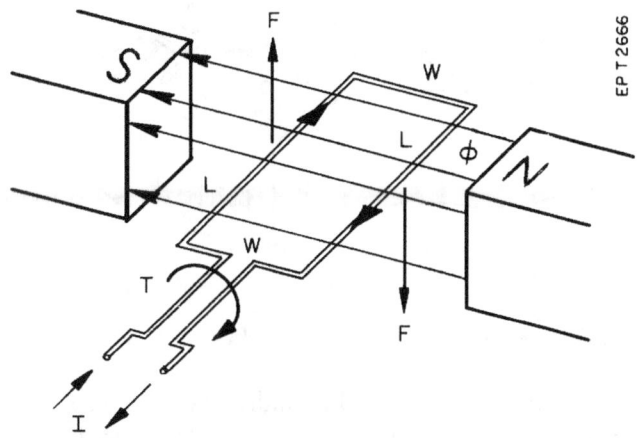

Exhibit 1-96 — Torque Produced By A Current-Carrying Coil In A Magnetic Field

$$T \propto \frac{W}{2} \times F$$

Since W/2 is constant and $F \propto I$, this can be simplified to

$$T \propto 1.$$

This principle is the basis of the most used meter movement, the permanent magnet-moving coil movement, also known as the D'Arsonval movement. Exhibit 1-97 illustrates a simplified view of the D'Arsonval movement. It consists of a coil carrying the measured current wound on a fixed, cylindrical, iron core mounted between the poles of a permanent magnet. The indicating pointer is secured to the coil, which is free to move within the air gap. Current to the coil is supplied through coiled springs mounted on each end of the shaft and electrically connected to the ends of the coil. In addition to providing a path for the coil current, the springs provide a force which holds the coil and pointer at the rest position when no current is flowing in the coil. The springs also provide a linear restoring force when they are deflected due to current flowing in the coil.

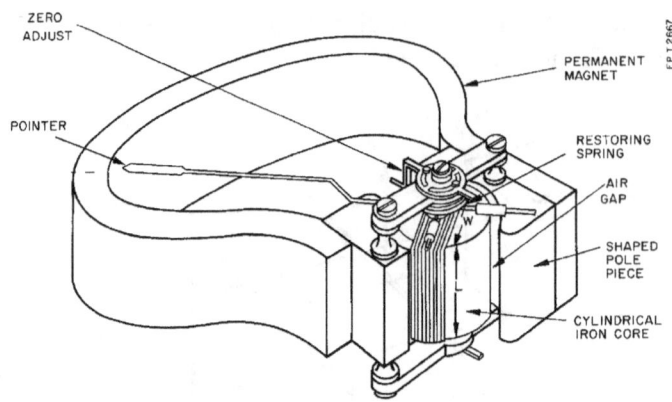

Exhibit 1-97 — Simplified View Of A Permanent-Magnet, Moving-Coil Meter Movement

As shown in Exhibit 1-98, the shaped pole pieces and cylindrical core shape the flux field so that the flux lines enter and leave the air gap at right angles to the curved surfaces. This provides a uniform flux field in the air gap and since the restoring spring is linear, the torque produced is directly proportional to the current in the coil. A multiturn coil is used to increase the active length of the conductor which increases the torque obtained with a given current, thus increasing the sensitivity.

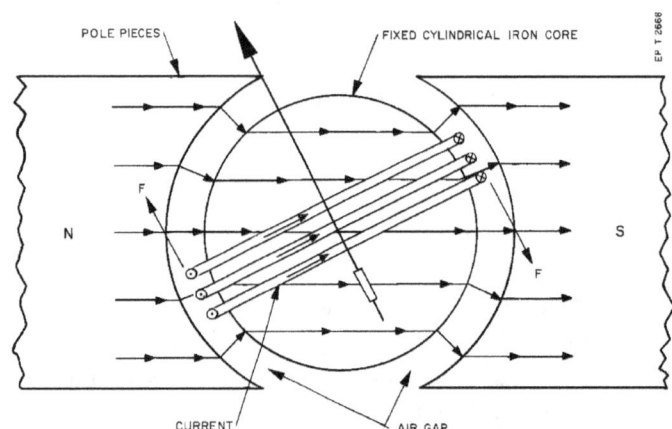

Exhibit 1-98 — Top View Of A Permanent-Magnet, Moving-Coil Meter Movement

Since an average torque can be produced only by a d-c current, the D'Arsonval movement does not respond to a-c. D'Arsonval movements are extremely sensitive, requiring only microwatts of power.

A second type of meter movement is the dynamometer movement shown in Exhibit 1-99. The dynamometer movement consists of two windings, one fixed and one movable. A magnetic field is established by the fixed-field coils which are normally in two sections to establish a more uniform field. Within this field, the movable coil is mounted on a pivoted shaft. Coiled springs are used to conduct current to the movable coil and to supply restoring force. Damping vanes rotate in close fitting air chambers to provide proper damping for the movement. If the instrument is not damped - that is, if friction or some other type of loss is not introduced to absorb the energy of the moving element - the pointer will oscillate for a long time about its final position before coming to rest. The movable coil may be connected in series with the field coil for measuring current or voltage, or the two windings may be supplied separately for measuring power. When the dynamometer is used as a d-c ammeter, the current to be measured flows through both windings. The field of each winding interacts with the current flowing in the other. Unlike the D'Arsonval movement, the field strength of the dynamometer is proportional to the current in the field coil instead of being constant as it was when established by a permanent magnet. Since the current flows in both windings, the torque is proportional to the square of the current, and hence the dynamometer is also a true RMS meter. Thus, for this application, the dynamometer is square-law and is compressed at the lower end.

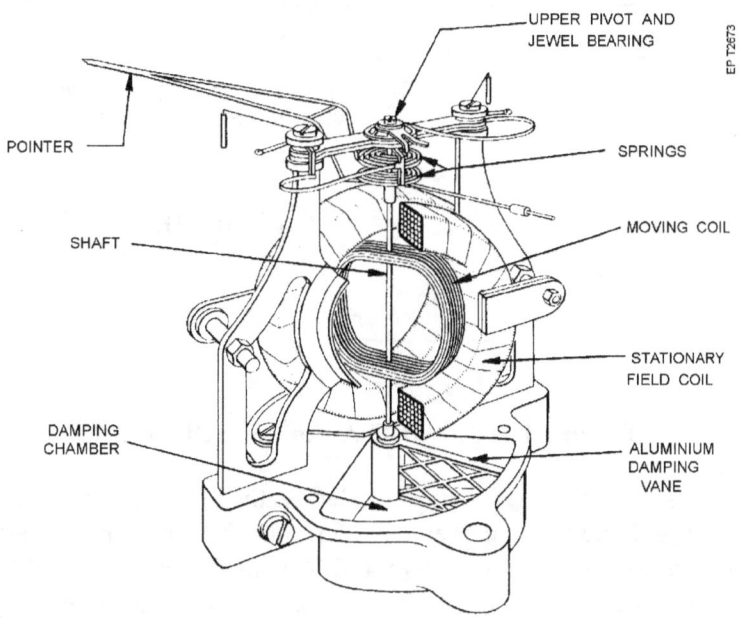

Exhibit 1-99 — Meter Movement Of A Dynamometer

One very important use for the dynamometer is based on its ability to measure power directly. In this application, the system current is circulated through the field winding and a current proportional to the system voltage is circulated through the movable coil. Since power is equal to the product of current and voltage, there is no square term in the proportionality and the scale for a dynamometer wattmeter is linear.

The dynamometer meter movement is inherently the most versatile movement. The absence of any magnetic materials, with their associated nonlinearities, makes the dynamometer a highly accurate movement. The dynamometer has the disadvantage that it is not as sensitive as the D'Arsonval movement, typically requiring 1 to 3 watts. The dynamometer will respond to d-c signals and a-c signals up to about 200 Hz for high accuracy.

The dynamometer movement is used in three-phase wattmeters to measure watts on electric plant SSTG and CTG sets. By using crossed-coil movements, it can also be used for power-factor, phase-angle, frequency, and capacitance measurement.

1.9.2 — D-C Ammeters

Exhibit 1-100 illustrates the schematic representation of a d-c ammeter. The meter movement could be either the D'Arsonval or the dynamometer type, depending upon the required sensitivity. For an ideal ammeter, the coil resistance, R_m, would be zero. For the actual ammeter to approach the ideal ammeter, the coil resistance should be much less than the resistance in the loop whose current is being measured. The sensitivity of the ammeter is the current required for full-scale deflection. The more sensitive the ammeter, the less current required for full-scale deflection.

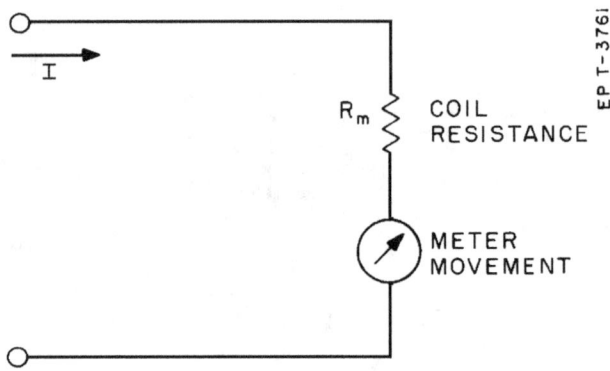

Exhibit 1-100 — D-C Ammeter Schematic

In circuits where the current is low and the voltage is moderate, the meter is often connected into the circuit so that the full current flows through the meter movement. Currents commonly measured directly by the meter movements are 5 microamperes to 5 milliamperes by the D'Arsonval movement and 50 to 200 milliamperes by the dynamometer movement. For currents above these ranges, shunts or current transformers are used.

A shunt is a low-value resistor that is connected in parallel with a meter movement. The theory and use of an ammeter shunt is the same regardless of the type of meter movement used. Since the shunt has a lower resistance than the meter movement, most of the circuit current is shunted around the meter and only a small part of the circuit current actually flows through the meter movement. From Exhibit 1-101, the meter current is related to the total current by:

$$I_m = \frac{IR_S}{R_m + R_S}$$

Since R_m and the maximum values of I_m and I are known, the required value of the shunt resistance, R_s, may be determined.

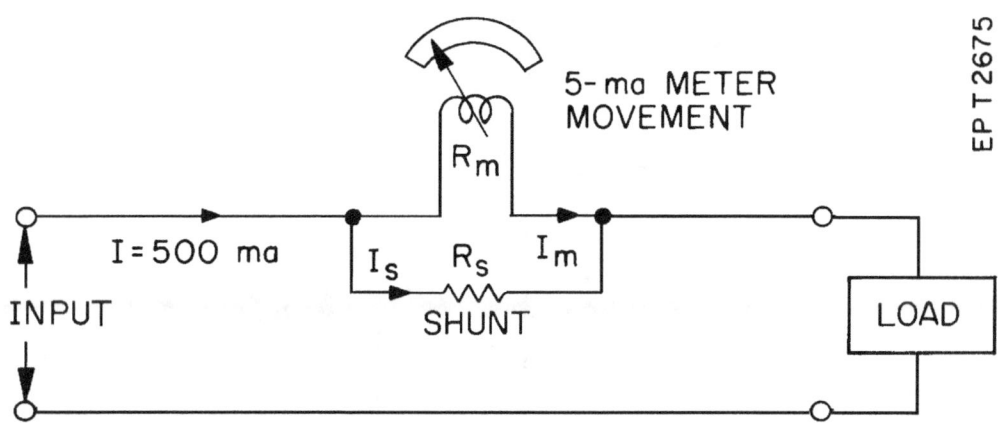

Exhibit 1-101 — Schematic Circuit Of A Shunt Used With A Milliammeter

For low currents, the shunt is usually mounted within the meter case. For large values of current where heating becomes significant, the shunt may be mounted external to the meter. By using several shunts and a selector switch as shown in Exhibit 1-102, the meter can be made to indicate a wide range of current. Although the circuit of Exhibit 1-102 appears to be the obvious way to connect the various shunts, it has disadvantages that limit its use. When the selector switch is moved from one shunt to another, the shunts are momentarily removed from the meter and the line current then flows through the meter movement. Even a momentary surge of current can damage the meter movement. Also, the contact resistance between the blades of the selector switch is in series with the shunt but not with the meter movement. Because of the variable nature of the contact resistance and the high currents carried by the shunts, the ammeter indication may be inaccurate. The selector switch problem can be solved simply by using a make-before-break switch. However, a more generally accepted method of range switching is shown in Exhibit 1-103. In this circuit, the selector switch contact resistance is external to the shunt and meter in each range position, and therefore has no effect on the accuracy of the current measurement.

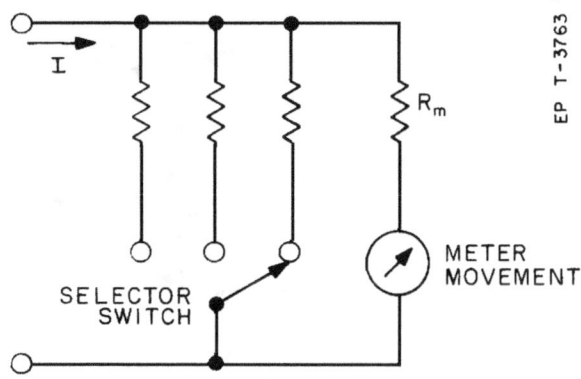

Exhibit 1-102 — Multirange Ammeter

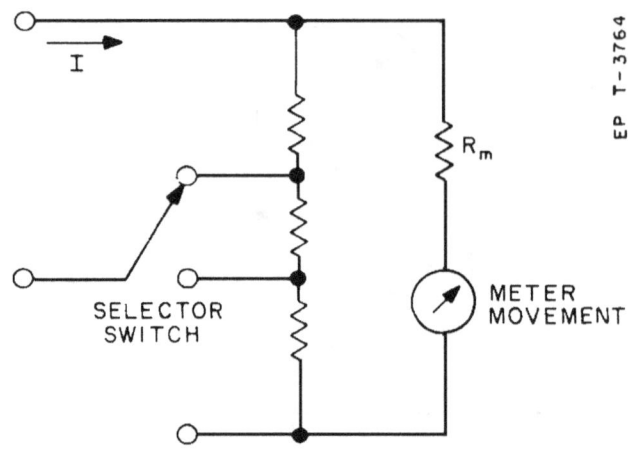

Exhibit 1-103 — Preferred Method Of Range Switching

1.9.3 — D-C Voltmeters

Any current meter can be conveniently used to measure voltage by connecting a large-value resistor in series with the meter movement as shown in Exhibit 1-104. For an ideal voltmeter, the combined resistance, $R + R_m$, would be infinite. For the actual voltmeter to approach the ideal voltmeter, the combined resistance, $R + R_m$, should be much larger than the resistances of the circuit being measured. Thus, the series resistor, R, is much larger than the coil resistance, R_m. The range of the voltmeter is the voltage required at the meter terminals for full-scale deflection. Another commonly used voltmeter sensitivity is the ohms/volt rating. The ohms/volt rating is obtained by dividing the series resistance $(R + R_m)$ of the voltmeter by the voltage required for full-scale deflection. As with the ammeter, a multirange voltmeter can be constructed by using several values of series resistance and a selector switch.

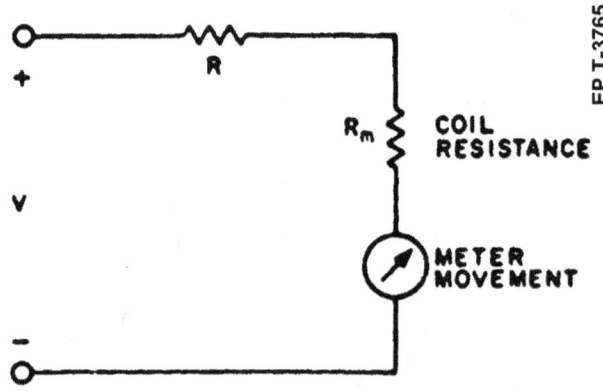

Exhibit 1-104 — D-C Voltmeter Schematic

1.9.4 — Ohmmeters

A current meter can be used as an ohmmeter by connecting it to a voltage source as shown in Exhibit 1-105, and remarking the scale to read in ohms. D'Arsonval meter movements are normally used in this application because of their greater sensitivity. The series resistor, R, is chosen so that the meter will read full scale when terminals A and B are connected together. The zero-adjust rheostat is provided to adjust the zero position of the pointer when the terminals are connected together. With an unknown resistor connected between terminals A and B, the meter will indicate some value less than full scale with a specific current being obtained for each specific value of resistance connected to terminals A and B. Ohmmeters may also be provided with selector switches to change the series resistance, R, and allow accurate resistance readings over a wider range.

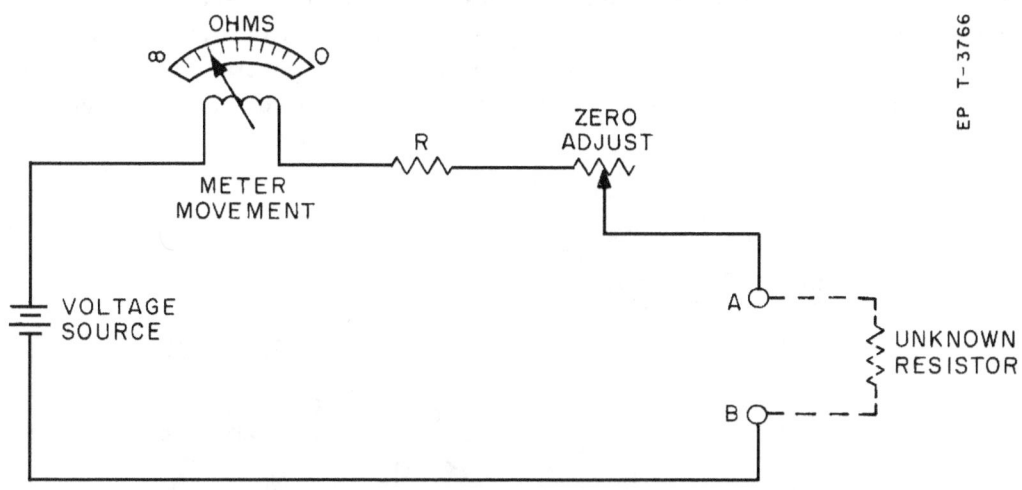

EP T-3766

Exhibit 1-105 — Ohmmeter Circuit Schematic

1.9.5 — Meggers

The measurement of resistances in the megohm and higher regions (such as conductor insulation) requires a much higher voltage to create a current flow which can be detected than is provided by the low voltage of batteries found in ordinary ohmmeters. An instrument called a **megger** (megohmmeter) is used to measure such high resistances. The megger is a portable instrument consisting of a hand-driven d-c generator and an instrument portion as shown in Exhibit 1-106. The instrument portion is of the opposed-coil type. Coils a and b are mounted on the movable member c with a fixed angular relationship to each other, and are free to turn as a unit in the magnetic field established by a permanent magnet. Coil b tends to move the pointer counterclockwise and coil a tends to move the pointer clockwise. There are no restraining springs on the movable member of the instrument portion of the megger. Therefore, when the generator is not operated, the pointer floats freely and may come to rest at any position on the scale. The guard ring intercepts leakage current. Any leakage currents intercepted are shunted to the negative side of the generator. They do not flow through coil a; therefore, they do not affect the meter reading.

If the test leads are open-circuited, no current flows in coil a. If the d-c generator is driven by the hand crank, current flows internally from the negative terminal of the generator through coil b, through R_2, and back to the positive terminal of the generator, deflecting the pointer to infinity. When a resistance such as R_x is connected between the test leads, current also flows from the negative terminal of the generator through coil a, through R_3, through R_x, and back to the positive terminal of the generator, tending to move the pointer clockwise. The moving element, composed of both coils and the pointer, comes to rest at a position at which the two forces balance. The position depends upon the magnitude of the current in coil a, which depends upon the value of resistance, R_x. Because changes in voltage of the hand-driven generator affect both coil a and coil b in the same proportion, the reading is independent of the voltage. If the test leads are short-circuited, the maximum current flows in coil a and the pointer rests at zero. Resistor R_3 limits the current to a safe value when the test leads are shorted.

Meggers provided aboard ship usually are dual range and are rated at 250 and 500 volts. To test conductors for insulation breakdown, the test leads are connected between the conductor and the outside surface of the insulation. To avoid excessive test voltages, most meggers are equipped with friction clutches. When the generator is cranked faster than rated speed, the clutch slips, limiting the generator voltage. Because of the high test voltage, all low-voltage equipment must be disconnected from any circuit which is to be meggered.

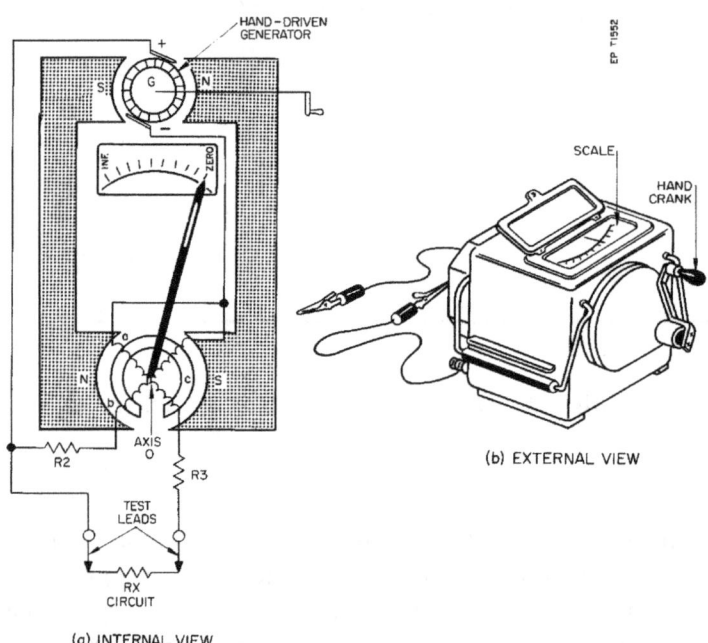

Exhibit 1-106 — Internal And External Views Of Megger

1.9.6 — A-C Measurements

Although several types of meter movements will respond to a-c directly, they are usually limited in frequency response to a few hundred Hz. Thus, the majority of a-c meters convert the a-c waveform to a d-c current which is then measured by a D'Arsonval movement. Exhibit 1-107 illustrates the use of a fullwave bridge rectifier to convert the a-c to a pulsating d-c. The d-c meter registers the average value of the rectified waveform. Although the meter is actually responding to a d-c current, the meter scale is calibrated in terms of an a-c quantity. Since the RMS value of a sine wave is frequently required, most a-c meters are calibrated to read the RMS values of sine waves directly. For example, the meter in Exhibit 1-107 registers the average value of a full-wave rectified sine wave which is equal to $2A/\pi$, where A is the peak value of the a-c sine wave. However, the RMS value of a sine wave is related to the peak value, A, by $A/\sqrt{2}$. Thus, the ratio of the RMS value to the average value is $\pi/2\sqrt{2}$ and this is the calibration factor used to mark the meter dial in RMS values. It is important to note that a particular meter calibrated to read the RMS value of a sine wave will not read the RMS value of an arbitrary waveform. This is because the ratio of the RMS value to the average value for an arbitrary waveform is not the same as the ratio for a sine wave. To use an a-c meter to read the RMS value of an arbitrary waveform, the type of a-c to d-c conversion employed by the meter must be known in order that the average value may be calculated. (The average value of a half-wave rectifier is not the same as the average value of a fullwave rectifier.) Once the average value is known, the RMS value is determined and the ratio of the RMS to average value is computed and compared with the same ratio for a sine wave to obtain a correction factor to be applied to the meter reading.

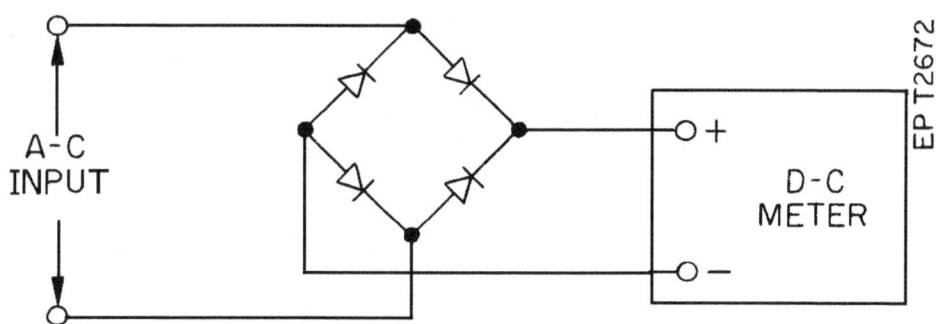

Exhibit 1-107 — Measuring An A-C Signal With A D-C Meter

1.9.7 — Multimeter

For testing purposes, it is usually inconvenient to have on hand all the different meters required to perform the various measurements required. To alleviate this problem, multimeters, combining the functions of ohmmeters, ammeters, and voltmeters, both a-c and d-c, are provided in one convenient enclosure. The multimeter contains only one D'Arsonval movement with function and range switches that permit the operator to switch to the proper circuitry to read the signal over the proper range.

In general, a very sensitive movement is used to permit low current, and thus small batteries, to be used. Care must be exercised to ensure that the multimeter switches are in the proper positions before connecting the meter to a voltage source, otherwise it may be damaged by overcurrent. It is good practice to set the range switch to the highest possible range when measuring unknown levels.

1.9.8 — Wattmeters

Because of its two-coil arrangement, the dynamometer movement is used to measure watts in an a-c circuit. A single-phase wattmeter consists of two coils: a fixed coil, made of very low resistance, heavy wire called the current coil, and a movable coil, made of a greater number of turns of high-resistance, fine wire called the voltage coil. The wattmeter is connected to a single-phase load so that the current flowing in the current coil is proportional to the current flowing into the load and the voltage across the voltage coil is proportional to the voltage across the load, as shown in Exhibit 1-108.

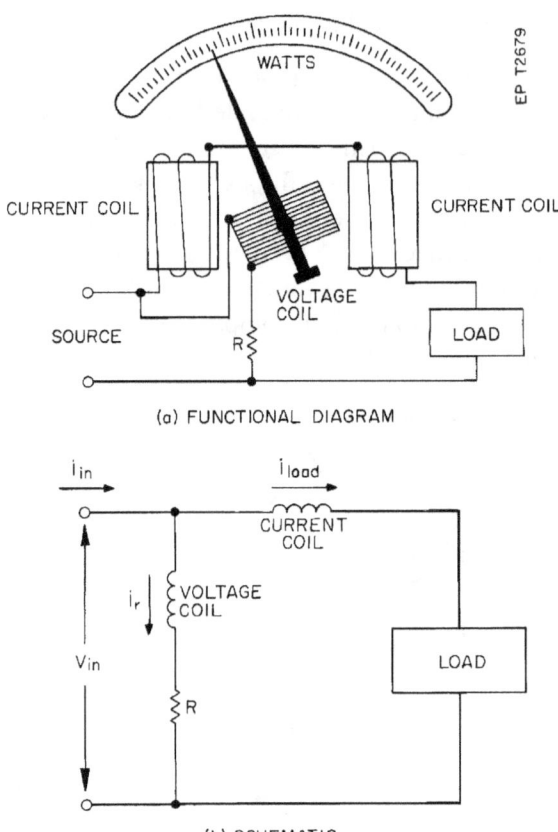

(a) FUNCTIONAL DIAGRAM

(b) SCHEMATIC

Exhibit 1-108 — Dynamometer Used As A Single-Phase Wattmeter

When connecting a wattmeter, care must be taken to avoid placing the current coil across the load as this could severely damage the meter. It is also important when using a wattmeter to remember that it measures average real power.

For a three-phase system, the total power can be expressed by:

$$P = \text{Ave} \left[V_a i_a \right] + \text{Ave} \left[V_b i_b \right] + \text{Ave} \left[V_c i_c \right].$$

This result indicates that the power in a three-phase system can be measured by three meters, with one meter sampling V_a and i_a, another sampling V_b and i_b, and the third sampling V_c and i_c. The total power is obtained by summing the readings of the three meters. Although three meters will give the correct result, it will now be shown that by proper connection only two meters are needed to measure three-phase watts for a three-wire load balanced or unbalanced and a four-wire load balanced. For such three-phase loads:

$$i_a + i_b + i_c = 0$$

$$i_a = -(i_b + i_c)$$

Substituting this result for i_a into the general power relationship:

$$P = \text{Ave} \left[V_a(-i_b - i_c) \right] + \text{Ave} \left[V_b i_b \right] + \text{Ave} \left[V_c i_c \right]$$

$$P = \text{Ave} \left[i(V_b - V_a) \right] + \text{Ave} \left[i_c (V_c - V_a) \right]$$

This result indicates that active power can be measured by two meters, with one sampling i_b and the voltage V_b - V_a and the other sampling i_c and the voltage V_c - V_a. The circuit of Exhibit 1-109 accomplishes this. A three-phase wattmeter may be constructed using four coils (two fixed and two moving) corresponding to two single-phase wattmeters connected as shown in Exhibit 1-109 but both mounted on a single shaft with a single pointer as shown in Exhibit 1-110. The position of the pointer depends upon the sum of the torques exerted by the two moving coils.

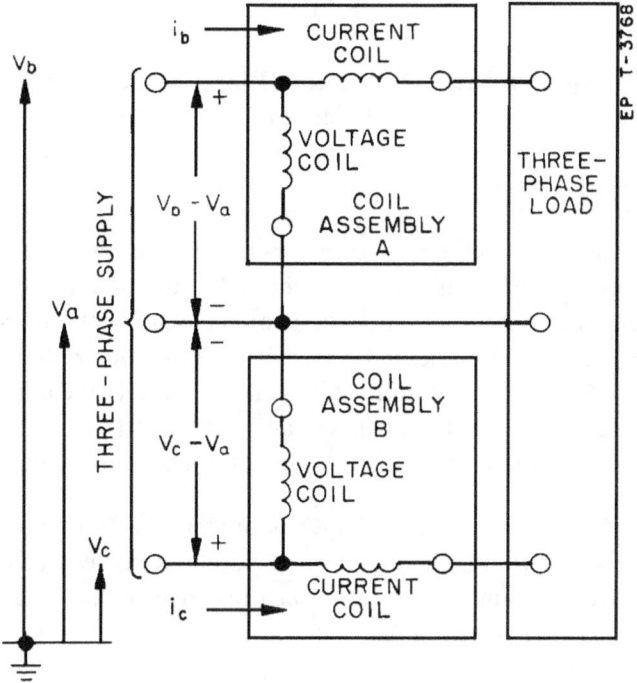

Exhibit 1-109 — Two Dynamometers Used As A Three-Phase Wattmeter

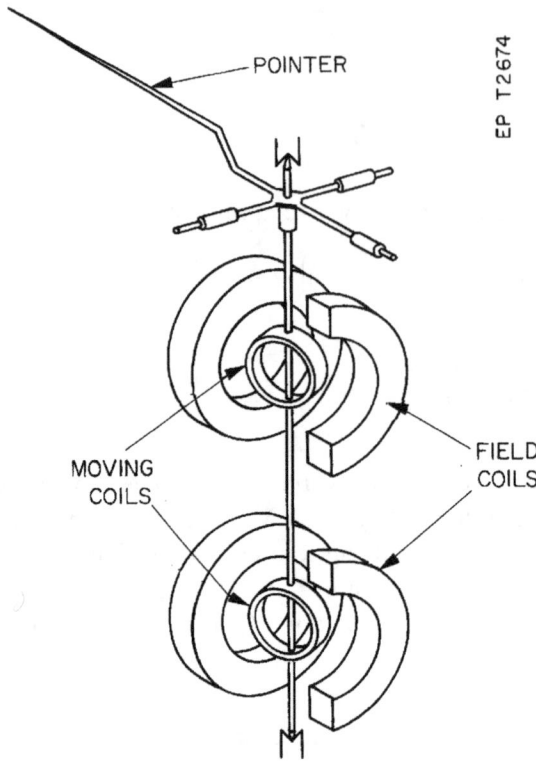

Exhibit 1-110 — Simplified View Of A Three-Phase Wattmeter With Dynamometer Movements

1.9.9 — Instrument Transformers

It is not usually practical to connect instruments and meters directly to high-voltage circuits. Unless the high-voltage circuit is grounded at the instrument, a dangerously high potential-to-ground may exist at the instrument. Further, instruments become inaccurate when connected directly to a high voltage, because of the electrostatic forces that act on the indicating element. By means of instrument transformers, instruments may be entirely insulated from the high-voltage circuit and yet indicate accurately the current, voltage, and power in the circuit. Low-voltage instruments having standard current and voltage ratings may be used for all high-voltage circuits, irrespective of the voltage and current ratings of the circuits, if instrument transformers are utilized.

Potential transformers are used to measure high voltages with standard low-voltage meters. Potential transformers do not differ in construction from the constant-potential power transformers with the exception that they need only a small power handling capability since only instruments, meters, and sometimes indicator lights are ordinarily connected to their secondaries.

Exhibit 1-111 shows schematically the connection of a potential transformer to measure a high voltage across a load. The secondary of the potential transformer should always be grounded at one point to eliminate static electricity from the instrument and to ensure personnel safety. Care must be taken that the secondary side of the potential transformer is never short-circuited, since this will produce excessively high currents in both the primary and secondary windings, resulting in damage to the transformer.

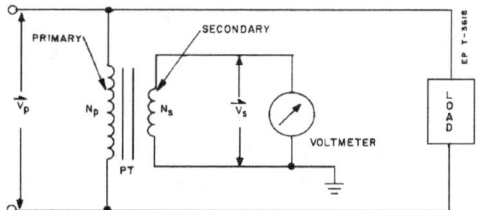

Exhibit 1-111 — Typical Potential Transformer Application

The actual load voltage ($\vec{V}_p$) is related to the meter voltage ($\vec{V}_s$) by the turns ratio of the potential transformer.

$$\vec{V}_p = \frac{N_p}{N_s} \vec{V}_s .$$

Since potential transformers are generally designed to utilize a 150-volt meter, the ratio of primary to secondary turns is quite high.

Current transformers are used to measure high currents with standard low amperage meters. They also provide isolation from high-voltage lines.

The current transformer has a primary winding of a relatively few turns of heavy wire connected in series with the line. The secondary winding of a standard current transformer is rated at 5 amperes. The connection of a current transformer to measure a large load current is shown in Exhibit 1-112. One side of the secondary is grounded for personnel safety.

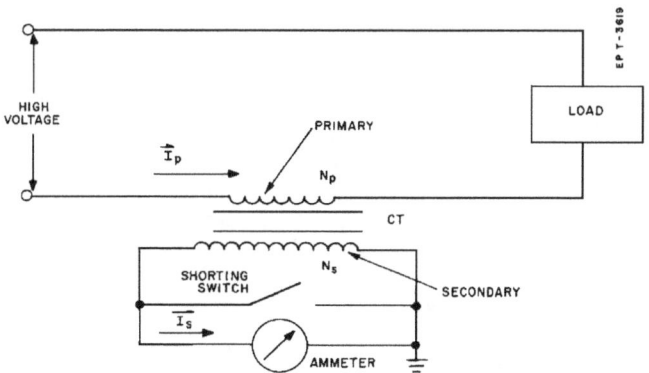

Exhibit 1-112 — Typical Current Transformer Application

The current transformer differs from the ordinary constant-potential transformer in that its primary current is determined entirely by the load and not by its own secondary load. If the secondary of a current transformer becomes open-circuited, a high voltage will exist across the secondary because the large ratio of secondary to primary turns causes the transformer to act as a step-up transformer. Also, since the effects of the opposing magneto-motive force of the secondary no longer exist, the magnetic flux in the core will depend on the total primary magneto-motive force acting alone. This creates a large increase in magnetic flux, producing excessive core loss and heating in addition to the dangerously high voltage across the secondary winding. For this reason, the secondary winding of a current transformer **should never be opened** while the primary winding carries current. As shown in Exhibit 1-112, a shorting switch is always provided with current transformer circuits with portable meters. If the meter is to be removed, the switch is always closed first. It is good practice to keep the current transformer secondary terminals shorted by a meter or switch even when the primary winding is not energized. This precludes an accident due to inadvertent energization of the primary winding before the secondary winding is connected.

The actual load current ($\vec{I}_p$) is related to the meter current ($\vec{I}_s$) by the turns ratio of the current transformer:

$$\vec{I}_p = \frac{N_s}{N_p}\,\vec{I}_s \, .$$

Chapter 2 — ELECTRONICS REVIEW

Section 2.1 — Solid State Devices

2.1.1 — Semiconductors

Semiconductors are a group of materials having electrical conductivities intermediate between conductors and insulators. The most important consideration in determining the electrical conductivity of a material is the availability of conduction or free electrons: that is electrons that are free to move within the material to constitute an electric current. In insulators, all electrons are bound tightly to individual atoms and are unavailable for conduction; in conductors there is a plentiful supply of free electrons. The vital feature of semiconductors is that the supply of free electrons can be readily controlled by any or all of several means: chemical, thermal, electrical, and others.

Chemical control is used during initial fabrication. The most popular electronic semiconductors are silicon and germanium. Each of these is a valence-4 material, normally arranged in a diamond-type covalent-bonded lattice, that provides a very small, but significant, supply of free electrons when at room temperature. By "doping" the pure material with a very small amount (a few parts per billion) of valence-5 material, such as arsenic, a much larger supply of conduction electrons is provided and increased conductivity results.

Doping with a valence-3 material (such as gallium) has a different effect. The gallium atom occupies a lattice site normally occupied by a silicon atom but lacks one electron needed to form all available covalent bonds with its neighbors. This "hole" in the bonding system is easily filled by an electron shifting (due to thermal agitation) from an adjacent site. Thus the hole is mobile, even among silicon atoms in the doped material. Since the hole carries an effective charge of +1, due to the one charge unit of the silicon atom that is not offset by an electron, a hole behaves much like a positive free electron.

Doped semiconductors in which the majority of carriers are electrons are called n-type; those in which the majority of carriers are holes are called p-type. Electronic semiconductor devices are fabricated as various combinations of p- and n-type materials to perform specific electrical functions.

2.1.2 — P-N Junction

Suppose that pieces of p-type and n-type material are brought together, forming a p-n junction while maintaining a continuous crystalline lattice structure. Electrons from the n-type material will migrate into the p-type material and drop into holes near the junction to complete the available covalent bonds. There is now an imbalance incharge in the materials. The p region has gained electrons and taken on a negative charge, while the n region has lost electrons and taken on a net positive charge.

The movement of electrons across the p-n junction to fill holes is called carrier migration and the carrier-free region thus developed is called a depletion zone. This zone is typically about 0.5 microns deep. The effects of carrier migration are illustrated in Exhibit 2-1.

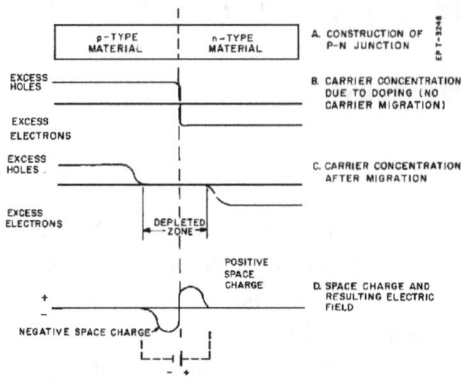

Exhibit 2-1 — Forward And Reverse Biased P-N Junctions

As previously mentioned, the migration of electrons from n- to p-type material also results in formation of regions of space charge near the junction. This space charge establishes an electric potential of a few tenths of a volt across the junction which, once established, opposes further migration of electrons from n- to p-type material. The depletion zone is maintained essentially free of carriers by this potential. Any electrons or holes that appear in the depletion zone (e.g., by thermal formation of electron-hole pairs) will be swept back to the n and p regions respectively. The presence of the space charge is denoted by the imaginary battery shown by dashed lines in Exhibit 2-1, Part (d).

If an external voltage is applied to the p-n junction and reinforces the space charge potential, the depletion zone remains free of carriers and acts as an insulator. No current flows through the junction. This is called reverse biasing and is illustrated in Exhibit 2-2, Part (b). The width of the depletion zone increases because electrons in the n-type material are attracted to the positive terminal and holes in the p-type material are attracted to the negative terminal.

When an externally applied potential opposes the space charge potential, the junction remains nonconducting until the external voltage is just sufficient to offset the space charge potential. At this point electrons and holes have sufficient energy to cross the depletion zone (which will have become quite narrow) and a current flows. The current increases very rapidly with increasing voltage and, for moderately large applied voltage, the junction will offer almost no resistance to current flow. This is called forward biasing and is illustrated in Exhibit 2-2, Part (a).

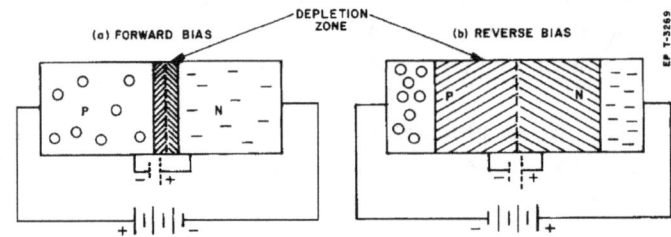

Exhibit 2-2 — Formation Of Depletion Zone At P-N Junction

A simple p-n junction thus acts as a one-way device, allowing current flow in the forward direction and no current flow in the reverse direction. A device that uses the simple p-n junction in this way is called a diode.

2.1.3 — Diodes

2.1.3.1 — Ideal Diodes

An ideal diode acts like a short circuit when forward biased and acts like an open circuit when it is reverse biased. In other words, when an ideal diode is conducting, there is no resistance across the p-n junction and therefore there is no associated voltage drop. When an ideal diode is non-conducting, no current flows through the diode. The schematic symbol and the voltage-current (v-i) characteristic for an ideal diode are shown in Exhibit 2-3. The arrowhead, known as the <u>anode</u>, points in the direction of current flow when the diode is conducting. The other terminal is called the <u>cathode</u>.

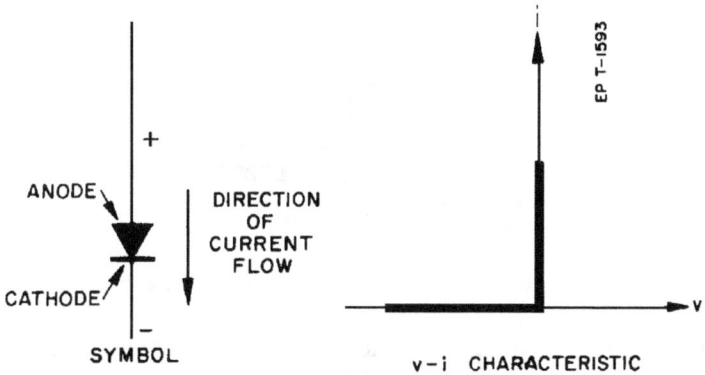

Exhibit 2-3 — The Ideal Diode

Exhibit 2-4 shows a simple half-wave rectifier using an ideal diode. When the input signal (v_{in}) is on the positive half-cycle, the diode is conducting and the output signal (v_{out}) is equal to the input signal. When v_{in} is on the negative half-cycle, the diode is non-conducting and v_{out} is zero.

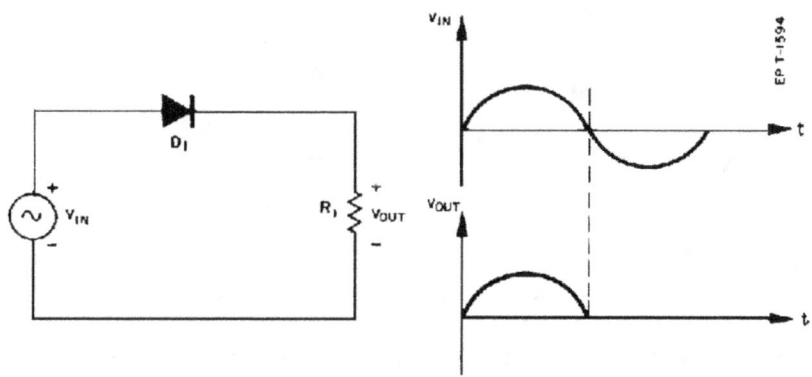

Exhibit 2-4 — The Half-Wave Rectifier

Exhibit 2-5, (a) shows a <u>full-wave</u> or <u>bridge rectifier</u>. Exhibit 2-5, (b) and Exhibit 2-5, (c) show the full-wave rectifier circuits for the positive and negative half-cycles of v_{in}. The output of the full-wave rectifier is shown in Exhibit 2-5, (d).

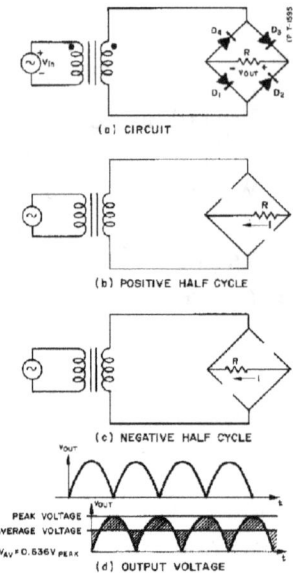

Exhibit 2-5 — Full-Wave Rectifier

Diodes are also be used in logic circuit applications such as the "OR" gate and the "AND" gate shown in Exhibit 2-6. The output of the OR gate assumes the voltage of the highest input voltage, where the AND gate output assumes the voltage of the lowest input voltage. Logic circuits will be discussed in more detail in Section 2.9.

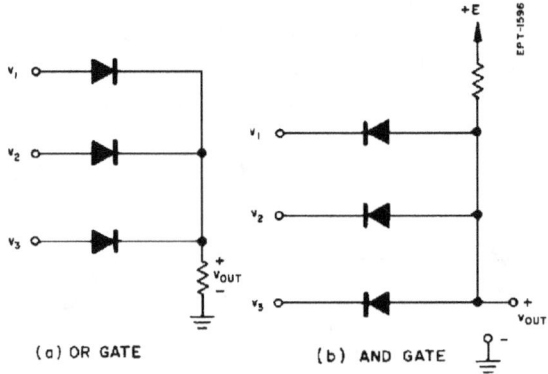

Exhibit 2-6 — Logic Circuit Applications

Diodes are also used in various wave-shaping circuits. Two such circuits, the clipper and the clamper circuit are shown in Exhibit 2-7. In the clipper circuit, when the input voltage is less than E1, the diode is reverse biased allowing no current to pass and thus the output voltage (v_{out}) is equal to the input voltage. When the input voltage is greater than E1, the diode is forward biased and acts like a short circuit forcing v_{out} to the value of

E1, therefore clipping off any part of the input signal with a voltage greater than E1. The clamper circuit is used to shift the DC level of the input signal. The output, taken across the diode, is

$$V_{out}(t) = V_{in}(t) - V_{max}$$

As Exhibit 2-7 shows, the maximum value of the output voltage is zero and the circuit is said to clamp the output voltage to a maximum of zero.

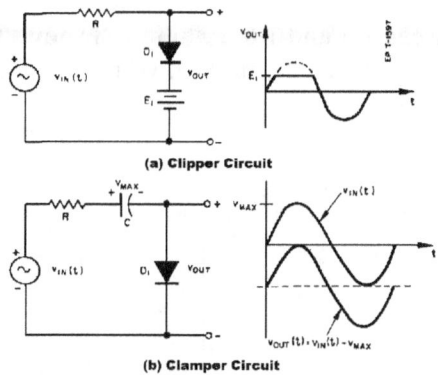

(a) Clipper Circuit

(b) Clamper Circuit

Exhibit 2-7 — Wave-Shaping Circuits

2.1.3.2 — Real Diodes

The actual or real diode has characteristics that are similar to those of an ideal diode. The relationship between applied voltage and current in a real diode is as follows:

$$I = I_o \left(e^{\frac{qV}{nkT}} - 1 \right)$$

I	= current through the diode
Io	= reverse saturation current
q	= electron charge (1.6 * 10E-19 coulombs)
V	= voltage across diode
n	= empirical constant related to material (between 1 and 2)
k	= Boltzmann's constant (1.38 * 10E-22 Joules/K)
T	= absolute temperature (K)

The above equation can be simplified by letting $V_T = kT/q = T/11600$. This results in the following equation:

$$I = I_o \left(e^{\frac{V}{nV_T}} - 1 \right)$$

The reverse saturation current is a function of material purity, doping, and geometry of the diode. Reverse current through the diode approaches this value as the reverse bias voltage is increased. The empirical constant n, sometimes known as the exponential ideality factor, varies with the current and voltage levels. However, some diodes operate over a large range of voltage and current levels with a fairly constant value of n. For germanium diodes, n is usually considered to be 1. Theoretically, n should be 2 for silicon diodes. However, in practice n is usually between 1.3 and 1.6. Exhibit 2-8 shows the typical operating characteristics of a typical diode. In the figure, V_γ is known as the turn-on voltage and is typically 0.7 volts for silicon diodes and 0.2 volts for germanium diodes.

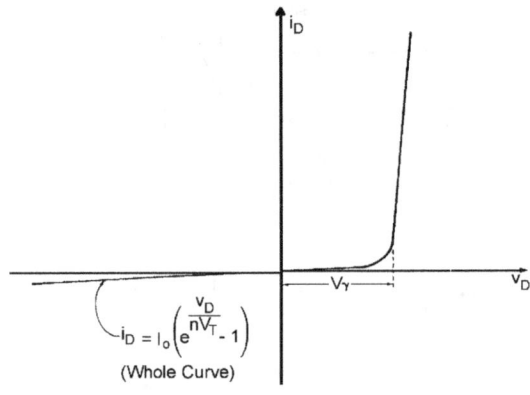

Exhibit 2-8 — Typical Diode Characteristics

2.1.3.3 — Zener Diodes

In all real diodes, there is a reverse bias voltage beyond which the reverse bias current increases greatly with little increase in reverse bias voltage. A <u>zener diode</u> is a device where doping is performed in such a way that the slope of the v-i characteristic curve past the reverse breakdown voltage is very steep. Exhibit 2-9 shows the schematic symbols a commonly used for zener diodes.

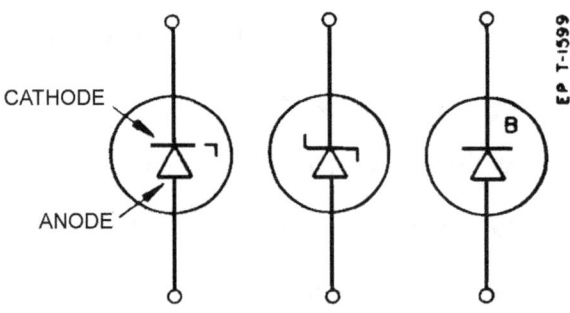

Exhibit 2-9 — Symbols For Zener Diodes

The v-i characteristic curve for a typical zener diode is shown in Exhibit 2-10. The maximum reverse current (I_{zmax}) that a zener diode can withstand is dependent on the design and construction of the diode. The minimum leakage current (I_{zmin}) below the knee of the characteristic curve is usually assumed to be 0.1 I_{zmax}. Use of I_{zmin} assures that the breakdown curve remains close to parallel with the I-axis. The amount of power that the zener diode can withstand is $P_{zmax} = I_{zmax} * V_z$.

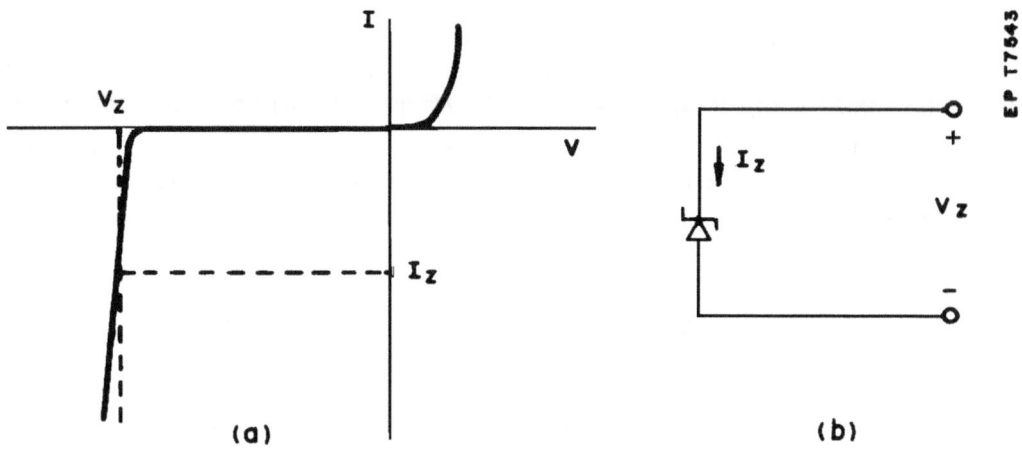

Exhibit 2-10 — Typical Zener Diode Characteristic Curve

Exhibit 2-11 shows a basic d-c voltage regulating circuit using a zener diode. In designing this regulating circuit, a zener diode with a breakdown voltage equal to the desired regulated output voltage is chosen. Next, a limiting resistor is chosen to drop the input voltage (E_{in}) to the desired level (E_{out}) while maintaining the current flow through the zener diode (I_z). The zener regulator operates as follows. When the input voltage increases or the load current decreases, current flowing through the zener diode increases, increasing the voltage drop across R1 so that the output remains constant. When the input voltage decreases or the load current increases, the diode current decreases and the voltage drop across R1 decreases to maintain a constant output voltage.

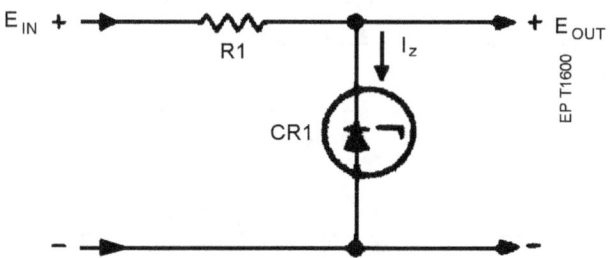

Exhibit 2-11 — Zener Diode Voltage Regulating Circuit

2.1.4 — Transistors

The transistor is a three terminal device consisting of two n-type materials separated by a p-type material (NPN transistor) or two p-type materials separated by an n-type material (PNP transistor). Transistors which can be idealized as two junctions placed back to back are known as junction transistors. Another group of transistors in which current is controlled by means of an electric field are known as field effect transistors.

2.1.4.1 — Bipolar Junction Transistors

The three different layers or sections are identified as the emitter, base, and collector and are shown in Exhibit 2-12. The emitter is a heavily doped, medium sized layer designed to inject or emit electrons. The base is a thin layer of medium doped material designed to pass electrons. The collector is a wide layer of lightly doped material designed to collect electrons. Because current flow exists as both electron and hole flow, these devices are commonly known as bipolar junction transistors (BJTs). The principles of operation for both types of BJTs are the same, with the only difference in circuit operation being the direction of current flow. For this reason, the operation of a PNP transistor is described here and the principle will be extended to the NPN transistor in a later paragraph.

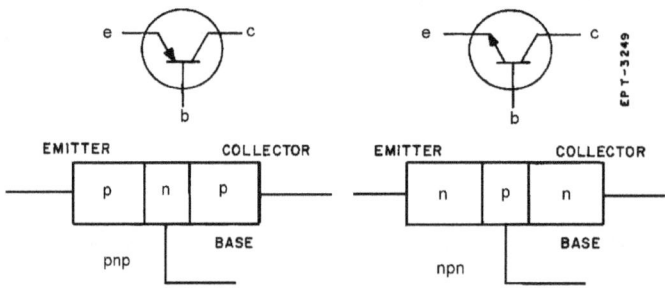

Exhibit 2-12 — Transistor Construction Symbols

A PNP transistor is operated with the collector negative with respect to the emitter. The collector bias, shown as a battery in Exhibit 2-13, ranges from 3 to 100 volts depending on the transistor and application. If the base is held at the same potential as the emitter, the base-collector p-n junction is reverse-biased and no current flows through the collector.

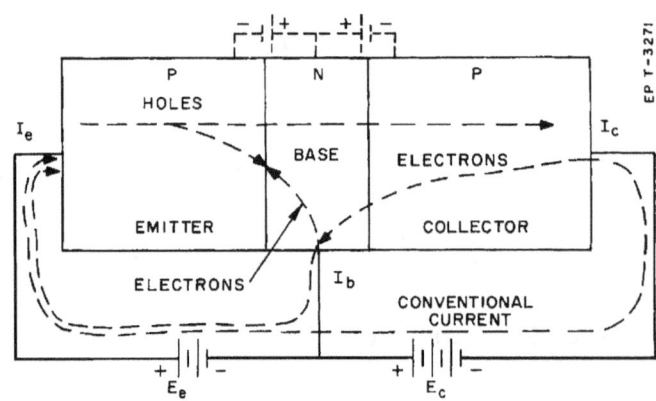

Exhibit 2-13 — Operation Of A PNP Transistor

If the base is biased a few tenths of a volt negative with respect to the emitter, the emitter-base p-n junction becomes forward-biased and a current flows through that junction. This current consists of electrons flowing from n- to p-type material and also of holes flowing from p- to n-type material. The electrons that cross into the p-type emitter quickly combine with holes. The holes entering the n-type base, however, have a high probability of continuing through to the collector before they recombine with electrons in the base. This occurs because the base is thin and lightly doped (only a slight excess of electrons). In a properly built transistor, only a small fraction of the holes entering the base recombine or are picked up by the base lead. Most of them (sometimes over 99 percent) continue through to the collector. These holes constitute an emitter-to-collector current many times larger than the emitter-to-base current. The basic operation of a transistor is then as follows: if a small current is drawn from emitter to base, a proportionately larger current will flow from emitter to collector.

An NPN transistor operates in much the same manner as a PNP transistor (Exhibit 2-14). The chief differences are that the collector is positive with respect to the emitter, and that electrons, rather than holes, flow from emitter to base and from emitter to collector. However, once the transistor is built and connected into a circuit, the type of carrier involved is essentially an internal property of the transistor. The user is concerned only with the amount and direction of bias required to make the transistor operate. The following rules apply:

1. The collector-base junction is reverse biased.
2. The emitter-base junction is forward biased.
3. Emitter-to-collector current is proportionately larger than emitter-to-base current.

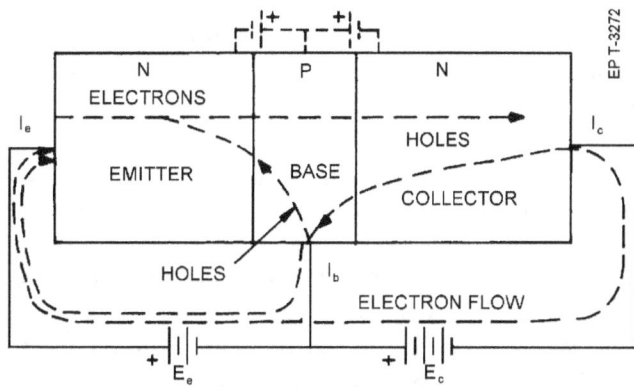

Exhibit 2-14 — Operation Of An NPN Transistor

In transistor applications the base current I_b is usually varied and the collector current I_c follows the time dependence of I_b. Normally I_c is much larger than I_b. The ratio $\beta = I_c/I_b$, called the current gain, ranges from 10 to several hundred. A transistor is a current-actuated device. That is, **no** collector current will flow **unless** a base current is introduced.

2.1.4.2 — BJT Circuits

Transistor circuits are classified according to which lead is common to both input and output circuits. There are three leads to a transistor; hence, three possible circuits: common emitter, common base, and common collector. "Common" as used here simply means that the particular lead is a part of both the input and output circuits: the lead is not necessarily grounded.

2.1.4.2.1 — Common Emitter

The common emitter configuration is the most widely used transistor amplifier circuit. Schematics of PNP and NPN circuits are shown in Exhibit 2-15. The input circuit consists of the emitter-base junction the bias components V_{BB} and R_B, and the input voltage V_S. The purpose of V_{BB} and R_B is to ensure that the emitter-base junction is biased in the forward direction and carrying some current for either a positive or negative input voltage V_S. If the input bias were not used the emitter-base junction would act as a rectifier for the input signal and only half of an a-c waveform would be amplified.

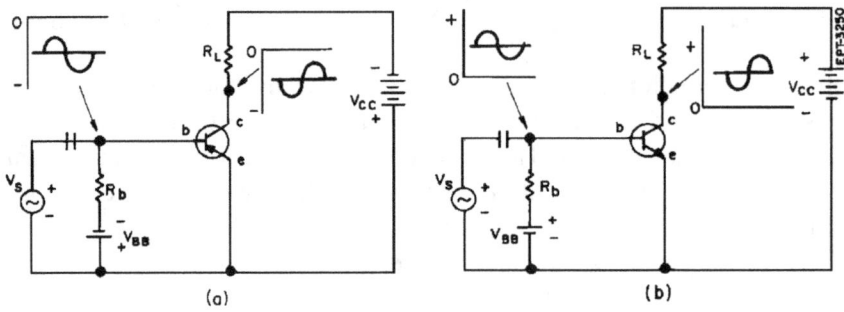

(a)   (b)

Exhibit 2-15 — Common Emitter Amplifier: (a) PNP Type And (b) NPN Type

The input signal V_S is connected between the emitter and base by means of a capacitor. The capacitor is used in this example to ensure that only the a-c portion of V_S is amplified. The d-c bias of the transistor input is established independently of V_S by V_{BB} and R_B.

The output circuit consists of the bias supply VCC, the load resistance RL, and the collector-base and emitter-base junctions of the transistor. The output voltage is measured between the collector and the emitter.

The common emitter amplifier is an inverting amplifier because a positive-going input signal results in a negative-going output signal and vice-versa. Consider the PNP circuit of Exhibit 2-15. A positive value of V_S will cause the voltage at the base to be less negative, decreasing the emitter-to-base current. The emitter-to-collector current will be decreased by a proportionately larger amount. The reduced current through R_L will result in a reduced voltage difference across R_L; thus the collector will be more negative. For the NPN circuit, a positive input voltage will increase the base current causing an increased collector current. The higher collector current will increase the voltage drop across R_L, and the collector voltage will be less negative.

The common-emitter amplifier has the following properties:
1. Moderate voltage gain
2. Inverting output
3. Intermediate input impedance, compared to other amplifiers
4. Intermediate output impedance, compared to other amplifiers.

2.1.4.2.2 — Common Base

The common base amplifier is shown schematically for PNP and NPN transistors in Exhibit 2-16. The input circuit consists of the bias supply V_{EE}, resistor R_E, the emitter-base junction, and the input signal V_s. The function of the bias components V_{EE} and R_E is the same as for the common emitter circuit. The input voltage is applied between emitter and base.

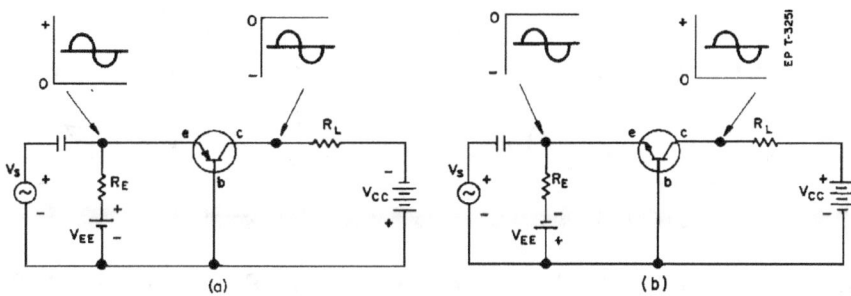

Exhibit 2-16 — Common Base Amplifier: (a) PNP Type And (b) NPN Type

The output circuit consists of the load resistance R_L, bias supply VCC, and the collector-base junction. Output voltage is measured between collector and base.

The common-base amplifier does not invert the signal. This is shown for the PNP case as follows. An increase in V_S reinforces V_{EE} and increases the emitter-to-base current, leading to a much larger increase in emitter-to-collector current. The collector voltage V_{out} becomes less negative because of the increased voltage drop across R_L. In the NPN circuit a positive V_S decreases the emitter-to-base current, leading to a larger decrease in emitter-to-collector current, a decreased voltage drop across R_L, and thus a more positive collector voltage.

The common base amplifier is not used as often as the common emitter type. Its principal use is to couple a low-impedance signal (low voltage and high current) to a higher-impedance amplification stage without inverting the polarity of the signal.

The common base amplifier has the following properties:
1. Moderate voltage gain (about the same as common emitter
2. Non-inverting output
3. Low input impedance
4. High input impedance.

2.1.4.2.3 — Common Collector

The common collector amplifier is also called the emitter follower. Circuits for PNP and NPN transistors are shown in Exhibit 2-17. In this configuration, the load is connected between the emitter and the common lead and the output voltage is measured across the load. The common collector amplifier does not invert the input signal. In the PNP case, a positive input signal drives the base less negative and reduces the base current. This results in a reduced emitter-to-collector current and thus a less negative output voltage at the emitter. In the NPN case, a positive-going input signal drives the base more positive with respect to the emitter, causing the base current to increase. This results in an increase to the emitter current causing a more positive output voltage at the emitter.

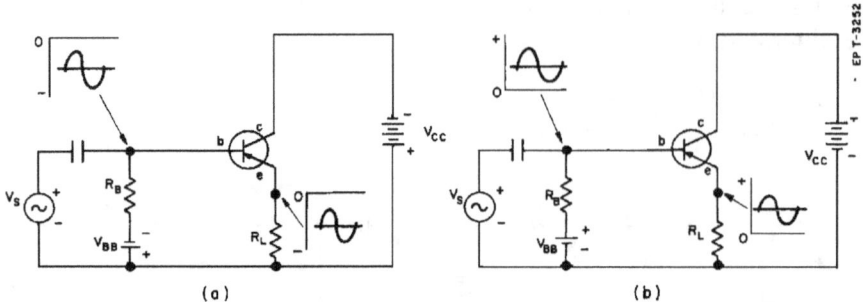

Exhibit 2-17 — Common Collector Amplifier: (a) PNP Type And (b) NPN Type

Since the voltage drop across the forward-biased emitter-base junction never exceeds a few tenths of a volt, the output voltage closely follows the input voltage; hence, we derive the alternate name of emitter follower for this amplifier. For the same reason, the emitter voltage must always be slightly less than the base voltage, and the voltage gain of the emitter follower can never exceed unity.

The common collector amplifier has the following properties:

1. Voltage gain slightly less than unity
2. Non-inverting output
3. High input impedance
4. Low input impedance.

The most frequent use of the common collector amplifier is to provide high output current to a low-impedance load.

2.1.4.3 — Field Effect Transistors

In addition to the bipolar junction transistors, there is another type of transistor known as the field effect transistor (FET). There are two types of FETs, the junction FET (JFET) and the insulated gate FET more commonly known as the metal-oxide semiconductor FET (MOSFET). The FET is a unipolar device since current exists either as electron flow or hole flow. The controlling parameter for a FET is voltage instead of current as with a BJT.

2.1.4.3.1 — MOSFETs

Exhibit 2-18 shows the connections for a typical MOSFET. The current between the source and drain terminals is controlled by the voltage present between the gate and the source terminals. The gate terminal is insulated from the main channel, preventing current flow through the gate. As a result, the potential between the gate and source causes an electric field whose intensity is a function of the potential difference. This electric field exerts a force on charges moving between the source and drain terminals and thus controls the current flow.

If the voltage between the gate and source (V_{gs}) is negative, the region where electrons can move is limited. This is shown as the cross-hatched region in Exhibit 2-18. If V_{gs} is made more negative, the force acting on the charges is increased and the region where electrons can move (and current can flow) becomes narrower, decreasing the source-drain current. If V_{gs} is made less negative, the channel becomes wider, allowing more current to flow.

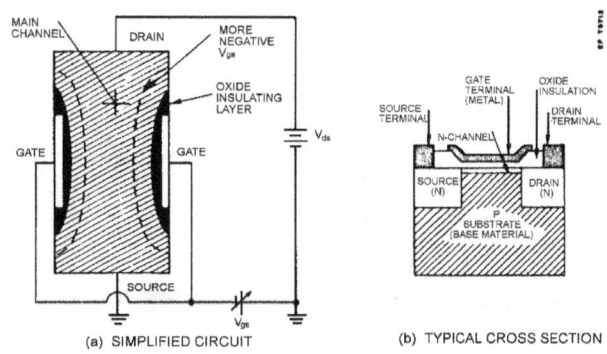

(a) SIMPLIFIED CIRCUIT   (b) TYPICAL CROSS SECTION

Exhibit 2-18 — Field Effect Transistor (MOSFET)

Exhibit 2-19 shows a typical MOSFET amplifier circuit. The capacitor (C) prevents any d-c component of the input signal from affecting the bias set by R_1 and R_2. The bias network consisting of R_d and R_s establishes the operating conditions so that the source-drain current varies linearly with V_{gs}. An increase in gate voltage caused by the input signal causes the current through R_d to increase. This increases the voltage drop across R_d and causes the output voltage (V_o) to drop. The change in output voltage is greater than the change in gate voltage, so the stage amplifies. MOSFET amplifiers typically have moderate gain (<10) and input impedances as high as 10^{14} ohms.

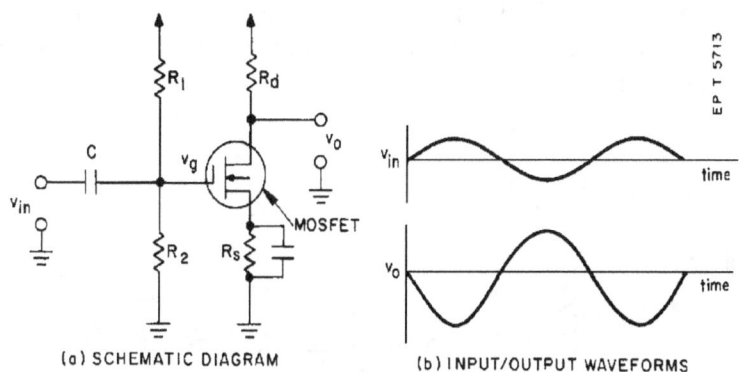

(a) SCHEMATIC DIAGRAM   (b) INPUT/OUTPUT WAVEFORMS

Exhibit 2-19 — MOSFET Amplifier Circuit

2.1.4.3.2 — Junction Field Effect Transistors

The junction field effect transistor or JFET is a unipolar device connected as shown in Exhibit 2-20. The JFET and MOSFET are very similar except that the gate terminal of the JFET is not insulated from the main conduction channel. Like the MOSFET, the drain-source current in the JFET is controlled by the voltage potential across the gate and source terminals. Applying a negative voltage between the gate and source terminals will reverse bias their junction, restricting current between the drain and the source. JFET amplifiers have a high input impedance, allowing some current to flow into the gate, and moderate gain compared to BJT amplifiers.

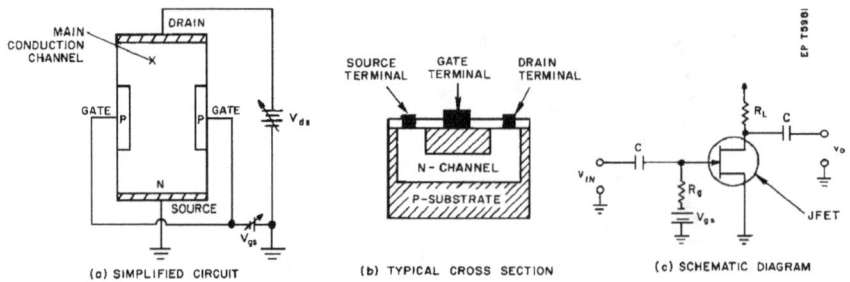

Exhibit 2-20 — Field Effect Transistor (JFET)

2.1.4.4 — Four-layer Devices

In previous sections, both two-layer devices (diodes) and three-layer devices (transistors) have been discussed. This section discusses the following four-layer devices: thyristors, silicon controlled rectifiers (SCRs), and gate turn-off thyristors (GTOs).

2.1.4.4.1 — Thyristors

The four-layer diode, commonly known as the thyristor, is shown in Exhibit 2-21. The anode is at the outside p-type region and the cathode is at the outside n-type region. When the anode is biased positively with respect to the cathode (positive v), the device is forward biased. As shown in the v-i characteristic in Exhibit 2-21, the forward biased condition has two different states, the high impedance, forward-blocking state and the low impedance, forward-conducting state.

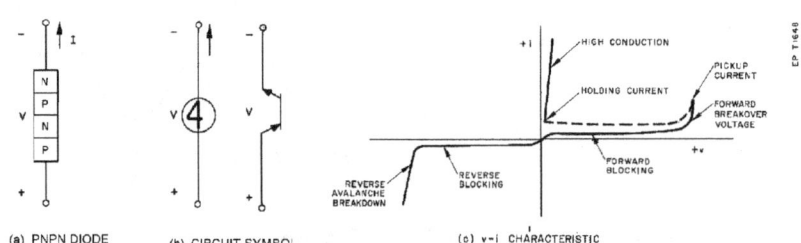

Exhibit 2-21 — Thyristor

The operation of a thyristor can best be described with the aid of the two-transistor analogy as shown in Exhibit 2-22. For simplicity, both transistors will be assumed to have identical characteristics. From the symmetry of the thyristor's structure, it is apparent that the base current of the PNP transistor is the same as the collector current of the NPN transistor, and vice versa. This condition implies that each transistor is driven with a base current equal to its own collector current. The voltage drop across the saturated base-collector junctions (the middle p-n junctions of the device) essentially cancels the voltage drop across one of the outer junctions, and the entire structure behaves like a single forward- biased diode. The voltage drop of the device is therefore reduced to a few tenths of a volt. This is the "on" state and there are several ways of forcing the thyristor into this state, but each method involves changing the β of the transistors to make them both greater than 1.0.

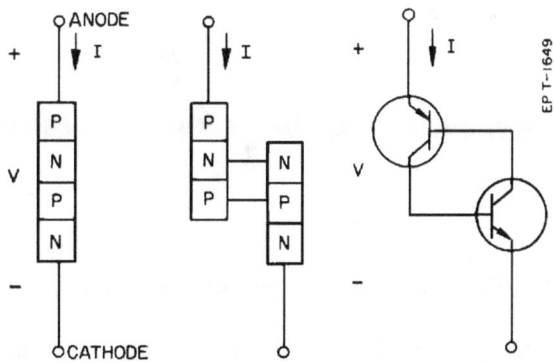

Exhibit 2-22 — Two Transistor Analog For A Thyristor

Operation of a thyristor is as follows. As the voltage across the thyristor is increased, the current increases only slightly since the center junction of the thyristor is reverse-biased. As the voltage is increased, avalanche breakdown occurs at this junction causing an increase in β. When β becomes greater than one, both transistors are driven into saturation and the device is in the "on" state. The thyristor will remain in the "on" state until the current through it is reduced to the point where β becomes less than 1. The current at this point is known as the holding current (I_H).

2.1.4.4.2 — Silicon-Controlled Rectifier

A silicon-controlled rectifier (SCR) is a four-layer device with a PNPN arrangement of doped layers similar to a thyristor. Exhibit 2-23 shows the construction of an SCR, a typical SCR, and its schematic symbol. An SCR is a rectifier which conducts in the forward direction only after a positive voltage of greater than .6 volts is applied to its gate terminal. The positive voltage need not remain for conduction to continue and can therefore be a pulse. When an SCR begins to conduct, the gate loses control and the device continues to conduct until the current is stopped by external means. Stopping the current flow, known as commutation, occurs either by removing the power source from the anode (natural commutation) or by forcing the current to zero by the application of a positive voltage to the cathode (forced commutation). Forced commutation requires a current of equal or greater magnitude than the anode current be applied at the cathode with a cathode voltage equal to or greater than the anode voltage.

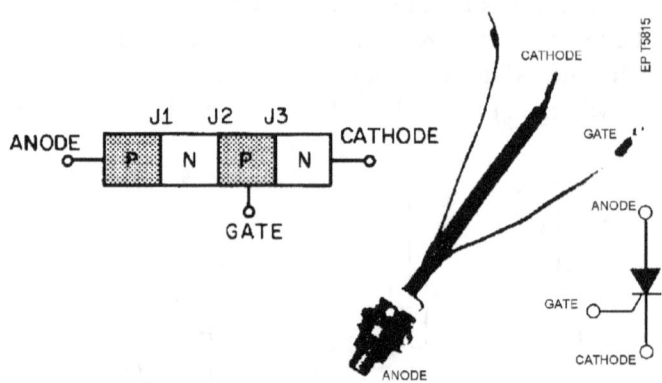

Exhibit 2-23 — Silicon-Controlled Rectifier

As shown in Exhibit 2-24, the holding current (I_H) is the minimum forward current at which the SCR will be able to maintain the "on" state. The breakover voltage (V_{BO}) is the minimum voltage needed to turn the SCR on when sufficient gate voltage is applied. The forward voltage drop across the SCR when it is in the conducting state is approximately 1.4 volts. The forward current rating of an SCR is primarily dependent on the power dissipation of the device and the thermal efficiency of the device mounting.

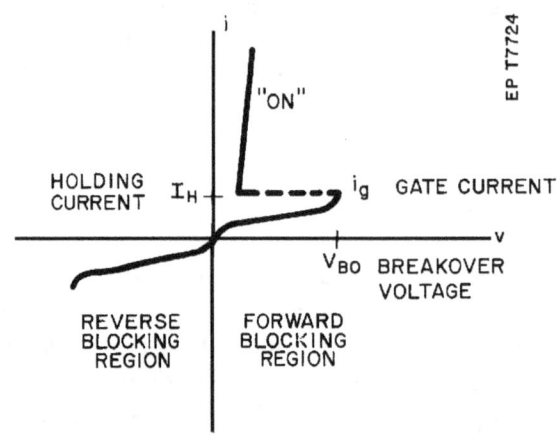

Exhibit 2-24 — SCR v-i Curve

2.1.4.4.3 — Gate Turn-off Thyristors

One of the major drawbacks of an SCR is that commutation is required to turn it off. The gate turn-off thyristor (GTO) operates in a similar manner to an SCR, but it can be turned off by applying a negative voltage to the gate terminal. The gate current required to turn off a GTO is only a fraction of the anode current (typically about one fifth) at a voltage no greater than the gate-cathode reverse breakdown voltage of about 20 volts. Conversely, the SCR required an auxiliary commutation circuit to provide full anode current at the peak anode voltage. Therefore, the GTO requires much less power than the SCR to turn it off. The use of GTOs instead of SCRs in equipment operating from d-c power supplies eliminates the need for bulky commutation circuits, reducing the weight and size of the equipment. GTOs however, are generally not capable of handling as high a current as SCRs.

2.1.5 — Operational Amplifiers

Operational amplifiers (op-amps) are one of the most widely used integrated circuits used in analog circuitry. They are so common because they are a building block that can be used to form circuits that perform a wide variety of functions, including inverting and non-inverting amplifiers, summing and subtracting amplifiers, buffers, comparators, differentiators, integrators, and signal filters. This section will describe the operation of the ideal op-amp and its use in several applications.

2.1.5.1 — Ideal Op-amp

The symbol for the ideal op-amp is shown in Exhibit 2-25. Connections consist of inverting (-) and non-inverting (+) inputs, an output, and the IC's power supply V^+ and V^-. The ideal op-amp has the following characteristics:

1. Infinite open loop gain, meaning any input (a voltage difference between the inputs) results in an infinite output.
2. Infinite input impedance, meaning the device draws no current from the outside circuit. This also implies that the + and - terminals are always at the same potential since they are connected and no current flows between them.
3. Zero output impedance. This makes the device look like an ideal voltage source, not affected by any external circuitry.
4. Infinite frequency response, meaning that the gain of the op-amp is constant at any frequency of operation.

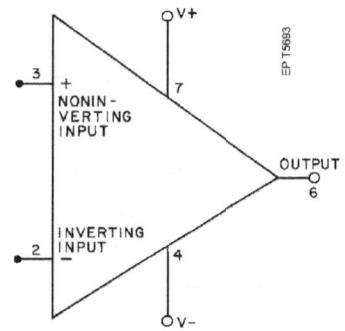

Exhibit 2-25 — Ideal Op-amp

It should be noted, that even though the gain is infinite, the output voltage cannot exceed the bounds of the supply voltage.

Modern op-amps closely approximate the characteristics of an ideal op-amp. This allows one to determine the operation of an real op-amp circuit by treating the op-amp as ideal. However, there are some differences between ideal and real op-amps. Characteristics of typical op-amps are as follows:

1. Gain is high (10^4-10^5), but not infinite.
2. Input impedances, although very high (>100 kohm), are not infinite.

3.   Output impedances are low (<100 ohm) but not zero.
4.   Frequency response deteriorates at very high frequencies.

2.1.5.2 — Analysis Method

To analyze an op-amp circuit, we rely on two properties of the ideal op-amp. One is the property that the current into both the + and - terminals is zero. The other property is that the voltage between the + and - terminals is zero, implying that the voltage at the + and - terminal is equal.

The following is a step-by-step method of analyzing any ideal op-amp circuit:
1.   Write the Kirchhoff node equation (i. e. the sum of the current entering and leaving a node must equal zero) at the noninverting terminal.
2.   Write the Kirchhoff node equation at the inverting terminal.
3.   Set the $v_+ = v_-$ and solve for the desired close loop gain.

Exhibit 2-26 shows the circuit diagram for an inverting amplifier.

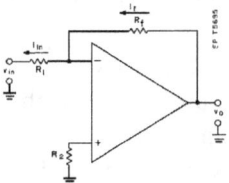

Exhibit 2-26 — Inverting Amplifier

The following demonstrates the method of analysis:

Step (1)

$$v_+ = 0 \, v$$

Step (2)

$$\frac{v_{in} - v_-}{R_1} = \frac{v_- - v_o}{R_f}$$

Step (3)

$$\frac{v_i}{R_1} = \frac{-v_o}{R_f}$$

Now, rearranging terms, the gain (A) is equal to:

$$A = \frac{v_o}{v_i} = -\frac{R_f}{R_1}$$

This analysis method is applicable to solving all ideal op-amp circuits. The following table shows the circuit diagrams for several op-amp circuits and their input-output relationships.

OTPT-0162

Circuit	Circuit Diagram	Equation	Assumptions
Voltage Follower (Buffer)	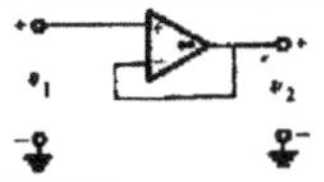	$v_2 = v_1$	
Inverting	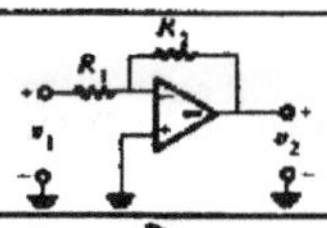	$v_2 = -Kv_1$	$K = R_2/R_1$
Noninverting	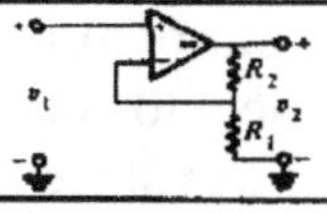	$v_2 = Kv_1$	$K = 1 + R_2/R_1$
Voltage Adder	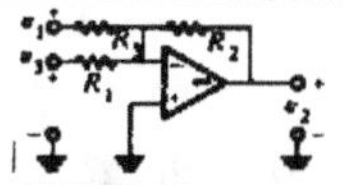	$v_2 = -(K_1v_1 + K_3v_3)$	$K_1 = R_2/R_3$ $K_3 = R_2/R_1$
Voltage Subtractor	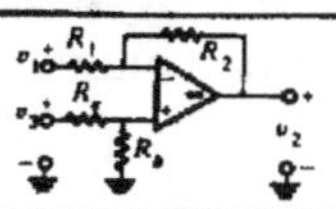	$v_2 = K_3v_3 - K_1v_1$	$K_1 = R_2/R_1$ $K_3 = (1+R_2/R_1)/(1+R_a/R_b)$
Difference Amplifier	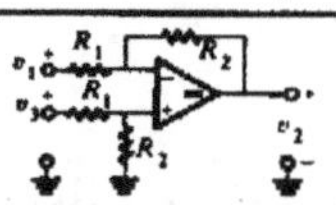	$v_2 = K(v_3 - v_1)$	$K = R_2/R_1$
Differentiator	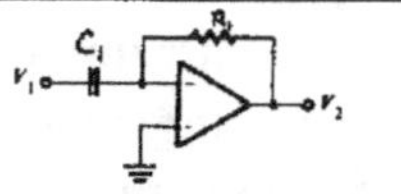	$v_2 = -K(dv_1/dt)$	$K = R_1C_1$
Integrator	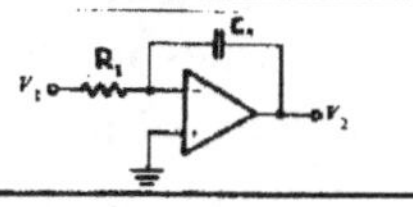	$v_2 = -K\int v_1 dt$	$K = 1/(R_1C_1)$
Comparator	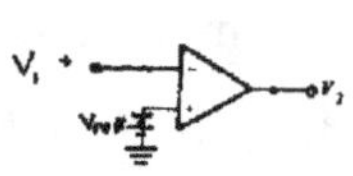	$v_2 = V+$	$v_1 > V_{ref}$
		$v_2 = V-$	$v_1 < V_{ref}$

Exhibit 2-27 — Circuit Diagrams for Op-Amp Circuits And Associated Input-Output Relationships

Section 2.2 — Magnetic Amplifiers

2.2.1 — Introduction

A magnetic amplifier (mag-amp) consists primarily of a variable impedance placed in series between a power source and a load, by which a small current controls the impedance and effects a large change in the load current. The devices are amplifiers because the expenditure of relatively small power in the control windings permits control over a relatively large amount of power in the output windings.

The mag-amp has certain advantages over other types of amplifiers. These include:
- high efficiency (90 percent)
- reliability (long life, freedom from maintenance, reduction of spare parts inventory)
- ruggedness (shock and vibration resistance, overload capability, freedom from effects of moisture)
- no warmup time.

The mag-amp also has certain disadvantages. These include:
- poor transient response for high gain applications
- limited frequency response
- output is a chopped sine wave and not an amplified reproduction of the input.

Because of its very high power gain, the mag-amp is used for control of large blocks of power, such as theater lights and heavy industrial loads. The mag-amp is used extensively to control actuating clevises such as valves, clutches, relays, and solenoids. It is also very often used as the last stage in a control system to provide power to a two-phase induction motor.

Nuclear plants use the mag-amp in the conventional aspects of the electrical system as well as in the reactor control circuitry. Typical circuits which employ mag-amps include voltage regulators, speed regulators, sensing devices and control systems.

2.2.2 — Review Of Magnetic Principles

In the region about a current-carrying conductor, effects are produced like those in the region about a permanent magnet. Forces act on magnets, iron, or other current carrying conductors which are in the region. Exhibit 2-28 illustrates the magnetic field about a conductor carrying a steady d-c current. The magnetic field is "mapped" by drawing magnetic flux lines. The strength of the magnetic field is indicated by the density of the magnetic flux lines, while the direction of the magnetic flux lines indicates the force that would be exerted on an isolated north pole placed within the magnetic field. Because of the potential ability to produce forces and do work, the magnetic field stores energy. The number of magnetic flux lines within a specified area is termed the magnetic flux. Magnetic flux is symbolized by ϕ and is measured in webers. At any point within a

magnetic field, there is a specific magnetic flux density. Magnetic flux density, symbolized by $\vec{B}$, is a vector quantity whose magnitude and direction indicate the magnitude and direction of the magnetic force on an isolated north magnetic pole at a point. Magnetic flux density has the units of webers/(meter)^2. Magnetic flux and magnetic flux density are related by

$$\phi = \int \vec{B} \cdot d\vec{A}$$

where the dot product operation is indicated. For a uniform magnetic flux density passing perpendicularly through a uniform area, this result reduces to

$$\phi = BA.$$

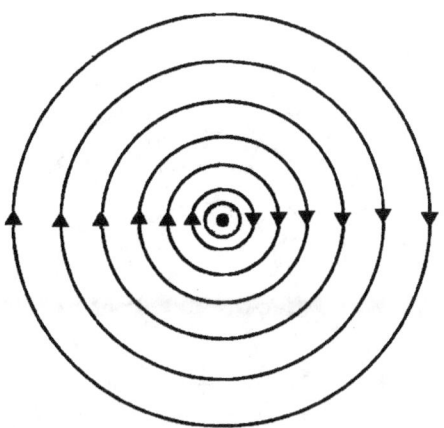

● DENOTES CURRENT FLOW INTO THE PAGE

Exhibit 2-28 — Magnetic Field About A Current-Carrying Wire

The magnetic flux density, $\vec{B}$, established at a point by a current is a function of the current and the particular material. The current is considered to set up a <u>magnetizing force</u> ($\vec{H}$) at a point; which is independent of the medium. Magnetizing force and magnetic flux density are related by

$$\vec{B} = \mu \vec{H}$$

where μ is an experimentally determined property of the material, called the <u>permeability</u>. Magnetizing force, $\vec{H}$, has the units of ampere-turns/meter and thus permeability, μ, has the units of webers/ampere-turn-meter. Like magnetic flux density, $\vec{B}$, magnetizing force, $\vec{H}$, is also a vector quantity. $\vec{B}$ and $\vec{H}$ must have the same direction with magnitudes differing by the permeability of the material.

For a given magnetizing force, $\vec{B}$, the resulting magnetic flux density, $\vec{H}$, depends upon the permeability of the material at that point. Permeability may be thought of as a measure of the ease of establishing flux within a material. Materials used in practical electromagnetic devices have values of permeability ranging from 2000 to 5000 times the permeability of air. Thus, by winding a coil on a material of high permeability, as illustrated in Exhibit 2-29, the magnetic flux will be constrained within the magnetic material. The design of electromagnetic devices is such that magnetic materials are shaped to control the magnetic flux path in the same way conductors control current flow. Although magnetic materials are not as good conductors of magnetic flux as low resistance materials are conductors of current, they are good enough that for simplified analysis it is sufficient to assume that all of the magnetic flux is confined within the magnetic material. When this assumption is made, the problem becomes a magnetic circuit problem and many analogies to an electrical circuit may be drawn. Exhibit 2-30 illustrates the magnetic circuit of Exhibit 2-29 redrawn alongside an analogous electrical circuit.

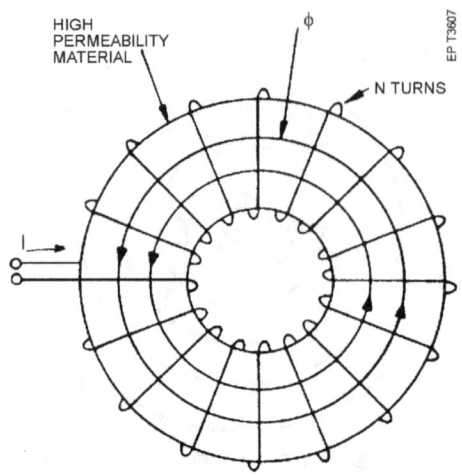

Exhibit 2-29 — Using Magnetic Materials To Confine Flux To A Desired Path

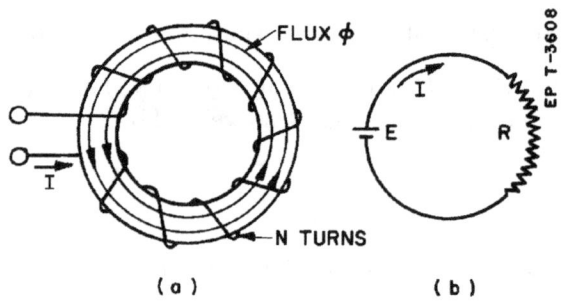

Exhibit 2-30 — Analogous Magnetic And Electric Circuit

The magnetic flux produced within the magnetic material is proportional to the number of turns of the winding, N, and the magnitude of the current, I. The product, NI, is termed the <u>magnetomotive force</u>, abbreviated mmf, and measured in ampere-turns. Magnetomotive force is related to magnetizing force by

$$NI = \int \vec{H} \; c \; d\vec{\ell}$$

where the dot product operation is indicated. For a magnetic circuit such as Exhibit 2-29, where the mmf is uniformly applied through a uniform magnetic material, this result simplifies to

$$NI = H\ell$$

where ℓ is the length of the magnetic flux path.

The production of a magnetic flux within the core by an applied mmf in the magnetic circuit is analogous to the production of current through an electrical circuit by an applied emf (voltage). Thus, Ohm's law for magnetic circuits can be stated as

$$NI = R\phi$$

where R is termed the <u>reluctance</u>. Table 2-1 summarizes the analogy between electric and magnetic circuits.

Table 2-1 — Analogous Electric And Magnetic Circuit Quantities

Electric Circuit		Magnetic Circuit	
Electric Field Density	$\vec{E}$	Magnetizing force	$\vec{H}$
Current density	J	Magnetic flux density	$\vec{B}$
emf (voltage)	V	mmf (magnetomotive force)	NI
Current	I	Magnetic flux	ϕ
Resistance	R	Reluctance	R
Conductivity	σ	Permeability	μ
$\vec{J} = \sigma\vec{E}$		$\vec{B} = \mu\vec{H}$	
$V = IR$		$NI = \phi R$	
$R = \dfrac{\ell}{\sigma A}$		$R = \dfrac{\ell}{\mu A}$	

2.2.3 — Properties Of Magnetic Materials

The flow of current through an air-core inductor creates a magnetizing force, $\vec{H}$, which establishes a magnetic flux density, $\vec{B}$, within the core. As the magnetizing force is increased, the magnetic flux density increases linearly in proportion to the permeability. If the same magnetizing force is applied to an inductor of the same size, but wound on a ferromagnetic core, the resulting magnetic flux is greatly increased. Exhibit 2-31 illustrates the $\vec{B} - \vec{H}$ curves for these two situations. The slope of the $\vec{B} - \vec{H}$ curve at a particular point yields the incremental permeability. Note that for the air-core inductor, the permeability is constant while for the ferromagnetic-core inductor, the permeability is highly nonlinear. The $\vec{B} - \vec{H}$ curve for the ferromagnetic-core inductor may be explained by the domain theory of magnetism. Ferromagnetic materials may be considered to be composed of small irregularly shaped regions, in which the magnetic moments of the individual atoms combine to create a net magnetic moment. These small regions, called <u>domains</u>, act like small permanent bar magnets. With no applied magnetizing force the domains are randomly aligned and the material exhibits no net magnetic field. As a magnetizing force is applied, the individual domains are brought into alignment with the applied field and the magnetic flux density is greatly increased. The magnetic flux density increase continues until all the domains are aligned with the applied field. At this point, the material is saturated and further increases in magnetizing force will only result in the small increase of magnetic flux density which occurs in the air core.

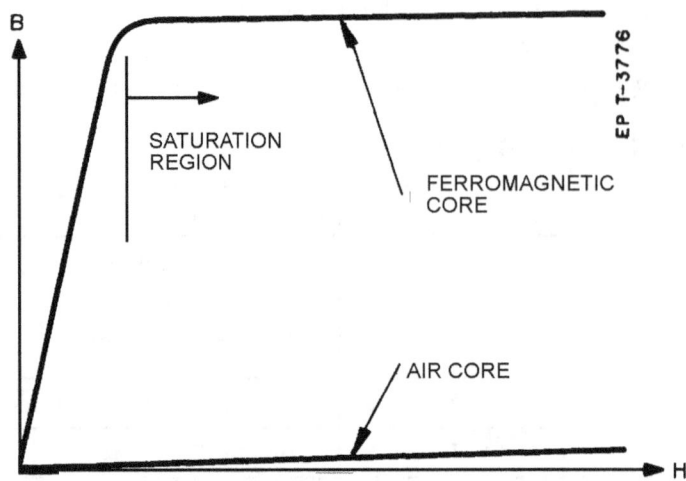

Exhibit 2-31 — Magnetization Curves For An Air-Core And A Ferromagnetic-Core Inductor

The application of a magnetizing force to a coil wound on a core of an initially unmagnetized ferromagnetic material causes an increase in magnetic flux density as discussed above. This action is shown by curve O-A on the B-H curve of Exhibit 2-32. If the magnetizing force is now reduced to zero, the magnetic flux density follows curve A-B. Even though the magnetizing force is reduced to zero, the ferromagnetic material remains magnetized. This effect is known as <u>residual magnetism</u> and the magnetic flux density existing at point B is the <u>residual magnetic flux density</u>. The reason for residual magnetism is that there is a frictional lag tending to keep the domains aligned. To return the magnetic flux density to zero, the magnetizing force must be increased in the opposite direction along curve B-C. The magnetizing force at point C is termed the <u>coercive force</u>. If the magnetizing force is increased further, the magnetic flux density follows curve C-D, which drives the core into saturation again. However, now all the domains are aligned in the opposite direction from the previous saturation condition. If the magnetizing force is again decreased, the magnetic flux density will follow curve D-E. The loop may be closed by increasing the magnetizing force in the original direction, causing the magnetic flux density to follow curve E-F-A.

The curve A-B-C-D-E-F-A is called the <u>hysteresis loop</u>. The area within the hysteresis loop represents the amount of energy lost in the core as heat resulting from the frictional opposition to realignment provided by the domains, and is termed the <u>hysteresis loss</u>. The hysteresis loop of Exhibit 2-32 was formed by a slowly varying magnetizing force and hence it is called a <u>static loop</u>. A <u>dynamic loop</u>, formed by an a-c magnetizing force, will always be wider than a static loop because the effect of domain friction is greater, resulting in increased hysteresis loss.

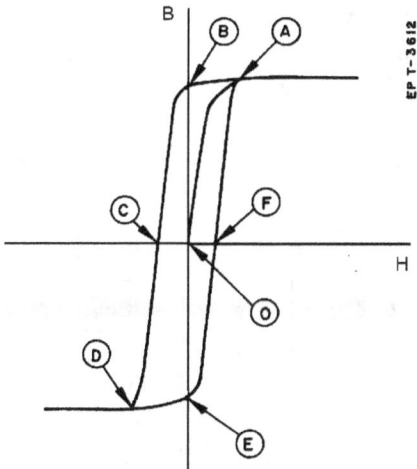

Exhibit 2-32 — Hysteresis Loop For A Ferromagnetic Material

2.2.4 — The Double-Core SRMA

Exhibit 2-33 illustrates schematically a double-core saturable reactor mag-amp (SRMA), which eliminates these problems. The control windings are connected series-opposing so that the voltage transformed by both cores to the control circuit cancels.

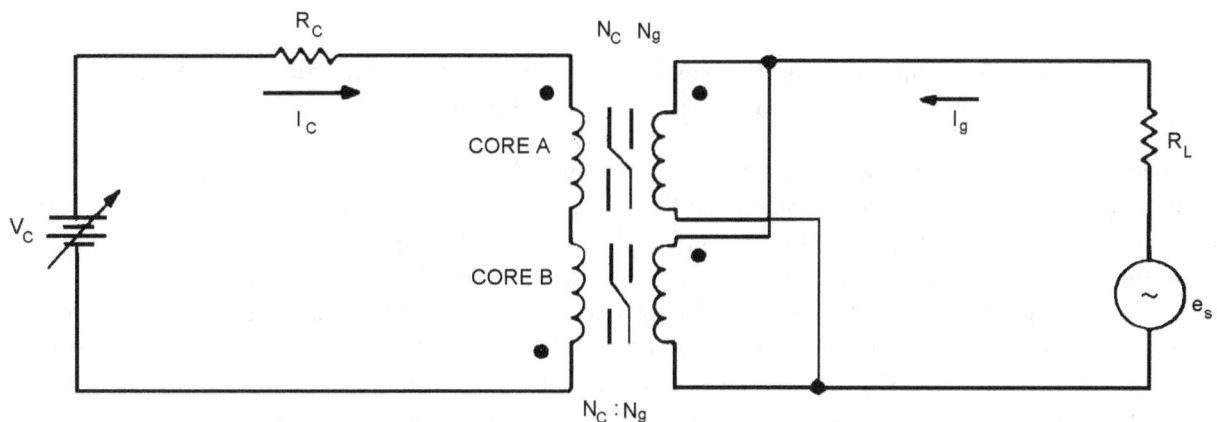

Exhibit 2-33 — Schematic Diagram Of A Double-Core SRMA With A Parallel-Connected Gate

Exhibit 2-34 illustrates the operation of the double-core SRMA when the magnitude of the applied source voltage, e_s, is such that the peak value is of the exact magnitude to require a magnetic flux within the cores to vary sinusoidally between $-\phi_{sat}$ and $+\phi_{sat}$. Note that the B-H curves have been relabeled as ϕ-Ni curves for ease of discussing circuit voltages and currents. Exhibit 2-34 illustrates the situation when there is no control current applied. Since the saturation magnetic flux values are only reached at one instant during each half-cycle of e_s, the cores remain unsaturated over the entire cycle of e_s. Corresponding instants in time are indicated by the circled letters. Note that the gate current waveform obtained from the φ-Ni curve is drawn to a exaggerated scale. For practical purposes the gate current can be considered to be zero for this case. Since the gate windings of the two cores are connected in parallel, the gate voltage and magnetic flux within each core are identical.

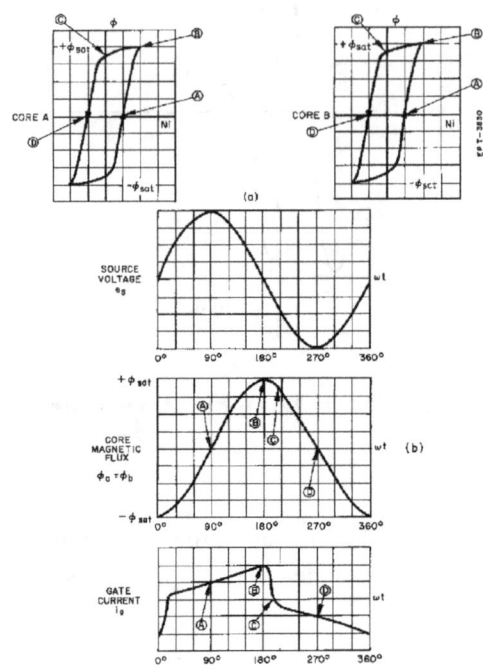

Exhibit 2-34 — Magnetization Curves For The Cores Of The Double-Core SRMA Of Exhibit 2-33 And Resulting Waveforms When Ic = 0

Now consider the situation when a medium value of d-c control current, I_c, is applied to the cores. Exhibit 2-35 illustrates the operation of the double-core SRMA for this case. Due to the orientation of the control windings (see Exhibit 2-33), the constant d-c mmf applied as a result of the d-c control current will produce magnetic fluxes in opposite directions within the two cores. Thus, the operation of core A is shifted towards positive saturation, while the operation of core B is shifted towards negative saturation. On positive half-cycles of e_s, core A is driven to saturation and on negative half-cycles of e_s core B is driven to saturation. Note that Exhibit 2-35 assumes the same sinusoidal e_s depicted in Exhibit 2-34. Since the gate windings are connected in parallel, when one core saturates, it shorts the other core, maintaining its magnetic flux constant at the level reached even though it is below saturation. Although not shown, the gate current, when the core magnetic fluxes remain constant, will be of the identical form as the load voltage with a magnitude given instantaneously by $i_g = e_s/R_L$. If the magnitude of the control current is further increased, the effects on the loops traversed for the cores is shown in Exhibit 2-36. The sequence of events is identical to Exhibit 2-35, the difference being that the magnitude of the mmf supplied by the control current is much greater causing the cores to saturate earlier during each alternate half-cycle and thus increasing the power delivered to the load.

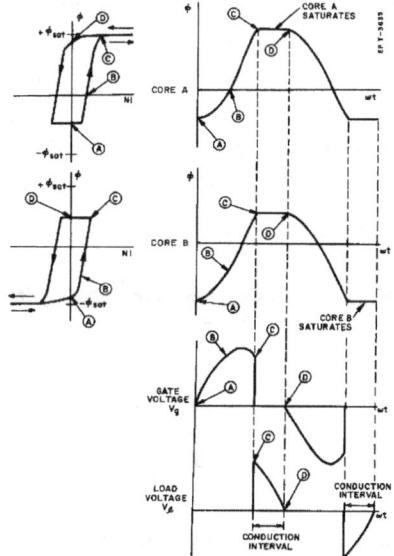

Exhibit 2-35 — Magnetization Curves For The Cores Of The Double-Core SRMA Of Exhibit 2-30 And Resulting Waveforms With A Small I_c

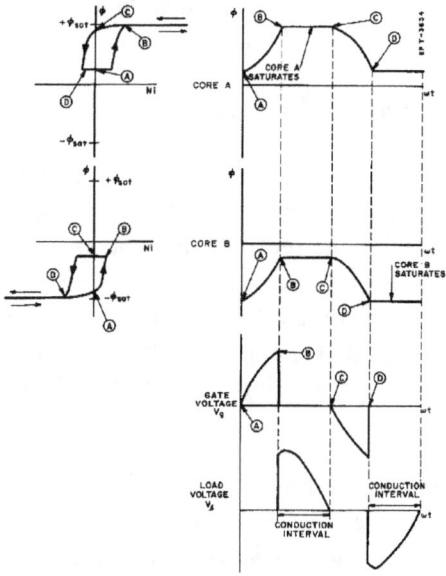

Exhibit 2-36 — Magnetization Curves For The Cores Of The Double-Core SRMA Of Exhibit 2-33 And Resulting Waveforms With A Large I_c

It is also possible to connect the gate windings of a double-core SRMA in series as shown in Exhibit 2-37. Note that the control windings remain connected in series opposition to avoid induced voltages in the control circuit. As with the parallel connected gate, the condition of both cores unsaturated causes the gate windings to look like an open circuit and no power is delivered to the load. When one core saturates, its windings appear as a short circuit. Though the other core is not saturated, only a small voltage will appear across its windings in this condition. Since the sum of the voltages across the gate windings is small in this condition, they behave essentially as a short circuit. By transformer action, the unsaturated core and its control circuit load may be represented by a reflected impedance $(N_g/N_c)^2 R_c$, where the gate winding acts as the primary and the control winding acts as the secondary. Since the turns ratio is large in a properly designed series-connected double-core SRMA, the reflected impedance in the gate winding will be small compared with R_L. Therefore, the voltage drop across the gate winding of the unsaturated core will be small, when the other core is saturated.

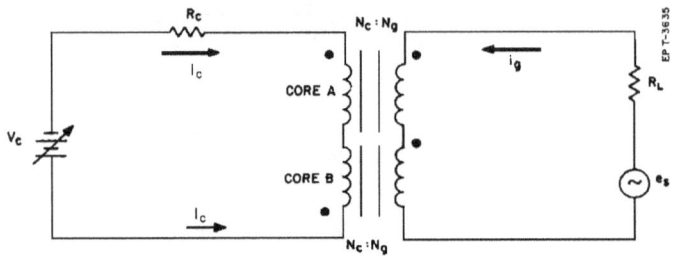

Exhibit 2-37 — Schematic Diagram Of A Double-Core SRMA With A Series-Connected Gate

Except for the operational difference when one core is saturated and the other core is unsaturated, no distinction as to the connections of the gate windings is necessary in the discussion of the double-core SRMA. The parallel-connected gate winding has the operational advantage of a larger current-handling capability and it is usually employed only when this requirement must be satisfied.

2.2.5 — Transfer Characteristics And Gain

It is desirable to be able to compare the amplification properties obtained for different values of control current. To do this, a means of relating the d-c control current (I_C) to the chopped sine-wave gate current (i_g) is needed. A simple comparison is made by comparing the rectified average values of the control and gate current wave forms. The rectified average value of a periodic waveform (no matter what its shape) is directly proportional to the absolute area enclosed by the wave form and the horizontal axis. To obtain a physical significance for an average value, it should be remembered that it corresponds to a measure of d-c which would enclose the same absolute area as the original waveform. Since the input to the double-core SRMA is already d-c, its average value is the d-c value itself.

Graphs relating output quantities to input quantities for a device are called <u>transfer characteristics</u>. For the mag-amp, the quantities to be plotted are the average value of the control current (I_C) and the average value of the load (gate) current (I_L). The transfer characteristic for a double-core SRMA is shown in Exhibit 2-38. Note that the transfer characteristic is composed of a series of linear portions. The nonzero slopes are called the <u>linear regions</u> of the transfer characteristics; the flat portions are called the <u>resistance limited regions</u> (because the current is limited solely by R_L during the entire cycle of e_s). With no applied control current ($I_C = 0$) only the negligibly small magnetizing current flows in the gate circuit and this current may be assumed to be zero ($I_L = 0$). As the control current is increased, the cores spend a greater portion of each cycle in saturation and the average value of the load current is increased. However, a limit is reached when the cores remain saturated for the full cycle of e_s. At this limit, the SRMA is acting like a short circuit for the entire cycle, and the full value of e_s is dropped across the load. The maximum average load current occurs in this resistance limited region, and is determined by Ohm's law:

$$I_{L_{max}} = \frac{E_s}{R_L}$$

where both I_L and E_s are average values.

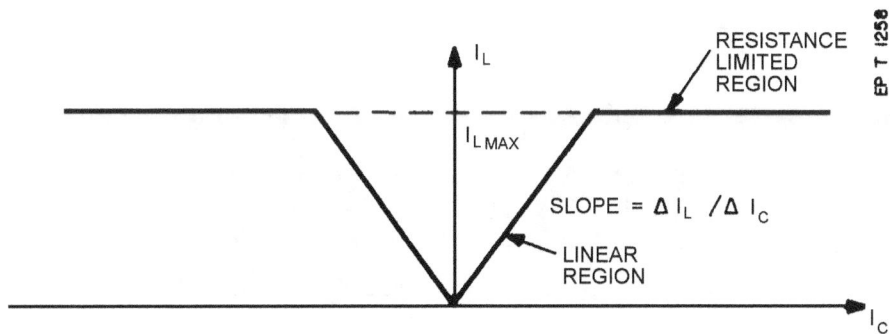

Exhibit 2-38 — Transfer Characteristic For An SRMA

Reversing the polarity of I_C has no effect on the value of the average load current (the cores simply interchange roles); thus, the transfer characteristic is symmetrical about $I_C = 0$.

The current gain of the mag-amp is defined as the absolute slope of the linear region of the transfer characteristic;

$$G_I = \left| \frac{\Delta I_L}{\Delta I_C} \right|$$

where:

    ΔI_L = the change in the average value of load current due to a change in control current

    ΔI_C = the change in d-c control current

2.2.6 — Bias And Feedback

There are many SRMA applications where the basic transfer characteristic does not satisfy the particular requirements at hand. The basic transfer characteristic may be altered by adding additional windings to the cores of the SRMA. One set of windings energized by a d-c source provides a <u>bias</u> which shifts the entire curve point-by-point to the left or right, depending upon whether the bias magnetic flux aids or opposes the control magnetic flux.

A set of bias windings supplied with direct current causes a fixed quantity of magnetic flux to be added to or subtracted from the total magnetic flux within the amplifier cores. With bias present, the SRMA transfer characteristic is no longer symmetric about the origin. Exhibit 2-39 illustrates schematically the cases of an SRMA with aiding and opposing bias; in addition, the effects on the transfer characteristics are shown. In general, an SRMA is said to have <u>aiding bias</u> if the bias current and the positive control current flow into corresponding terminals dotwise and is said to have <u>opposing bias</u> if the currents flow into noncorresponding terminals dotwise. The amount of shift of the transfer characteristic varies directly with the bias current (I_B) which in turn varies inversely with the value of the bias resistor (R_B) according to Ohm's law:

$$I_B = \frac{V_B}{R_B}$$

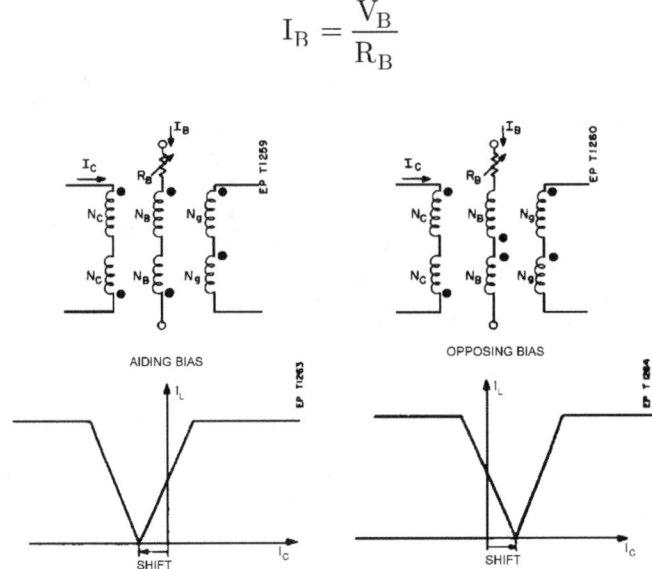

Exhibit 2-39 — Effect Of Adding Bias Windings To A Double-Core SRMA

While bias acts to shift the transfer characteristic to the right or left of the origin, feedback may be added to change the slope of the linear regions of the transfer characteristic. With feedback, control of the transfer characteristic is effected by supplying a portion of the output current to a set of feedback windings added to the cores of the SRMA. Exhibit 2-40 illustrates schematically the addition of feedback windings and the effects upon the transfer characteristic for both positive and negative feedback. Observe that the gate circuit of the SRMA has been modified to allow a portion of the load current to flow through the feedback windings. A bridge rectifier is inserted in this circuit so that the gate current in part of its journey is forced to be unidirectional. The current through the gate windings is bidirectional. The current through the parallel paths containing the load and the feedback windings is unidirectional. The bridge acts to split the gate current with a portion going through the load (I_L) and a portion going through the feedback windings (I_F). It is necessary for the feedback current to be unidirectional, as with the control current, so that the magnetic flux produced by the feedback windings will properly combine with the magnetic flux produced by the control windings. Note that the bridge rectifier is only one means of supplying a unidirectional current proportional to the load current to the feedback windings.

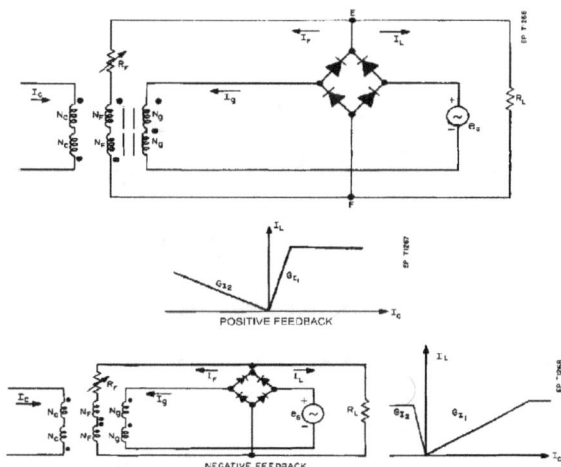

Exhibit 2-40 — Effect Of Adding Feedback Windings To A Double-Core SRMA

Feedback current can either aid or oppose the effect of control current on both cores. An amplifier is said to have positive feedback if the feedback current and positive control current flow into corresponding terminals dotwise. Negative feedback will result when the feedback current and positive control current do not flow into corresponding terminals dotwise.

Attention is now given to a qualitative look at the effect of positive feedback on the transfer characteristic. First consider the case when I_C is flowing in the positive direction. Under this condition, an increase in I_C will cause an increase in gate current (i_g). This in turn causes an increase in I_F, a change which, due to the winding orientation, aids the original change in I_C. Thus, with positive feedback, the gain (G_{I1}) is increased with the consequent steepening of the slope of the right branch of the transfer characteristic. Next, consider the case when I_C is reversed and made to flow in the negative direction. An increase in I_C still causes an increase in i_g. However, since the direction of I_F is unchanged, the resulting change in I_F opposes the original increase in I_C. Therefore, the gain (G_{I2}) is decreased in magnitude and the slope of the left branch of the transfer characteristic will be decreased. Notice that the only difference between the circuit for negative and positive feedback is the reversal of the feedback windings. The reversing of all the diodes in the bridge will accomplish the same thing as reversing the orientation of the feedback windings. With negative feedback, the feedback current opposes positive control current, decreasing G_{I1}, while the feedback current aids negative control current, increasing G_{I2}. Since the magnitude of the feedback current can be controlled by an adjustable feedback resistor, R_F, gain is thus adjustable.

Exhibit 2-41 illustrates one very important application of positive feedback. Note the effect of increasing the feedback; that is, decreasing R_F. The right branch gets steeper and steeper, and the left branch gets less and less steep, as shown in part A of Exhibit 2-41. Eventually, a point will be reached when the right branch is vertical. The right-hand gain region is said to undergo critical feedback. When the positive feedback is increased beyond this point, the right-hand gain region undergoes supercritical feedback and the transfer characteristic of part B of Exhibit 2-41 results.

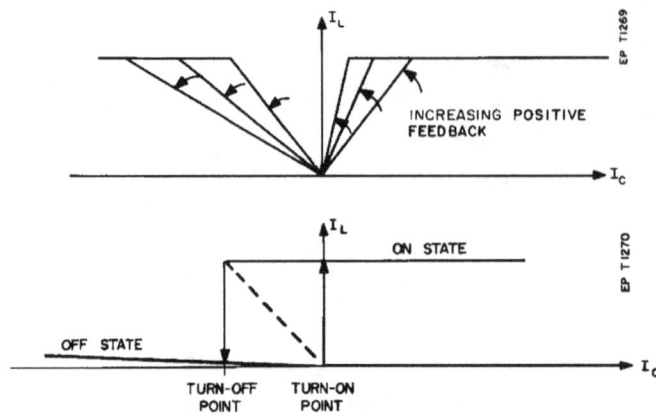

Exhibit 2-41 — Effect Of Increasing Positive Feedback On SRMA Transfer Characteristic

The double-core SRMA with supercritical feedback is termed a <u>bistable</u>, because its output is bivalued for certain inputs. The right-hand gain region is now an unstable region. Physically, the feedback current has been made so large that an incrementally small positive control current will result in core magnetic saturation with the corresponding step change in load current resulting. By examining part B of Exhibit 2-41, the following observations can be made. As I_C is decreased from a large negative value towards zero, the SRMA is considered in the OFF state. When the control current reaches zero, the output will jump to the maximum I_L value. The point at which I_L jumps from zero to its maximum value is called the <u>turn-on point</u>. As the control current is increased in the negative direction, the amplifier will remain ON until I_C reaches the left side of the bistable region; then it will turn OFF. The point at which the load current drops from its maximum value to zero is called its <u>turn-off point</u>. The difference in I_C between the turn-on point and the turn-off point is defined as the <u>bandwidth</u>.

It is possible, and actually the most common application, to employ both bias and feedback to the same amplifier. The effect of bias is independent of the presence of feedback. The bias with or without feedback will shift the transfer characteristic to the left for aiding bias and to the right for opposing bias with no effect on gain. Using both bias and feedback allows the amplifier to be adjusted to a desired gain by adjusting the amount of positive or negative feedback and then by adjusting the bias, the region of desired operation can be centered over the values of control current available. For bistable operation, the turn-on point is set directly by bias adjustment. The turn-off point is a result of both the bias and feedback. After setting the bias for the proper turn-on point, the feedback is adjusted to yield the proper turn-off point.

2.2.7 — Self-Saturation Magnetic Amplifiers

In general, the SRMA has inherent negative feedback due to the load current flowing in both directions through the gate windings. In each core, alternate half-cycles of supply voltage tends to desaturate the core, so that the control current or the feedback current has to overcome this effect on the following half-cycle to place the core in a saturated state, making an output possible. By use of diodes in the gate circuit, it is possible to eliminate the desaturating effect of the gate current. A diode in series with each gate winding is an identifying characteristic of the self-saturating mag-amp (SSMA). The double-core SSMA is shown schematically in

Exhibit 2-42. The presence of the diodes allows the cores to saturate without the aid of control current. Thus, the name self-saturating comes from the ability of the amplifier to saturate on gate current alone.

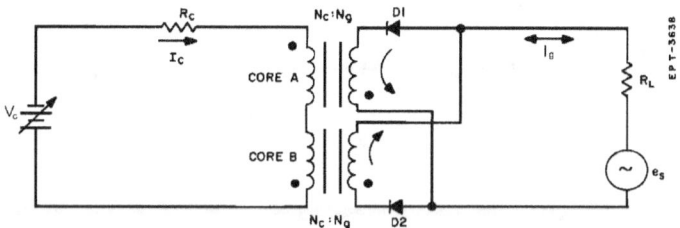

Exhibit 2-42 — Schematic Diagram Of A Double-Core Self-Saturating Magnetic Amplifier (SSMA)

Consider operation of the circuit of Exhibit 2-42 without the presence of control current ($I_C = 0$). On the positive half-cycles of e_s, diode D1 conducts and diode D2 blocks. On the negative half-cycles of e_s, diode D1 blocks and diode D2 conducts. Due to the diode action, a pulsating gate current will flow only in one direction within each gate winding as indicated by the unmarked arrows adjacent to the windings in Exhibit 2-42. On the positive half-cycles of e_s, a positive pulse flows through the gate winding of core A. On the negative half-cycles of e_s, a negative pulse flows through the gate winding of core B. For initially unmagnetized cores, both gate windings present high inductances. Each succeeding pulse of gate current creates a magnetizing force that builds up the magnetic flux within the cores. As the magnetic flux builds up, the inductance of the gate winding decreases, allowing the magnitude of the gate current to increase and accelerating the magnetic flux buildup. In core A, the current flows into the dotted winding, tending to drive core A to positive saturation. In core B, the current flows into the undotted winding tending to drive core B to negative saturation. After a few pulses of gate current, the cores saturate and further pulses just move the point of operation back and forth within the saturation region, with no appreciable change in magnetic flux. Thus, after the initial magnetic flux buildup, the cores remain saturated over the entire cycle of e_s. Both gate windings act as short circuits over the entire cycle of e_s; thus, no voltage is developed across the gate windings, the maximum load current is obtained $i_g = e_s/R_L$, and conduction is over the entire cycle of e_s. Exhibit 2-43 illustrates the operation of the SSMA with no applied control current.

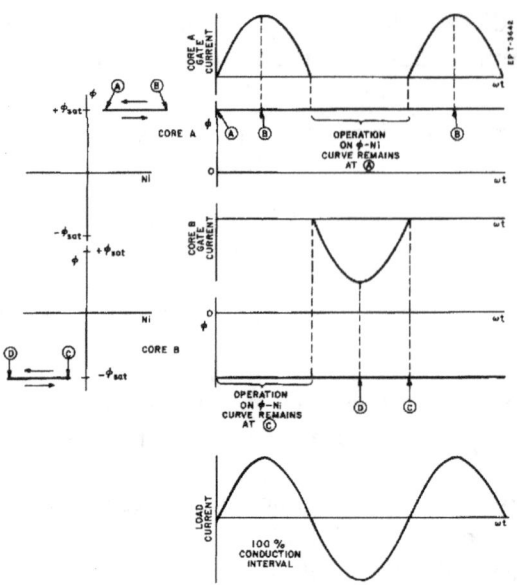

Exhibit 2-43 — Magnetization Curves For The Cores Of The Double-Core SSMA Of Exhibit 2-42 And Resulting Waveforms With No I$_c$

Now consider the effect of control current on the operation of the SSMA. In core A, both the control current and the gate current flow into the dotted terminal of their respective windings. In core B, both the control and the gate current flow into the undotted terminals. Thus, a positive control current produces an aiding magnetic flux within both cores. However, since both cores saturate on gate current alone, control current simply drives the cores deeper into saturation and has no effect on the output since the SSMA was already delivering maximum output when no control current was applied.

If a negative control current is applied, opposing magnetic fluxes are produced in both cores. The effect of negative control current is to alternately bring the cores out of saturation during the time that the diode in series with the gate winding blocks. This in turn requires the gate current pulses to spend the first portion of each alternate half-cycle bringing a core back into saturation, with the result that the load current is correspondingly reduced. As the control current is made more negative, the conduction interval of the amplifier is reduced and less power is delivered to the load. Exhibit 2-44 illustrates the operation of the SSMA with negative control current.

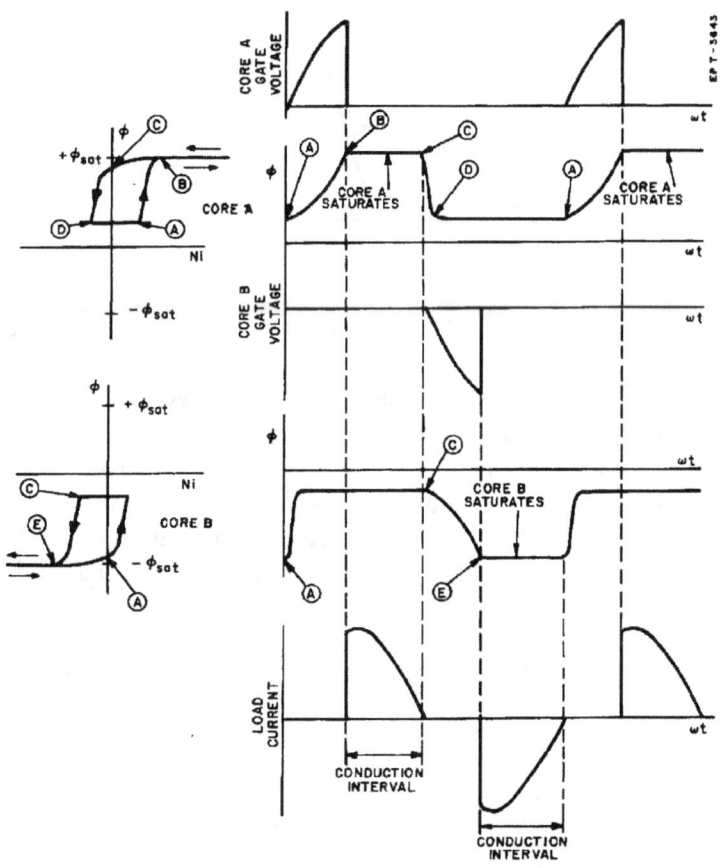

Exhibit 2-44 — Magnetization Curves For The Cores Of The Double-Core SSMA Of Exhibit 2-42 And Resulting Waveforms With A Negative I_C

The characteristic curve for the SSMA is shown in Exhibit 2-45. For $I_C \geq 0$, the cores are always in saturation; therefore, i_g is a maximum. As I_C is made more negative, less of each half-cycle is in saturation and the average gate current decreases. This is the linear region of the characteristic. When I_C equals a value such that the cores no longer go into saturation during the entire cycle of e_s, the gate current has its minimum value. This value of I_C is called $I_{C\ cutoff}$. As I_C decreases further, the cores never saturate, but there is a slight increase in gate current due to the decrease in inductance of the cores caused by the additional control flux within the cores. One advantage of the SSMA over the SRMA is that it has a much higher gain, as seen by the steeper slope of the linear region.

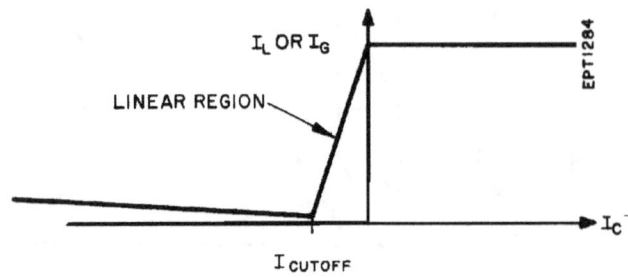

Exhibit 2-45 — Transfer Curve For An SSMA With No Bias And No Feedback

As discussed for the SRMA, bias and feedback windings may also be added to the SSMA as shown in Exhibit 2-46. The effects on the transfer characteristic of the SSMA are shown in Exhibit 2-47. Unlike the SRMA, instead of choosing a direction for the control current and calling that positive, the gate current must now be used as the reference, for due to the diodes, the gate current is allowed to flow in only one direction through the gate windings. With reference to the direction of gate current through the gate windings, the control current is referred to as either saturating or desaturating, just as this bias is either aiding or opposing, and the feedback is positive or negative.

For the SSMA, the gain is so high in the linear region that the addition of any amount of positive feedback results in bistable operation.

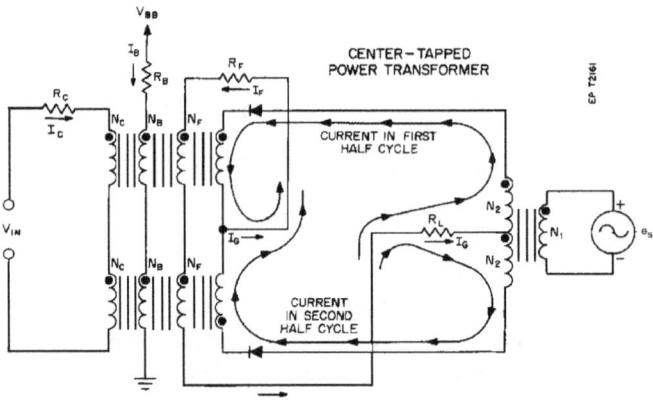

Exhibit 2-46 — Typical Schematic For An SSMA With Bias And Feedback

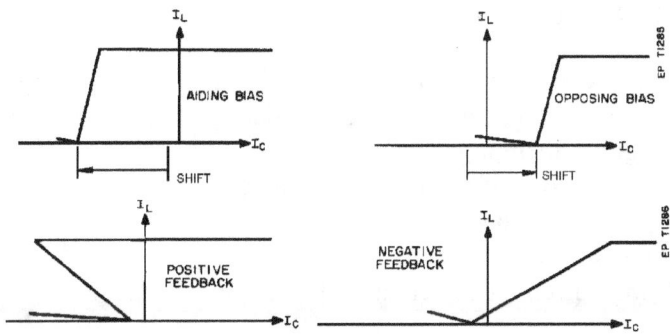

Exhibit 2-47 — Effects of Bias And Feedback On SSMA Transfer Characteristic

Another advantage of the SSMA over the SRMA is that its control circuit can be driven by a-c voltages. The only requirement is that the a-c control voltage and the a-c gate voltage must be operating in synchronism. Exhibit 2-48 illustrates schematically a possible arrangement using a-c control. The operation of the gate circuit is the same as discussed above. The control circuit is arranged similar to the gate circuit, employing diodes so that a pulse of current flows through one of the control windings on alternate half-cycles. However, the control circuit is oriented so that a pulse of control current flows through the control winding of the appropriate core when the diode in the gate circuit is blocking the flow of gate current in the corresponding gate winding. In the first half-cycle, core A is brought into saturation by the gate and core B is brought out of saturation by the control circuit. In the second half-cycle, the control circuit brings core A out of saturation and the gate saturates core B. The rheostat (R_C) controls the amount of control mmf added to the core. With R_C at its minimum value, the control mmf is equal to the gate mmf and both windings move the point of operation an equal, but opposite, amount on the B-H curve. In this situation, the operation remains in the vertical regions, the cores remain unsaturated, and the minimum (ideally zero) gate current flows. As the value of R_C is increased, the control mmf decreases and the gate mmf saturates the cores. Higher values of R_C decrease the control mmf further until R_C is infinite and the cores remain saturated over the entire cycle of source voltage.

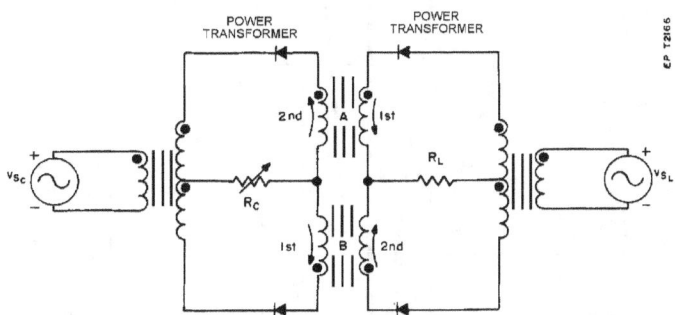

Exhibit 2-48 — Double-Core SSMA With A-C Control Circuit

Instead of using a rheostat, the control mmf may be adjusted by using a variable d-c voltage source in series with the a-c control voltage or an a-c control voltage with a varying amplitude may be used.

2.2.8 — Classification Of Magnetic Amplifiers

Mag-amps are found in many different forms and are used for a multitude of purposes. In order that the basic operating principles of a mag-amp be determined, no matter what its form or purpose, six different classifications will be used to group them together.

1. Type: Saturable reactor or self-saturating mag-amp
2. Bias: Aiding opposing, or none
3. Feedback: Positive negative, or none
4. Output: Alternating current or direct current through the load
5. Operation: Linear or bistable
6. Control: Saturable or desaturable (applies only to an SSMA)

Section 2.3 — Thermocouples

2.3.1 — Thermocouple Effect

An electromotive force (emf) is developed between two dissimilar metals that are held in contact with one another. The magnitude of the emf is 0 to 40 millivolts, and depends on two conditions: (a) the compositions of the two metals and (b) the temperature at the junction (point of contact). A graph of emf versus junction temperature is shown in Exhibit 2-49 for some combinations of metals.

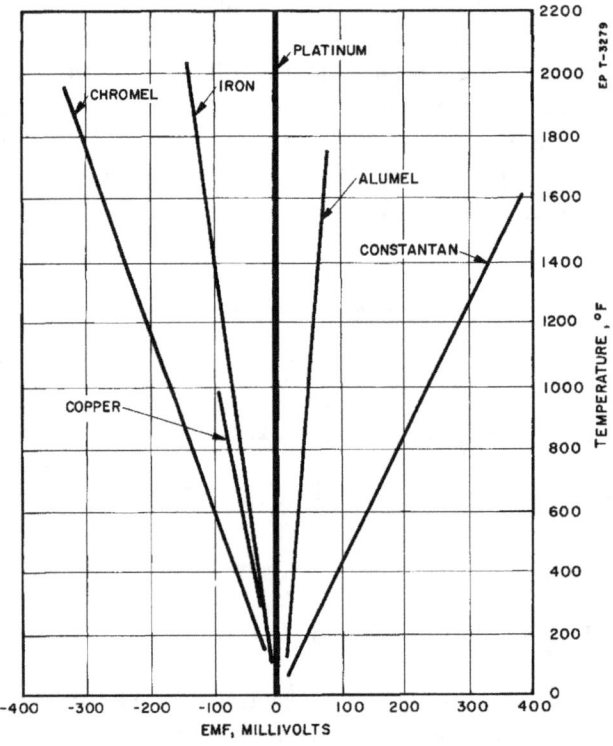

Exhibit 2-49 — Thermal EMF's Of Some Metals Versus Platinum

2.3.2 — Thermocouple Measurement Circuits

A thermocouple is simply a junction of two different metal wires, which is designed to serve as a temperature measuring device. Careful measurement of the emf between the two wires will allow the temperature at the junction to be determined.

Exhibit 2-50 shows a possible thermocouple arrangement. The thermocouple is a small junction at the temperature (T) that is to be measured. The thermocouple leads are chromel and alumel in this example; however, many other combinations of metals are possible. The leads are insulated from one another over their entire lengths except for the junctions. The emf developed at each junction in the circuit is represented by a dashed battery. The second junction at room temperature (T_r) is called the reference junction. A thermal emf is also developed at the reference junction, and this emf opposes the emf of the measurement junction. Since the reference emf is also temperature dependent, the temperature at this junction must be accurately known. The difference between the two emf's is measured by a suitably accurate instrument such as a potentiometer.

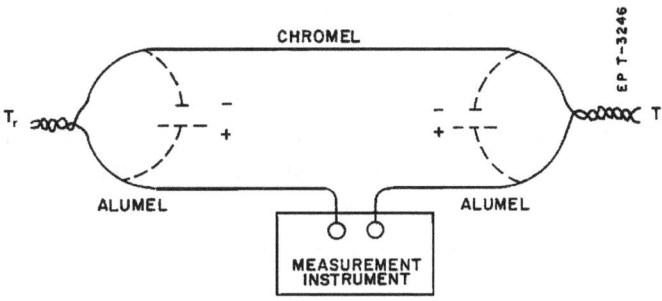

Exhibit 2-50 — A Thermocouple Circuit

In practice the location where temperature is to be measured may be remote from the location of the instrument. In this case the arrangement of Exhibit 2-51 may be more practical. Here copper wire conducts the emf (and associated current) to the instrument. The use of copper wire has no effect on the measurement provided the following conditions are satisfied:

1. The same intermediate material is used for each side of the circuit (other than at the measurement point, T). For example, copper is used for both sides of the circuit in Exhibit 2-51.
2. Each "pair" of junctions (i.e., the connection points at T_r and those at T_m) is at the same temperature.

These conditions apply because of the following laws of thermocouple behavior:

1. A wire of uniform composition will not develop a thermal emf along its length even if the size and temperature along the length vary.
2. Thermal emf's at a given temperature are algebraically additive among a series of metals. For example, the emf at the chromel-copper junction in Exhibit 2-51 is identical to the algebraic sum of emf's of a chromel-alumel junction plus an alumel-copper junction. Thus the two junctions at T_r in Exhibit 2-51 are effectively identical to the single reference junction in Exhibit 2-50.

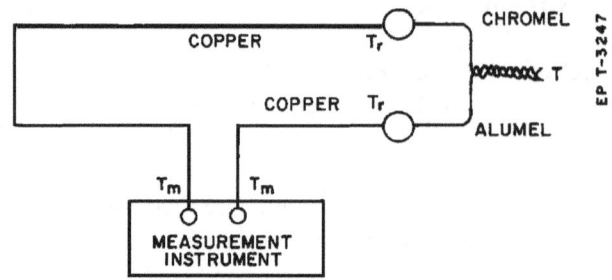

Exhibit 2-51 — A Practical Thermocouple Circuit

The effect of the two laws just stated is that the net thermal emf around a circuit is equal to the algebraic sum of the thermal emf's at each junction in the circuit.

2.3.3 — Thermocouple Design Considerations

2.3.3.1 — Composition

One normally chooses the combination of metals giving the highest thermal emf after considering the temperature range, cost, availability, and corrosion resistance of the medium in which it is used. The melting point of the metals used must be above the application range. The emf of the thermocouple must increase continuously with increasing temperature over the application range.

2.3.3.2 — Size

For a given composition, the smaller the thermocouple wire and the junction, the faster the response to changes in temperature. The larger the wires, the greater the effect of heat conduction on the measuring junction.

2.3.3.3 — Construction

Junctions may be made by means of solder, hard solder, or by welding. Acid flux or any other material that would have a corrosive effect on the wires should be avoided. Wires may be twisted together before soldering or welding or may be butt welded. The junction length should be restricted to a distance equal to the wire diameter. The twist may be omitted if it results in a long junction in which a temperature gradient might exist. The temperature measured is that at the last point of electrical contact along the junction. The thermocouple wires should be insulated from each other and brought out to a terminal block where extension leads are attached. The terminals should be at equal temperatures to avoid errors.

2.3.4 — Installation Of Thermocouples

2.3.4.1 — Location

The measuring junction should be as near as possible to the temperature to be measured. If a large temperature gradient exists near the measuring junction, an error due to heat conduction may exist unless a sufficient length of the insulated wire is provided at the measuring temperature.

2.3.4.2 — Cold Junction Compensation

Inasmuch as the emf developed by the thermocouple circuit depends upon the temperature of both the measuring and reference junctions, a knowledge of the following quantities is required for determining temperatures with thermocouples:

1.   the calibration data for the thermocouple
2.   the measured emf
3.   the temperature of the reference junction.

(Calibration data for thermocouple wire are usually supplied by the wire vendor, who guarantees that the wire matches the reference curves.)

2.3.4.3 — Intermediate Junctions

Intermediate junctions should generally be made in pairs. The temperatures of each junction pair should be equal to avoid errors.

Thermocouple circuits vary widely, depending upon the particular application. The indicating instrument may be a millivoltmeter, a potentiometer, an electronic temperature recorder or controller, or any other device suitable for detecting and indicating small direct-current emf's. The input impedance of the instrument should be as high as practical to avoid errors in measurement. The magnitude of these errors is a direct relationship between the resistance of the measuring instrument and the other resistances in the thermocouple circuit.

2.3.5 — Thermocouple Calibration

Thermocouples may be calibrated by comparing with (1) a standard thermocouple or (2) a standard resistance thermometer. The accuracy of the thermocouple is on the order of $11°C$. The accuracy of the thermometer is on the order of $10.5°C$.

Section 2.4 — Resistance Thermometry

2.4.1 — Introduction

The resistance thermometer depends upon the inherent characteristics of metals to change in electrical resistance when they undergo a change in temperature. The change in electrical resistance of a material with a change in temperature is termed the "temperature coefficient of resistance" for the material (usually expressed in ohms per degree at a specified temperature).

For pure metals, the change in resistance with temperature is practically linear, at least over a large portion of the resistance-temperature curve.

2.4.2 — Design Factors

2.4.2.1 — Size And Shape

Size and shape are dependent to a large degree upon the application.

2.4.2.2 — Material Used In Resistance Winding

The required characteristics of the material used in resistance windings or coils include:
1. Relatively high temperature coefficient of resistance in order to obtain good sensitivity.
2. High resistivity: The higher the resistivity of the material, the more resistance there is available for a given length of wire and, consequently, for a given space. Also, the higher the resistivity the greater the resistance change per degree of temperature. This factor also contributes to the sensitivity.
3. Stability: The material must be stable over a long period of time and over a wide range of temperatures without changing its electrical characteristics.
4. Linearity of resistance-temperature relationship: This is a desirable characteristic since it results in a linear temperature scale.
5. Ductility and strength: The material must be readily drawn into wire or ribbon of small size and yet possess sufficient strength for use in the winding. The smaller the wire cross section, the faster the speed of response.
6. High thermal conductivity and low specific heat, if high speed of response is desired.

Platinum and molybdenum are the two common metals used for resistance thermometers in naval reactor plants. Platinum is generally used in the slow-speed well-type thermometers, and molybdenum in high-speed thermometers. The following explains why "moly-type" resistance thermometers are used instead of platinum, in most applications:
1. It is necessary to braze the platinum in the fabrication of the nonseparable-type thermometer. This process changes the characteristics of the platinum at the braze joint and seriously affects the resistance temperature relationship. Platinum cannot stand the change from an oxidizing to reduction process of brazing. Welding has been shown to be difficult and virtually impossible to inspect to ascertain whether a good joint was made. Welding has been abandoned as a fabrication method.
2. The linear temperature resistance characteristic of platinum is not as good as that of molybdenum. New, well-type thermometers are platinum types and the nonlinearities are compensated for in the receivers.

2.4.2.3 — Fabrication

Meticulous fabrication techniques are required to provide a resistance thermometer with the response, accuracy, and degree of reliability required.
1. Resistance element and circuit must be insulated to prevent parallel resistance paths.
2. The material may be formed as wire or strips, and is generally wound around an insulating arbor or imbedded within a ceramic tube or rod.
3. Resistance thermometer winding configurations and bulb configurations are generally dependent upon the intended application.

2.4.2.4 — Resistance Thermometer Circuit Design Considerations

The following listing points out some of the important considerations in resistance thermometer circuit design.
1. Stray thermal emf problem: Ensure that all pairs of junctions are comprised of dissimilar metals and are at equal temperatures.

2. Self-heating: The effect is proportional to the amount of current flowing through the resistance temperature detector (RTD), the resistance of the RTD and the medium being measured by the RTD (cooling effect). Example: Still air versus flowing water.

3. Parallel resistance errors: All connecting resistance thermometer wires must be insulated to prevent parallel low-resistance paths.

4. Lead resistance errors:
   a. Two-wire system. In a two-wire system, the lead resistance and variations in lead resistance add to the reading.
   b. Three-wire system. In a three-wire system, a measuring lead is in each of two legs of the bridge. If their leads are initially balanced and are run in the same cable, initial resistance and changes in resistance as a function of temperature will not generally introduce error.
   c. Four-wire system. In a four-wire system, three leads of a four-lead thermometer are switched so that the effect of lead resistance is canceled out. The disadvantage of a four-wire system, however lies in the switching. Errors will be introduced if switch contact resistance is not repeatable when switching leads from one side of the bridge to the other.

2.4.3 — Resistance Thermometer Measuring Circuits

Resistance thermometer measuring circuits are generally one form or another of a null-bridge type of instrument. In the case of plant instruments, the RTD forms one leg of a bridge; changes in resistance change the balance of the bridge and provide an error signal for operation of plant instruments.

Recording type instruments, such as the Leeds and Northrup shield recorder utilize the bridge error (unbalance) signal for actuating a servomotor which positions a slidewire slide to balance the bridge. A pen mechanism attached to the slidewire shaft is calibrated to give an indication of the temperature.

2.4.4 — Some Advantages And Disadvantages Of Resistance Thermometers

Comparison of a resistance thermometer with a liquid-in-glass thermometer reveals the following advantages and disadvantages.
1. The resistance thermometer is usable over a wider range of temperatures.
2. Is less fragile.
3. Can be installed in relatively inaccessible locations and read at a distance.
4. Is adaptable to recording and control.
5. Is usually a more costly method of measurement.
6. Is a more complex instrument.

Comparison of a resistance thermometer with a thermocouple reveals the following advantages and disadvantages:
1. The resistance thermometer has higher sensitivity to a small change in temperature.
2. Is more readily adapted to measurement of the mean temperature of a space.
3. Does not require a reference junction.
4. Has a greater time lag (usually).
5. Can neither be intimately attached to a small body whose temperature is to be measured, nor inserted in as small a cavity as a thermocouple requires.
6. Is, in general, more subject to injury by vibration.

7.   Presents a greater problem in avoiding thermal radiation errors.

Section 2.5 — Nuclear Radiation Detectors

Nuclear radiation detectors can be characterized by the method in which the detector functions. The three principal methods used for detecting radiation are ionization in a gas-filled chamber, scintillation, where light is emitted because of ionization in a phosphor, and thermoluminescent dosimetry, where energy resulting from ionization is stored in a phosphor and is released as light when the phosphor is heated. This section discusses the first two detection methods listed above.

2.5.1 — Ionization Detectors

Many instruments for the detection of nuclear radiations are dependent upon the behavior of the ion pairs, formed by the ionizing particles in their passage through a gas, in the presence of an electric field. This behavior varies with the potential gradient of the electric field and can best be explained by considering the potential difference established across two electrodes, one a hollow metal cylinder and the other a wire running along the cylinder's axis (Exhibit 2-52), placed in a vessel of gas. As a general rule, the central wire is the positive electrode (anode), and the cylinder is the negative electrode (cathode). Negative ions, usually electrons, are attracted to the wire, and positive ions to the cylinder.

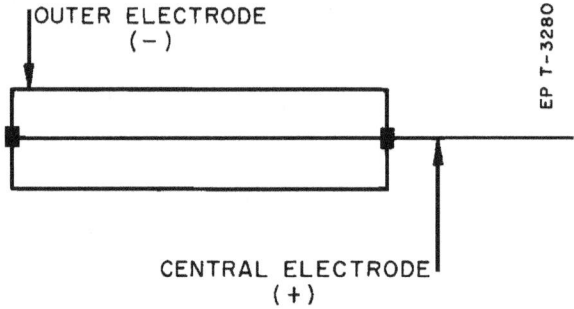

Exhibit 2-52 — Electrical Field For Study Of Ion Parts

Suppose a burst (or pulse) of ionizing radiation enters the gas between the electrodes so that a definite number of ion pairs is formed. Under the influence of the electric field, the positive ions move toward the negatively charged electrode (cylinder), and the negative ions migrate toward the positive electrode (wire). The magnitude of the charge collected on the electrodes depends on the applied voltage, and the general nature of the variation is indicated in Exhibit 2-53. In this figure, the charge collected is shown (on a logarithmic scale) as a function of the voltage. Curve 1 refers to a case in which the radiation pulse produces 10 ion pairs, and Curve 2 to a larger pulse producing 1000 ion pairs. It can be seen that the curves can be divided into six fairly distinct regions, marked A through F. Three of these regions (B, C, and E) are used in various types of nuclear radiation instruments and each of them will be considered in turn.

The charge, collected at very low voltages, increases at first with the applied voltage (Region A) and then attains a constant value (Region B). At very low potentials, the ions move slowly in the electric field, with the result that many of them recombine before they reach the electrodes. With increasing voltage, the ions migrate more rapidly toward the respective electrodes; thus, the extent of recombination is decreased and the charge collected on the electrodes increases. This accounts for the behavior in Region A, the <u>recombination region</u>.

Above a certain voltage, the ions move so fast that the recombination is negligible and essentially every ion produced by the nuclear radiation reaches the electrode. A further increase in the applied voltage does not result in any increase in the number of ion pairs collected. Hence, for Region B, referred to as the <u>ionization chamber</u> (or ion chamber) region, the charge remains constant as the voltage is increased. The charge collected is then equal to the number of ion pairs produced in the gas (ionization chamber) by the nuclear radiation. This effect is shown for Region B in Exhibit 2-53. (Note that the ordinate scale in Exhibit 2-53 is logarithmic.)

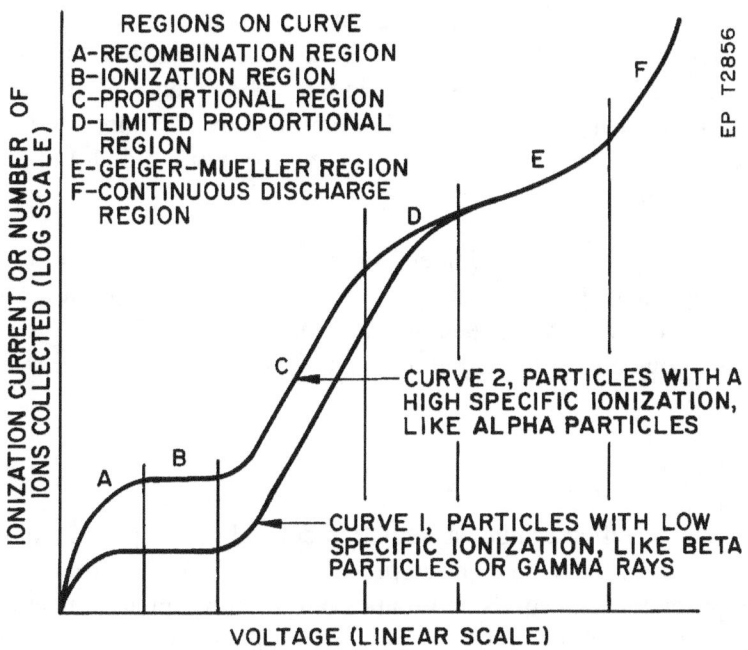

Exhibit 2-53 — Variation Of Charge Collected With Applied Voltage

As the voltage across the counter increases through Region C, the <u>proportional region</u>, the ions moving toward the electrodes are sufficiently accelerated between collisions with gas molecules to cause additional ionization. The resulting ion pairs are accelerated and collected by the electrodes. Thus, a single primary ion pair created by incoming radiation can result in the collection of a large number of ions at each electrode. This effect is called an ionization avalanche and the average number of ion pairs actually collected for each primary ion pair formed is called the gas amplification factor.

The gas amplification factor is unity in Region B since avalanche does not occur. In Region C, however, the avalanche effect becomes increasingly significant with increasing voltage, and may become as high as 104 near Region D. The avalanche effect is most significant near the center electrode, where the electric field intensity is the highest.

As the voltage is increased still further, other factors arise which limit the production of secondary ion pairs, so that the increase in the gas amplification factor does not continue indefinitely. The electrons (negative ions) are much lighter than the positive ions; thus, they are drawn toward the positive central electrode much faster than the positive ions are drawn to the cylinder. The result is a "cloud" of excess positive ions, which forms a space charge that masks the negative center electrode, reduces the electric field intensity in part of the counter volume, and thus tends to limit the extent of gas amplification. At sufficiently high voltage, in Region D, the amplification factor consequently approaches a limit, and the charge collected is no longer proportional to the initial ionization. Region D is therefore referred to as the limited proportional region. At the upper end of this region the space charge effect is effectively the only factor determining the amount of charge collected. The discharge in the tube has become space-charge limited, and the same charge is collected regardless of the number of primary ion pairs formed. Note that the gas amplification factor no longer has a definite value under these conditions.

The space charge limitation is effective across Region E, which is called the Geiger-Mueller region. Almost every particle entering the chamber operating within the Geiger-Mueller region is now recorded, whether it causes little or much primary ionization. By suitable design, the dependence of collected charge on applied voltage can be made quite small, such that for ionizing radiation above the threshold necessary to produce primary ionization, the pulse count rate is independent of the energy of incident radiation and the voltage applied to the ionization chamber.

If the applied voltage is increased beyond the Geiger-Mueller region, there is a rapid increase in the charge collected; this condition is represented by Region F of Exhibit 2-53. The potential is so high, then, that once a discharge is initiated, others follow in rapid succession so that there is effectively a continuous discharge. This region, called the <u>continuous discharge region</u>, is not used for the detection or measurement of ionizing radiations.

2.5.2 — Scintillation Counters

Most instruments used for the detection of nuclear radiation depend on the production of ion pairs in a gas, followed by their collection and recording in an appropriate manner. Nuclear radiations also produce ionization (and excitation) in their passage through solids and liquids but the subsequent collection and measurement of the ion pairs have proved difficult. The restriction to gaseous media is not serious, except in connection with the measurement of gamma radiation. For this latter purpose, an ionization medium of fairly high density and high atomic number is desirable if the detection efficiency is to be good, since the probability of interaction of the radiation with the medium is thereby increased. In the scintillation counter, a solid or liquid is used as the medium with which the radiation interacts.

It has long been known that certain solids and liquids, referred to as phosphors, emit light when exposed to nuclear radiations. In traversing the phosphors, the ionizing particles, or radiation, cause the molecules of the material to become ionized or excited. These molecules then emit their excess energy in the form of light, each interacting particle giving rise to a flash, generally called a scintillation. This is the basic principle used in the scintillation counter.

The most widely used phosphor or scintillation material is sodium iodide in the crystalline form with a trace of the element thallium also present in the crystal as an activator. Activators create fluorescent centers in the crystal which assist in the light emission process. There are other materials which scintillate when exposed to radiation, including solid zinc sulfate with silver.

The scintillator is optically coupled to a photomultiplier tube which converts the tiny flashes of light to an electronic signal and amplifies this signal by many orders of magnitude. The electronic pulses produced by the photomultiplier tube are then fed to an electronic circuit where the pulses are further amplified and then counted.

One method for combining the phosphor with a photomultiplier tube is shown in Exhibit 2-54. The aluminum foil acts as a reflector of the light flashes. A particle of nuclear radiation or a gamma ray photon entering the phosphor produces a flash of light. The intensity of the light is proportional to the energy of the ionizing radiation. This light falls upon the photocathode located on the inner surface of the glass envelope of the photomultiplier tube, where a number of photoelectrons are liberated. These are increased about a million-fold (or more) in passing through the tube, so that an appreciable voltage pulse is produced at the output stage.

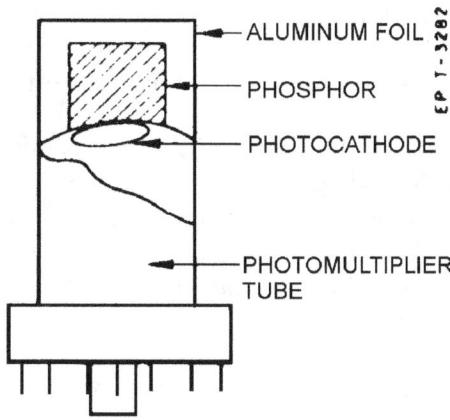

Exhibit 2-54 — Scintillation Counter

A schematic cross-section of one type of photomultiplier tube is shown in Exhibit 2-55. The photomultiplier is a vacuum tube with a glass envelope and containing a photocathode and a series of electrodes called dynodes. Light from a scintillation phosphor liberates electrons from the photocathode by the photoelectric effect. These electrons are not of sufficient number or energy to be detected reliably by conventional electronics. However, in the photomultiplier tube they are attracted by a voltage drop of about 50 volts to the nearest dynode. The photoelectrons strike the first dynode with sufficient energy to liberate several new electrons for each photoelectron. The second-generation electrons are in turn attracted to the second dynode where a larger third-generation group of electrons is emitted. This amplification continues through 10 to 12 stages. At the last dynode sufficient electrons are available to form a current pulse suitable for further amplification by transistor circuits. The voltage drops between dynodes are established by a single external bias on the order of 1000 volts d-c, and a network of external resistors to equalize the voltage drops.

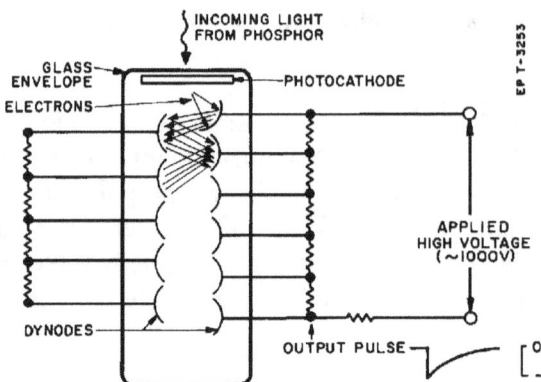

Exhibit 2-55 — Operation Of A Photomultiplier Tube

The chief advantages of the scintillation counter are its efficiency, resulting from the use of a relatively large mass of phosphor, and the high precision and counting rates that are possible. These latter attributes are a consequence of the extremely short duration of the light flashes, from about 10^{-9} to 10^{-6} seconds. The intensity of the light flash and the amplitude of the output voltage pulse are proportional to the energy of the particle responsible for the flash. Consequently, scintillation counters can be used to determine the energy, as well as the number of the exciting particles (or photons).

In connection with reactor operations, scintillation counters have previously been used mainly for detecting and measuring low-level gamma radiation. Formerly, Geiger counters have been largely employed for this purpose, but their efficiency is very low. On the other hand, scintillation counters are higher priced and require somewhat more complicated equipment than Geiger counters. By using phosphors containing lithium, scintillation counters can be adapted to neutron detection. Light flashes are produced by the helium and tritium nuclei resulting from the Li^6 (n, α)H^3 reaction.

2.5.3 — Counting Rate Meters

Most counting rate meters are arranged so that each input pulse feeds a known charge (q) into a tank capacitor that is shunted by a resistor, R. The voltage across the capacitor builds up to an equilibrium value at which the rate of loss-of-charge through the shunt resistor equals the rate of input-of-charge by the pulses. If the charge (q) per pulse is a constant, the equilibrium value of the output voltage (v) is:

Equation 2-1

$$v = rqR$$

where r is the average number of pulses per second. The voltage v is thus proportional to the counting rate r (q and R are constant) and can be read on a voltmeter.

The equilibrium voltage across the tank capacitor, as given by Equation 2-1 is seen to be independent of the size of the capacitor. However, the capacity does affect statistical accuracy and the time required to reach equilibrium.

The diode pump circuit which is shown in Exhibit 2-56, is a satisfactory method of feeding the charge into the tank capacitor. The generator (E), with internal resistance (R_f), produces a rectangular pulse of duration T and magnitude V. The capacitor (C_f) is charged, through the resistance R_f in series with D_1, up to nearly the pulse voltage (V) provided the pulse duration (T) is greater than about five time constants or $>5R_fC_f$. When the input pulse returns to zero, C_f discharges through D_2 and places a fixed charge (VC_f) per pulse on the tank capacitor (C_f) provided the following conditions hold:

$$C_f << C_t$$

$$v << V$$

and

$$\frac{1}{r} - T > 5R_fC_f.$$

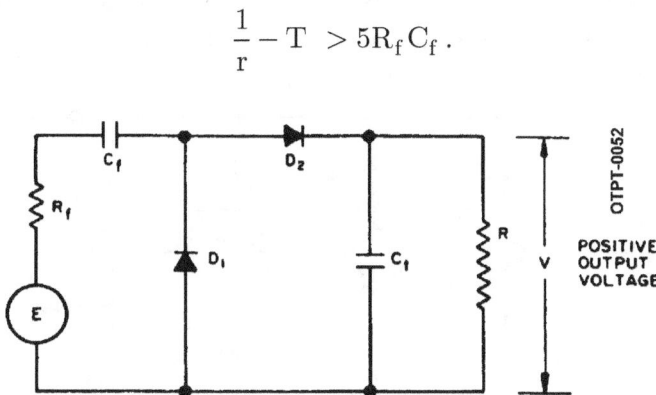

Exhibit 2-56 — Diode Pump Circuit For A Counting Rate Meter

The first two conditions ensure that negligible charge remains on C_f in equilibrium, while the third condition ensures that sufficient time elapses for equilibrium to be nearly reached before the next pulse occurs.

If the condition $v << V$ is not satisfied, the charge per pulse becomes $(V - v) C_f$; therefore, the relationship of output voltage to pulse rate becomes

Equation 2-2

$$v = \frac{VrC_fR}{1 + rC_fR}$$

Equation 2-2 indicates that, by proper choice of circuit components, relationships between voltage and counting rate other than linear can be obtained.

A logarithmic response can be obtained by placing a logarithmic element across the tank capacitor, C_t.

Section 2.6 — Nuclear Instrumentation Circuits

2.6.1 — D-C Amplifiers

Many reactor plant instrumentation applications require the amplification of d-c or slowly varying signals. In these applications, RC coupled amplifiers are unsuitable and it becomes necessary to direct-couple the amplifier stages. Exhibit 2-57 shows a basic direct-coupled transistor amplifier. When a signal is applied to the base of Q1, the amplified output is directly applied to the base of Q2 from the collector of Q1. The output is taken from load resistor R_L. Direct-coupling of transistor stages is limited to a few stages since all signals are amplified, including noise. Transistor bias shifts are due primarily to the change of transistor parameters (h_{FE}, I_{CO}, V_{BE}) with temperature and the variation of these parameters between transistors of the same type. When direct-coupled transistor stages are used, the magnitude of the quiescent collector voltages increases as one progresses through the circuit toward the load. With transistors, this problem can be solved by alternating complementary (PNP and NPN) transistors in cascade as shown in Exhibit 2-58.

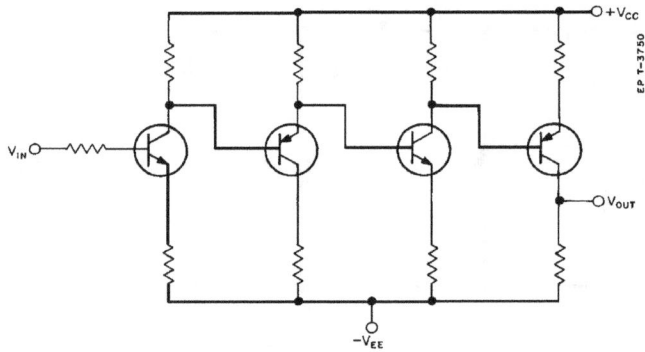

Exhibit 2-57 — Direct-Coupled Transistor Amplifier

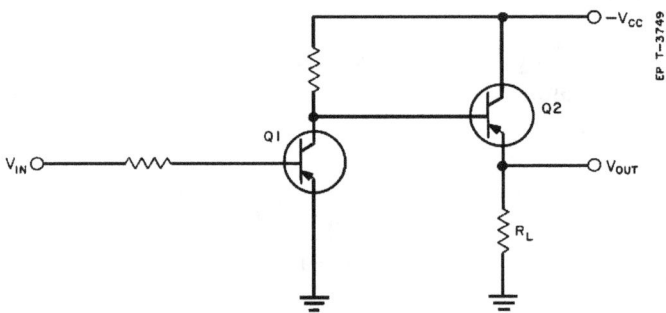

Exhibit 2-58 — Complementary Symmetrical, Direct-Coupled Transistor Amplifier

The d-c amplifier discussed above is acceptable in applications where the signal input level is high enough that drift considerations are not unduly severe. For lower input levels, down to around one microampere, differential amplifiers such as shown in Exhibit 2-59 are used. The circuit will amplify single-ended inputs (by setting V_2 = 0) or it will amplify the difference of two isolated inputs ($V_1 - V_2$). One feature of the differential amplifier is that it amplifies only the difference of the two input signals and reject the signal common to both inputs. The load is usually connected between the collectors of Q1 and Q2. A properly designed differential amplifier, with Q1 matched to Q2 and corresponding resistances, will provide stable d-c amplification even when the circuit is subject to temperature excursions. Any change in gain or leakage of Q1 is balanced out by an equal change in gain or leakage of Q2, due to the symmetry of the differential amplifier. Practical differential amplifiers will exhibit some drift, particularly when operated with a single-ended input. With a single-ended input, drift may be minimized by using several differential amplifiers connected in cascade with common-mode feedback.

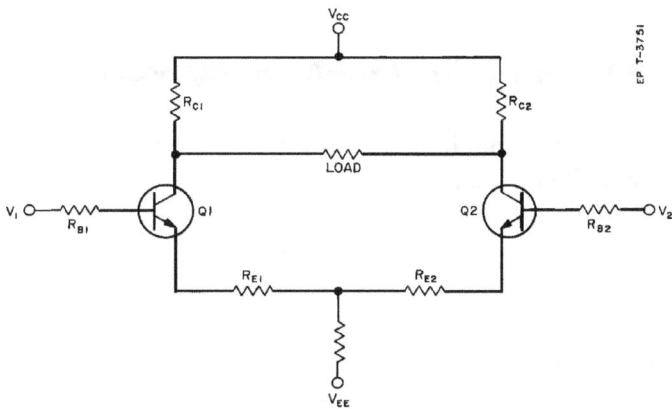

Exhibit 2-59 — Emitter-Coupled Differential Amplifier

The difficulties in designing a stable d-c amplifier are often overcome by using the d-c signal to modulate an a-c signal. This modulated a-c signal is then amplified by a conventional amplifier, after which it is converted back to d-c by demodulation.

2.6.2 — Differentiating Circuits

Our purpose is to obtain automatically a voltage proportional to the rate of change, or slope, of a given voltage. The output voltage will approximate the time derivative of the signal input voltage. A possible arrangement that satisfies this condition appears in Exhibit 2-60. We shall show that the voltage e_{out} is roughly proportional to the rate of change of the voltage e_{in} provided the time constant (RC) is suitably chosen. From Kirchhoff's Law,

$$e_{in} = e_{out} + e_c = iR + q/C$$

where q is the instantaneous charge stored in the capacitor, and e is the instantaneous voltage across C. Differentiating with respect to time, we obtain:

$$\frac{de_{in}}{dt} = R\frac{di}{dt} + \frac{i}{C}$$

or

$$C\frac{de_{in}}{dt} = RC\frac{di}{dt} + i.$$

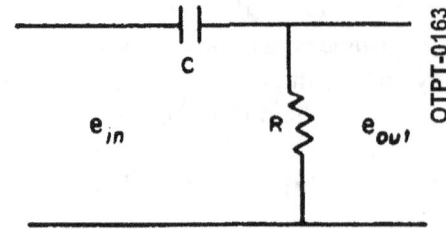

Exhibit 2-60 — Differentiating Circuit

If the time constant is sufficiently small and di/dt not too large, we can neglect the first term on the right-hand side of this equation in comparison with i and thus obtain:

$$i = C\frac{de_{in}}{dt}$$

Since $e_{out} = iR$, the output voltage is approximately:

Equation 2-3

$$e_{out} = RC\frac{de_{in}}{dt}$$

Therefore, the output voltage is roughly proportional to the time derivative of e_{in}. To see in more detail how this circuit acts consider an example (Exhibit 2-61) in which the signal is a square wave. At the instant t_o the source voltage suddenly jumps from zero to 100 volts. The transient is as follows: Initially all the source voltage appears across R, since the capacitor requires some time to charge. As the capacitor charges the current in the circuit and the voltage across R follow an exponential decay curve. If the time constant (RC) is small compared with the period (1/F) of the square wave, the output voltage drops to a negligible value in a small fraction of a cycle of the square wave. The voltage across R during this time is shown in Exhibit 2-61 (b).

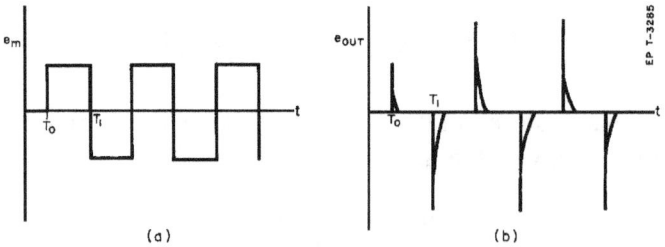

Exhibit 2-61 — Differentiation Of A Square Wave

At the instant t_1, the source voltage changes by 200 volts. The resultant transient is a pulse in the opposite direction. The differentiating circuit has transformed the square wave into a series of pulses. The pulses can be made extremely sharp if both the time constant and the rise time of the square wave are short.

The interpretation of a derivative as the slope of the input voltage furnishes a convenient method of predicting the output waveform for various inputs applied to differentiating circuits. As we have found, the derivative of a square wave is a series of pulses. When the slope of the input voltage waveform becomes zero, the output voltage vanishes. The greater the input slope (or rate of change), the greater is the output voltage at that time. For example, the derivative of the triangular wave of Exhibit 2-62 (a) is the square wave of Exhibit 2-62 (b).

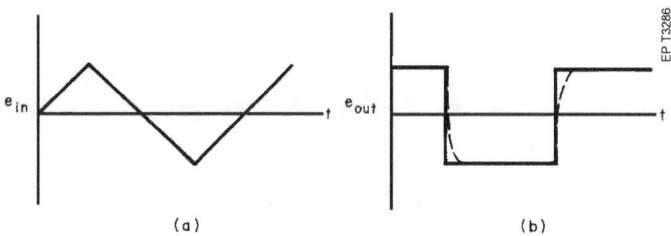

Exhibit 2-62 — Differentiation Of A Triangular Wave

The RC time constant is the time required for the capacitor to charge to 63 percent of the applied voltage. If 100 volts is applied, it takes one time constant for the capacitor to charge to 63 volts.

The RC time constant is found by multiplying the resistance in ohms by the capacitance in farads, and it is in seconds; or the resistance in megohms and the capacitance in microfarads, to give seconds; or the resistance in megohms and the capacitance in picofarads to give the time in microseconds.

A differentiating circuit has a short time constant if RC is one-tenth or less than the of the period (T) of the input cycle (T = 1/f), where f is frequency. The following terms are used to describe the length of the time constant:

Short = RC is 0.1T or less

Medium = RC is greater than 0.1T but less than 10T

Long = RC is 10T or more.

2.6.3 — Integrating Circuits

Suppose that we wish to shape a wave in the reverse manner from that of a differentiating circuit; that is, we wish to create a square wave from a series of uniform pulses or a triangular wave from a square wave. Since integration is the inverse of differentiation, it is reasonable to name this kind of circuit an integrator.

A series RC circuit with a comparatively long time constant yields an input across the capacitor that is approximately proportional to the integral of the input voltage. For Exhibit 2-63 we write:

$$e_{in} = e_R + e_{out} = IR + \frac{q}{C}$$

where q is the instantaneous charge stored in the capacitor. Multiply both sides of this equation by dt and integrate:

$$\int e_{in}dt = \int Ridt + \int \frac{qdt}{C}.$$

Since C and R are assumed constant,

$$C\int e_{in}dt = RC\int idt + \int qdt.$$

If the time constant RC is sufficiently long, we can neglect the last term on the right in comparison with the first term on the right. Thus:

$$C\int e_{in}dt = RC\int idt = RCq.$$

since $\int idt = q$. Replace q in the last term by its equivalent, which is Ce_{out}, and solve for e_{out}:

Equation 2-4

$$e_{out} = \frac{1}{RC}\int e_{in}dt.$$

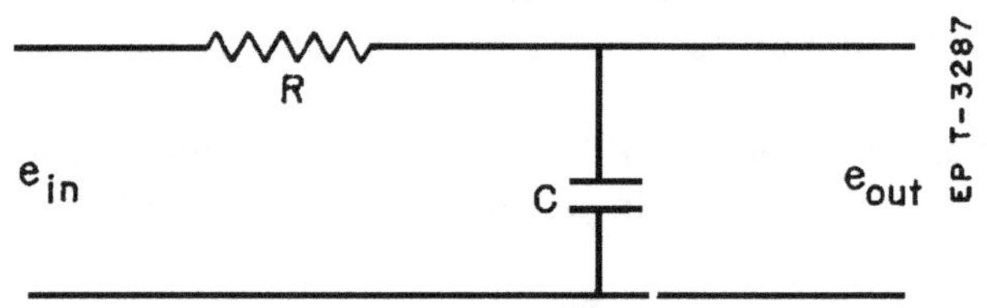

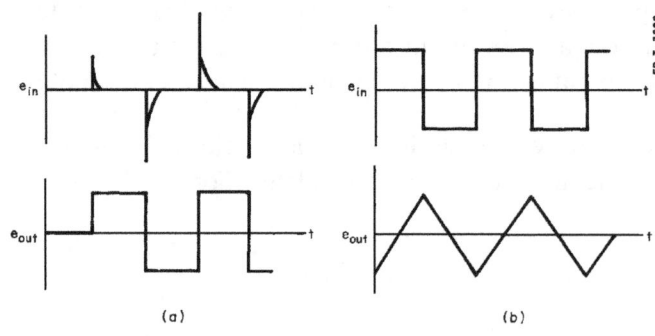

Exhibit 2-63 — Intergrating Circuit

From Equation 2-4, the output voltage is approximately proportional to the time integral of the input voltage, and this is the relationship that we wished to prove. The interpretation of the integral as the area under a curve affords a simple, graphic method of predicting the shape of the output voltage for various inputs (Exhibit 2-64).

Exhibit 2-64 — Integration Of A Series Of Pulses (a) And A Square Wave (b)

Section 2.7 — Differential Transformers

A differential transformer is essentially a multiwinding transformer with a movable ferromagnetic core used to develop an electrical signal which is proportional to a mechanical position. As shown in Exhibit 2-65, the primary winding is energized by an a-c source to provide a continuous magnetic field, and the two secondary windings are connected in series-opposition so that their induced voltages are 180 degrees out of phase. The position of the movable core within the winding assembly determines the relative flux distribution and hence the mutual coupling between the primary and each of the secondary windings.

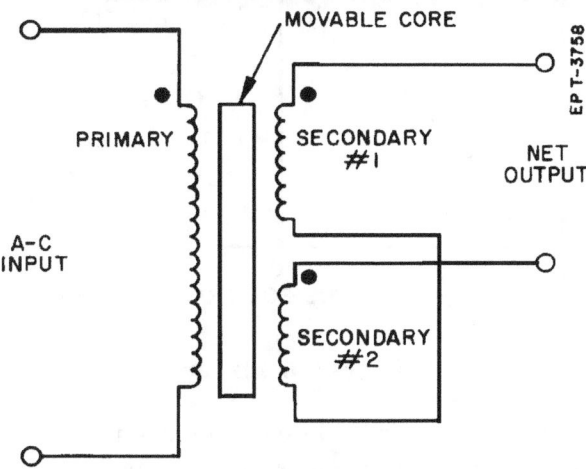

Exhibit 2-65 — Schematic Of Differential Transformer

When the movable core is positioned so that each secondary winding receives an equal amount of flux, the induced voltages will be equal. At this point, the induced voltages effectively cancel each other so that the net output voltage is zero. This condition denotes the electrical center of the device, referred to as the null position or balance point.

If the movable core is displaced by an actuating force to either side of the null position, the voltage induced in the secondary toward which the core is moved increases, whereas the voltage induced in the other secondary decreases. The difference between each secondary winding voltage results in an a-c output signal whose magnitude is proportional to the axial displacement of the core from the null position. Exhibit 2-66 shows the voltage curve of the secondary windings as the movable core is moved through its linear range of displacement.

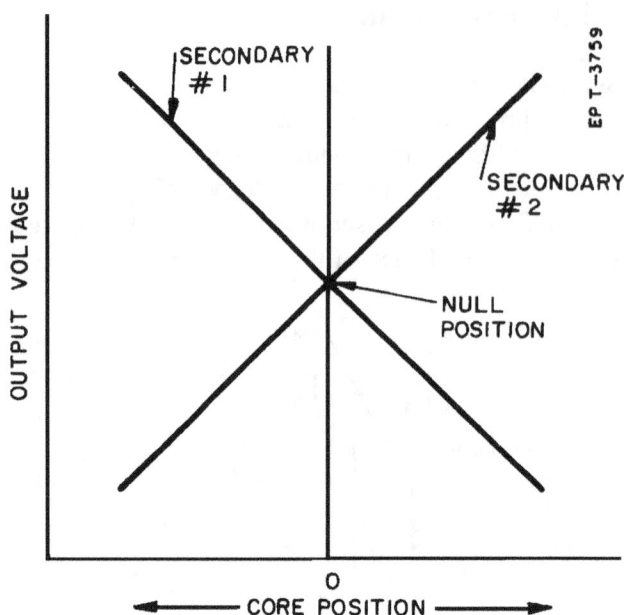

Exhibit 2-66 — Secondary Voltages For Linear Range Of Core Displacement

The primary and secondary windings are wound in close proximity on a tubular bobbin fabricated of a suitable nonmetallic material. To provide high resistance to environmental humidity, shock, and vibration, the entire winding assembly is encapsulated in a potting compound such as epoxy resin. The movable core is made of a high-permeability ferromagnetic alloy that is free to move axially within the bore of the cylinder without actually making mechanical contact. Since each core must be accurately matched to a particular transformer, the device must be precisely machined in order to obtain a linear magnetic characteristic. To further improve the magnetic properties, the core is heat-treated during the fabrication process to relieve mechanical stresses.

Differential transformers are particularly adaptable to the remote monitoring of system parameters necessary for the control and safe operation of a reactor plant. As a primary sensing element, they are used to accurately measure, control, and record such variables as temperature, pressure, and flow. A differential transformer is used for primary plant pressure detection. As the pressure changes, a Bourdon tube coils or uncoils and by mechanical linkage varies the mechanical position of the ferromagnetic plunger (core) .

Certain valve position indicators utilize "E core" differential transformers such as illustrated in Exhibit 2-67. In these devices, the center leg of the "E" is the primary and the outer legs are the secondaries. A ferromagnetic slug is attached to the valve stem and, as the valve moves, varies the coupling between the primary and the secondaries. The output voltages are rectified and their difference is taken.

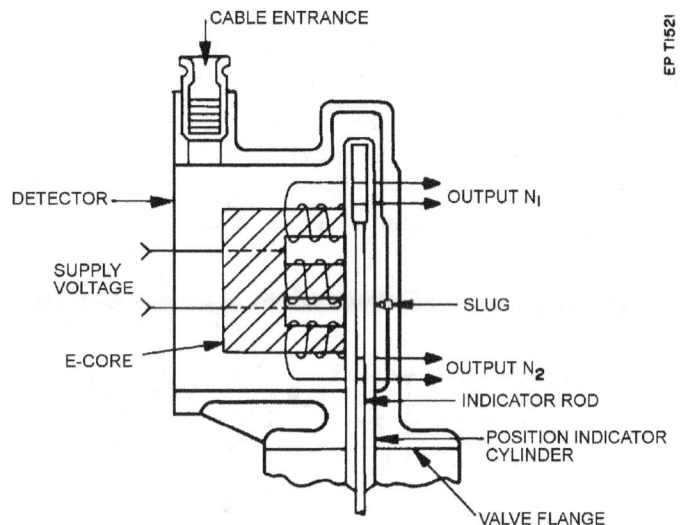

Exhibit 2-67 — "E" Core Position Indicator

The difference, which varies with the valve position, is amplified to give an indication of valve position. The "E" core detector is used to operate a relay and lights as this detector is designed to generate an abrupt change in voltage rather than the proportional change in voltage generated by most differential transformers.

Section 2.8 — D-C Power Supplies

2.8.1 — Introduction

A d-c power supply converts a-c line power to d-c power of various qualities by one or more of the following stages:

Transformer	= steps line voltage up or down to a convenient a-c voltage.
Rectifier	= converts a-c voltage to d-c voltage.
Filter	= reduces irregularities in rectified d-c voltage.
Regulator	= maintains a uniform d-c voltage.

The first stage of the power supply, the transformer, is used for one or both of the following reasons: either the available a-c line voltage is not appropriate, or d-c isolation from the line is desired.

2.8.2 — Rectifiers

All rectifiers in nuclear plants use silicon diodes. A simple rectifier circuit using a transformer and one diode is shown in Exhibit 2-68 (b). The diode passes current in only one direction, resulting in the waveform shown at the right. However, only one-half cycle of the a-c waveform is utilized.

The circuit in Exhibit 2-68 (a) uses both half-cycles of the a-c waveform but requires a center tap on the transformer. The bridge rectifier in Exhibit 2-68 (c) gives an identical output to the full-wave rectifier in Exhibit 2-68 (b). The only difference is the absence of a center tap and the fact that each diode encounters only half of the peak voltage when used in a bridge configuration.

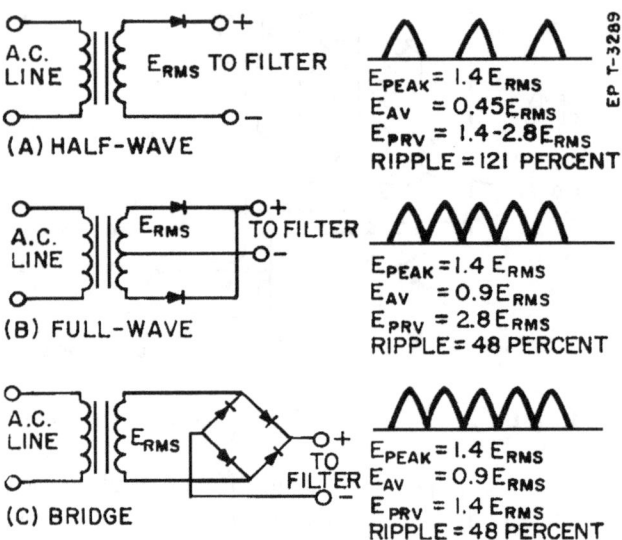

Exhibit 2-68 — Rectifier Circuits

2.8.3 — Filter Circuits

The amount of ripple (variation from a straight-line waveform) in the output of a rectifier circuit can be reduced by a filter circuit. Filter circuits depend on two types of circuit elements, capacitors and inductors, called <u>chokes</u>. Capacitors are connected across the output of the rectifier (in parallel with the load) to oppose changes in voltage. Chokes are connected in series with the load to oppose changes in load current.

Power supply filters are denoted as capacitor-input or choke-input according to whether the first element after the rectifier is a capacitor or a choke. Capacitor-input filters are characterized by relatively high output voltage: approaching the peak voltage of the rectifier output waveform for low load currents. They are most suitable for low-current loads: ripple increases significantly with load current. Choke-input filters produce a lower output voltage from a given transformer-rectifier system; the voltage is limited to the average of the rectifier output waveform for a properly designed filter. Intrinsically, a choke provides no filtering action unless a minimum current is drawn. Therefore, the output of the choke includes a "bleeder" resistance to ensure that the necessary current is drawn even if no load is connected. A bleeder resistance is also used for capacitor-input filters to reduce the effects of varying load currents on output voltage, and to ensure that a discharge path is available for the filter capacitor when the supply is shut off.

A filter, whether capacitor-input or choke-input, may include additional choke-capacitor pairs, called LC sections, for improved ripple reduction. Several combinations are shown in Exhibit 2-69. Although voltage ripple can be reduced to almost any degree, in principle, additional LC sections will not reduce the drop in output voltage with increasing load current. A voltage regulator circuit must be used to counter this effect.

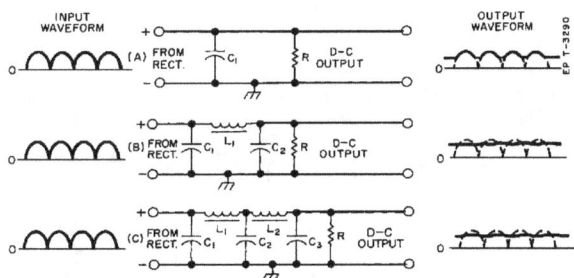

Exhibit 2-69 — Filter Circuits

2.8.4 — Voltage Regulator Circuits

A voltage regulator circuit is used to maintain a constant output voltage over wide variations in load current. Most regulating circuits make use of a zener diode. A simple regulating circuit is shown in Exhibit 2-70 (a). The zener diode maintains a constant output voltage, and the series resistor is chosen to limit the amount of current passed by the zener diode when no load is drawn.

The circuit of Exhibit 2-70 (a) will regulate voltage over a fairly wide current range at a single voltage. The more elaborate circuit shown in Exhibit 2-70 (b) permits improved regulation over a wider current range. In this circuit the transistor is used as a series regulating element. The base of the transistor is maintained at a constant voltage by the zener diode and R. Any increase in load current decreases the emitter voltage slightly, causing a slight increase in base-to-emitter current and consequently a larger increase in collector-to-emitter current. The advantage of this circuit (actually an emitter-follower amplifier) is that the range of currents that must be handled by the zener diode and its series resistor is much smaller than for the circuit of Exhibit 2-70 (a). Thus the reference voltage at the zener diode is maintained more accurately, and the series transistor (or transistors) can be chosen to handle a larger load current. The series-type regulator is used in most applications that require a regulated voltage supply. Note that for either of the regulator circuits in Exhibit 2-70, the d-c input voltage must exceed the desired output voltage by at least a few volts in order to allow a voltage drop across the regulator circuit.

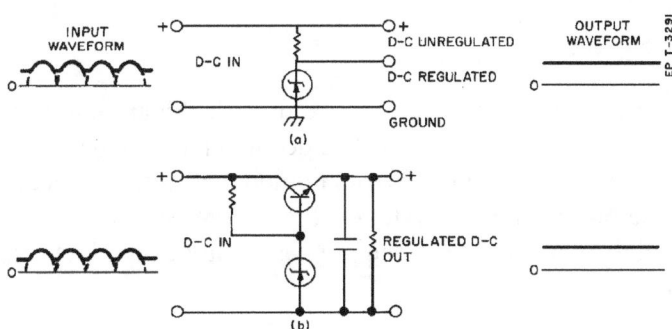

Exhibit 2-70 — Voltage Stabilization Circuits

Exhibit 2-71 shows a typical regulated d-c power supply that combines the rectifier, filter, and regulating circuits described above.

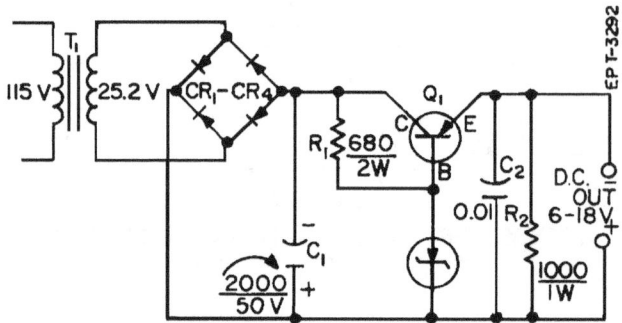

Exhibit 2-71 — Typical Regulated D-C Power Supply

Filtered, unregulated d-c power supplies are seldom used in nuclear plants, except in the d-c portions of some types of rod control systems. Filtered, regulated d-c power supplies are used extensively in primary plant and nuclear instrument systems where accurate signal processing is important. The output voltage of such power supplies is typically stable to within 1 or 2 percent.

Section 2.9 — Digital Integrated Circuit Devices

The use of digital equipment has expanded in recent years due to the decreased cost and increased reliability of modern integrated circuit (IC) devices. Digital ICs are the basic building blocks used to construct digital equipment. This section describes the binary number system and how it is applied in the building of digital IC devices.

In a digital circuit, signals exist as only one of two logic levels, or states, whereas analog signals vary continuously over a range of values. These states are usually referred to as 1 and 0, HIGH and LOW, or TRUE and FALSE. In actual digital circuits, these states are represented by two distinct voltage levels, most commonly 5 volts (HIGH) and 0 volts (LOW). The output is represented by a voltage, a logic state, or a binary symbol. This illustrates a type of digital circuit.

2.9.1 — Binary Number System

Understanding digital electronic circuits requires understanding the binary number system. A number system is a code that uses symbols to refer to a set of items. The decimal number system uses ten symbols (0, 1, 2, 3, 4, 5, 6, 7, 8, and 9), and is a base 10 system. The binary number system uses only two symbols (0 and 1), and is a base 2 system. The binary number system is the system best suited to digital circuits. The two symbols, 0 and 1, are referred to as binary digits and are commonly called bits (bits is a construction of the two words binary digits).

The decimal and binary number systems are positional or weighted number systems. This means that each position in a number carries a certain weight in determining the magnitude of the number. For example, the decimal number 659 has 6 hundreds, 5 tens, and 9 ones.

The binary number system also uses a weighted system. The position weights are increasing powers of the number system base (2) starting with 2^0 in the place on the right, as shown in the following table:

Table 2-2 — Demonstration Of Place Values In The Binary System

8	7	6	5	4	3	2	1	Place
2^7	2^6	2^5	2^4	2^3	2^2	2^1	2^0	Base 2 Weight
128	64	32	16	8	4	2	1	Base 10 Value

In order to convert a binary number to a decimal number, add the base 10 value for each place in the binary number that has a 1. For example, the binary number 1011 would be $8 + 0 + 2 + 1 = 11$.

To convert a decimal number to binary, repeatedly divide the decimal number by 2. The remainder (always a 1 or a 0) of the first division is the least significant bit and each following remainder is the next higher significant bit. For example:

Table 2-3 — Conversion Of Decimal Number 22 To Binary

	Remainder	Decimal Value
$22/2 = 11$	0	0
$11/2 = 5$	1	2
$5/2 = 2$	1	4
$2/2 = 1$	0	0
$1/2 = 0$	<u>1</u>	<u>16</u>
	10110	22

Other binary codes are used in digital electronics equipment, including binary coded decimal (BCD) and hexadecimal (HEX) codes. BCD code represents the decimal digits 0 through 9 with a 4-bit binary code. Note that in the BCD system, the numbers 1010 through 1111 are not defined. To represent a decimal number in its BCD form, replace each decimal digit with its binary equivalent, making sure to keep the same digit order. For example:

Table 2-4 — Conversion Of Decimal Number 312 To BCD Form

	3	1	2
	↓	↓	↓
BCD	0011	0001	0010

Note that the BCD system differs from the pure binary system in that the decimal number is converted to its 4-bit binary equivalent on a digit-by-digit basis. The decimal number 312 expressed in the pure binary system is 100111000 while in the BCD system it is 0011 0001 0010.

HEX (Base 16) code groups the binary code in 4-bit groups with a single digit symbol. The symbols 0 through 9, A, B, C, D, E, and F represent the 16 possible 4-bit groups. To convert a decimal number into HEX, the method of successive divisions is used as illustrated above. However, the decimal number is divided by 16 each time rather than dividing by 2. For example, the decimal number 4751 is equivalent to the hexadecimal 128F as shown below:

$$4751 \div 16 = 296 + \text{a remainder of (decimal 15) F}$$

$$296 \div 16 = 18 \div \text{a remainder of 8}$$

$$18 \div 16 = 1 + \text{a remainder of 2}$$

$$1 \div 16 = 0 + \text{a remainder of 1}$$

$$\text{Decimal } 4751 = \text{Hexadecimal } 128\,\text{F}$$

To convert a HEX number into a decimal number, multiply each digit by its place value and add the results. For example, HEX Number 2AB is $(2 \times 256) + (A(10) \times 16) + (B(11) \times 1) = 512 + 160 + 11 = 683$.

2.9.2 — Logic Gates

The basic building block of any digital circuit is the logic gate. Logic gates can be implemented using simple switches, relays, vacuum tubes, transistors, diodes, and integrated circuits (ICs). Because of their wide use, ICs are used in this discussion to illustrate the various logic gates. A logic gate is a circuit that makes the decision, "yes" or "no" (on or off; 1 or 0), at the output based on the inputs. The following discussion describes the various logic gates encountered in digital circuits.

2.9.2.1 — The Inverter

The inverter (or NOT gate) gives an output that is the opposite of the input. Sometimes the opposite is also called the complement or the inverse. The NOT gate is used so extensively that in some cases it is represented as a small circle at the inputs or output of another logic gate. The logic symbol for the NOT gate, as well as the other logic gates discussed in this section, are shown in Exhibit 2-72. When a binary 1 is applied to the input, the output is a binary 0 and when a binary 0 is applied to the input, the output is a binary 1.

Note the F = x' equation (read "F" equals "not x"). This is the Boolean algebraic expression of the NOT gate. Boolean algebra is a mathematical method of expressing, analyzing, and simplifying logic circuits. As each gate is discussed, its Boolean expression is also given. Another method of illustrating the various inputs and outputs of logic gates for all situations is by using a truth table. Exhibit 2-72 shows the truth table and expression for a NOT gate.

2.9.2.2 — The AND Gate

The AND gate is a circuit that has two or more inputs and one output. The output is binary 1 if and only if all of the inputs are a binary 1. When any input is a binary 0, the output of the AND gate is a binary 0. The AND gate encountered most is constructed from diodes and transistors, and is packaged in an IC. Exhibit 2-72 shows the logic symbol, Boolean expression, and truth table for a two-input AND gate.

Note the multiplication dot (•) can be used to symbolize the AND function in the Boolean expression. The AND gate may have more than two inputs, the exact number being determined by the application. If it had three inputs, the Boolean expression would be $F = x \bullet y \bullet z$.

2.9.2.3 — The OR Gate

Another basic logic gate is called the OR gate. Like the AND gate, it has two or more inputs and a single output. The output is a binary 1 when any input is a binary 1. The output is a binary 0 only when all the inputs are a binary 0. Exhibit 2-72 shows the logic symbol, Boolean expression, and truth table for a two-input OR gate. Note the addition symbol (+) in the Boolean expression. This is the Boolean symbol for the OR function.

2.9.2.4 — The NAND Gate

The AND, OR, and NOT gates are the three basic digital logic circuits. The NAND gate is a NOT AND or an inverted AND function. The logic symbol, Boolean expression, and truth table for the NAND gate are shown in Exhibit 2-72. The small circle on the output represents the inverter. The NAND gate is the most widely used gate in digital electronics due to circuit design considerations, and therefore it is sort of a "universal gate." Exhibit 2-73 gives some examples of how NAND gates are connected to perform the basic logic functions.

2.9.2.5 — The NOR Gate

The NOR gate is a NOT OR gate. The output of an OR gate is inverted to make a NOR gate. Exhibit 2-72 shows the logic symbol, Boolean expression, and truth table for the NOR gate.

2.9.2.6 — The Exclusive OR Gate

A combination of different gates form a widely used logic function known as the exclusive OR (XOR). The XOR gate outputs a binary 1 when only one of its inputs is a binary 1. Sometimes the XOR gate is called the "either but not both" gate. Exhibit 2-72 shows the logic symbol, Boolean expression, and truth table for the XOR gate.

2.9.2.7 — The Exclusive NOR Gate

A combination of logic gates form another logic function called the exclusive NOR gate. It, too, finds wide use in digital equipment. Its logic symbol, Boolean expression, and truth table are shown in Exhibit 2-72.

Example 2.9.2.7 — The Buffer Gate

The buffer does not produce any logic operation on the input, since the binary value of the output equals the binary value of the input. It is used simply for power amplification of the digital signal and is equivalent to two inverters cascaded.

Name	Graphic symbol	Algebraic function	Truth table

Name	Graphic symbol	Algebraic function	Truth table
AND		$F = xy$	$x\ y\ \vert\ F$ 0 0 | 0 0 1 | 0 1 0 | 0 1 1 | 1
OR		$F = x + y$	$x\ y\ \vert\ F$ 0 0 | 0 0 1 | 1 1 0 | 1 1 1 | 1
Inverter		$F = x'$	$x\ \vert\ F$ 0 | 1 1 | 0
Buffer		$F = x$	$x\ \vert\ F$ 0 | 0 1 | 1
NAND		$F = (xy)'$	$x\ y\ \vert\ F$ 0 0 | 1 0 1 | 1 1 0 | 1 1 1 | 0
NOR	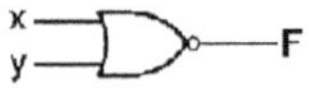	$F = (x + y)'$	$x\ y\ \vert\ F$ 0 0 | 1 0 1 | 0 1 0 | 0 1 1 | 0
Exclusive-OR (XOR)		$F = xy' + x'y$ $= x \oplus y$	$x\ y\ \vert\ F$ 0 0 | 0 0 1 | 1 1 0 | 1 1 1 | 0
Exclusive-NOR or equivalence	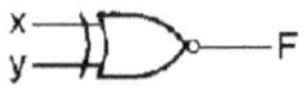	$F = xy + x'y'$ $= x \odot y$	$x\ y\ \vert\ F$ 0 0 | 1 0 1 | 0 1 0 | 0 1 1 | 1

Exhibit 2-72 — Digital Logic Gates

2.9.3 — Conversions Between Boolean Expressions And Logic Circuits

The Boolean expression is used to construct logic circuits. For example, the expression, $Y = A \cdot B + C \cdot D$, is two AND gates feeding an OR gate. Sometimes the dot symbol for the AND function is either left out or replaced with parentheses. For example, the preceding equation could also be expressed as $Y = (AB) + (CD)$ or $Y = AB + CD$.

The Boolean expression in the above example is known as a sum-of-products (SOP) form, because the various inputs are "ANDed" together and then "Ored" together (the AND function is Boolean multiplication; the OR function is Boolean addition).

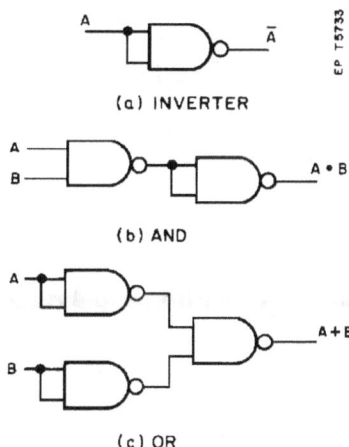

(a) INVERTER

(b) AND

(c) OR

Exhibit 2-73 — Implementing Logic Gates With NAND Gates

Another form for Boolean expressions is the product-of-sums (POS). The SOP and POS forms will permit writing the Boolean expression from any logic circuit and drawing the logic circuit from any Boolean expression. To write the expression of a given logic circuit, start at the inputs and write the output expression for each gate in the circuit from left to right until the output is reached.

2.9.4 — Flip-Flops

Logic circuits are placed in one of two categories, combinational logic circuits and sequential logic circuits. The logic gates described in Paragraph 2.9.2 (AND, OR, NOT, etc.) are called combinational logic circuits, because their output is exactly dependent on a particular combination of inputs. Flip-flops (FFs) are classified as sequential logic circuits, because their outputs depend not only on the present inputs, but also on input signals received in the past (that is, on a sequence of inputs).

The flip-flop is a basic digital circuit, as are the logic gate circuits. This section discusses five types of flip-flops used in digital equipment. Flip-flops are commonly used to construct memory storage devices. A flip-flop is capable of storing a single bit of binary data. It has two stable states, one representing a binary 1 and the other a binary 0. If a flip-flop is put into one of its stable states, it remains there until power is removed or until it is changed to another stable state by applying the proper sequence of input signals.

A flip-flop is triggered to change state either on the positive edge of the input pulse or on the negative edge. This is illustrated in Exhibit 2-74.

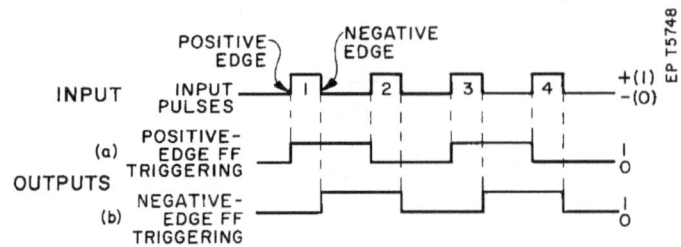

Exhibit 2-74 — Flip-Flop Triggering

2.9.4.1 — The Set-Reset Flip-Flop

The Set-Reset (RS) or latch flip-flop has two inputs, set and reset, and two outputs, Q and $\overline{Q}$ (the complement of Q is called Q-not). Exhibit 2-75 shows the logic symbol, logic diagram, and truth table for an RS flip-flop with NAND gates. The NAND flip-flop operates with both input normally at 1, unless the state of the flip-flop is to be changed. Assuming that this is the case, the momentary 0 applied to the set input causes output Q to go to 1 and $\overline{Q}$ to go 0, thus putting the flip-flop to the set state. After the set input returns to 1, a momentary 0 to the reset input causes a transition to the clear state, where Q is 0 and $\overline{Q}$ is 1.

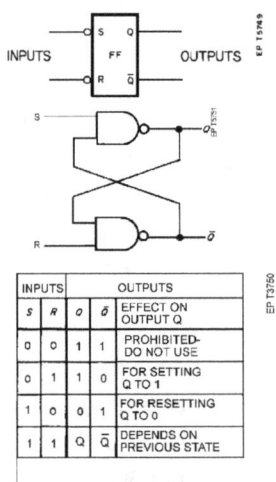

Exhibit 2-75 — RS Flip-Flop

2.9.4.2 — Clocked R-S Flip-Flop

The cross-coupled connection of the output of one gate to the input of the other gate constitutes a feedback path. For this reason, the flip-flop is classified as an asynchronous sequential circuit. It is also desirable to synchronize the operation of the flip-flop. A clocked (or synchronous) R-S flip-flop responds to an incoming signal at the R-S inputs only when a clock pulse of 1 is also at a third input, called the clock input. Exhibit 2-76 shows the logic symbol, logic diagram, and truth table for a clocked R-S flip-flop.

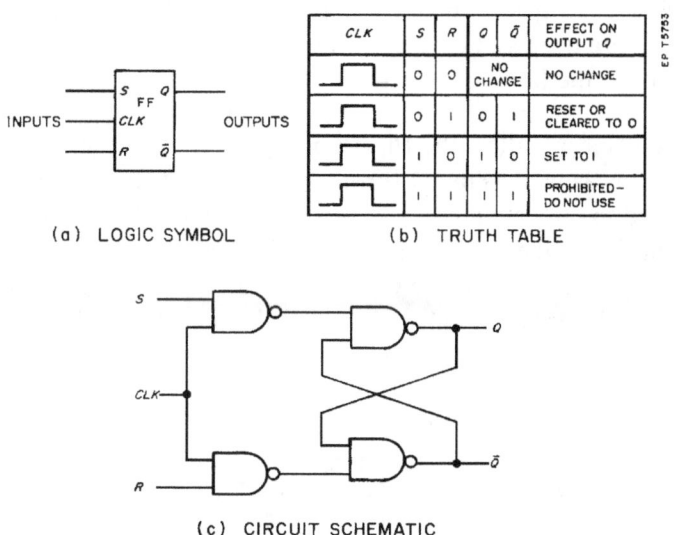

(a) LOGIC SYMBOL   (b) TRUTH TABLE

CLK	S	R	Q	Q̄	EFFECT ON OUTPUT Q
⎍	0	0	NO CHANGE		NO CHANGE
⎍	0	1	0	1	RESET OR CLEARED TO 0
⎍	1	0	1	0	SET TO 1
⎍	1	1	1	1	PROHIBITED — DO NOT USE

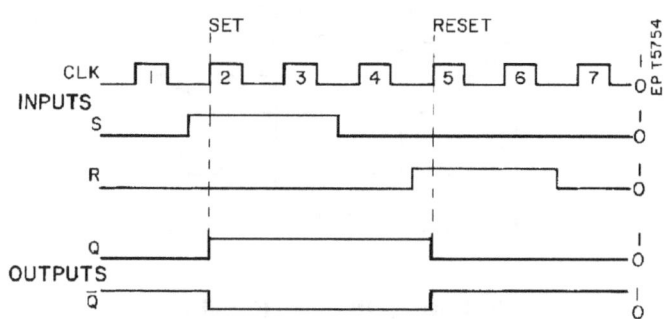

(c) CIRCUIT SCHEMATIC

Exhibit 2-76 — Clocked RS Flip-Flop

The flip-flop will not change states unless the clock (CLK) input is present at the same time as the other inputs (R and S). This gives the clocked flip-flop a characteristic that is valuable in counters and shift registers which will be discussed later. One noted difference from the asynchronous R-S flip-flop is that the prohibited state is where both the R and S inputs are 1 rather than 0. Exhibit 2-77 shows input and output waveforms for the clocked R-S flip-flop. The clock pulse has no effect on the output when both R and S inputs are 0. When the S input is 1, with R at 0, the next clock pulse changes the output (Q) to 1. Thus, the action of "set" occurs on the first clock pulse after S became 1.

Exhibit 2-77 — Waveforms For Clocked RS Flip-Flop

2.9.4.3 — The Delay Flip-Flop

A typical delay (D) flip-flop is shown in Exhibit 2-78. D flip-flops are widely used to form shift registers or storage registers, as presented in Paragraph 2.9.9. The D flip-flop has only one data input (D), a clock (CLK) input, and the outputs, Q and Q̄. The D flip-flop delays the data by one clock pulse from input to output. The truth table for the D flip-flop shows that no change in output occurs until the rising edge of the first clock pulse after the data input has changed from 0 to a logical 1. The arrowhead shown in Exhibit 2-78 at the D flip-flop clock input is used to specify that this particular flip-flop is triggered by the rising edge of a clock pulse. For flip-flops that are triggered by the falling edge of a clock pulse, a circle or bubble is added before the arrowhead but outside the box.

Two extra inputs, "preset" and "clear", allow the flip-flop to be "set" or "reset" by other logic signals. This feature is used in other flip-flops as well as the D flip-flop. The preset (PR) input sets the output to 1 (Q = 1) regardless of the signal applied to the data input. Likewise, the clear (CLR) input sets the output to 0 (Q = 0) regardless of the signal applied to the data input. This is illustrated in Exhibit 2-78. For this figure, both PR and CLR are normally 1. A change from 1 to 0 activates the preset and clear functions. The circle in Exhibit 2-78 (a) at the preset and clear terminals means that the preset and clear functions are triggered by a low pulse (logical state 0).

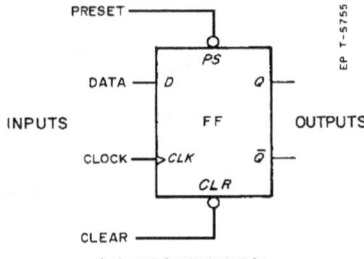

(a) CIRCUIT SYMBOL

INPUT		OUTPUT
CLK	D	Q
O	O	NO CHANGE
O	I	NO CHANGE
⌐	O	O
⌐	I	I

(b) TRUTH TABLE

PRESET (PRI)	CLEAR (CLR)	Q
1	1	NO CHANGE
1 0 ⎍	1	1
1	1 0 ⎍	0

Exhibit 2-78 — D Flip-Flop

2.9.4.4 — The J-K Flip-Flop

The J-K flip-flop is one of the most widely used flip-flops. Exhibit 2-79 shows the logic diagram, logic symbol, and the truth table for a J-K flip-flop. This flip-flop has two data inputs (J and K), and a clock (CLK) input. The operation of the J-K flip-flop is similar to the clocked R-S flip-flop, except when the inputs are all binary 1. For the clocked R-S flip-flop, the combination R = 1, S = 1 was prohibited. For the J-K flip-flop, when both data inputs J and K are binary 1 and repeated clock pulses are applied, the output will switch from 1 to 0 with each pulse. This alternating on-off action is called toggling. The J-K flip-flop could also have preset and clear inputs as discussed in the previous section on the D flip-flop. The J-K flip-flop is especially useful to form counters. Counters are discussed in Paragraph 2.9.8.

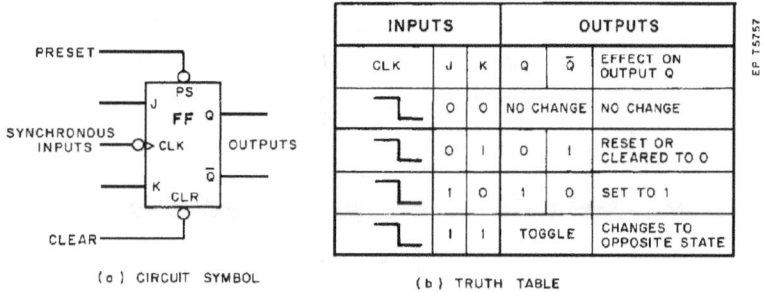

INPUTS			OUTPUTS		
CLK	J	K	Q	$\overline{Q}$	EFFECT ON OUTPUT Q
⌐_	0	0	NO CHANGE	NO CHANGE	
⌐_	0	1	0	1	RESET OR CLEARED TO 0
⌐_	1	0	1	0	SET TO 1
⌐_	1	1	TOGGLE		CHANGES TO OPPOSITE STATE

(a) CIRCUIT SYMBOL   (b) TRUTH TABLE

Exhibit 2-79 — JK Flip-Flop

2.9.4.5 — The Toggle Flip-Flop

The toggle (T) flip-flop is a single input version of a clocked JK flip-flop where the inputs J and K are tied together to form a single input T. The toggle flip-flop has a T input and a clock input. When the T input is at

logical 0, the flip-flop ignores incoming clock pulses. When the T input is at logical 1, the Q and $\overline{Q}$ outputs of the flip-flop cycle back and forth between 0 and 1 each time the rising edge of a clock pulse arrives.

2.9.5 — Clocks

A very important factor in the operation of digital circuits is that they are clocked. This means that there is some master clock sending out signals that are carefully regulated in time. These signals initiate the operations performed. Clock signals synchronize the changing and exchanging of data in sequential circuits (make things happen at the right time).

The clock signal, generated by a oscillator, is a series of pulses (this is typically a square wave) with a certain frequency and amplitude. It should be noted that most flip-flops now in use respond to either (but not both) the rising edge or falling edge of the square wave generated by the system clock and not the d-c level of the clock signal.

Some circuits require additional clock signals of a different frequency than the main clock signal; however, the clock signal must be related to the main clock signal. In this case, frequency dividers produce clock signals of different frequencies while maintaining the timing action synchronized to one master clock oscillator.

2.9.6 — Multivibrators

A multivibrator is a circuit that has no, one, or two stable states. If it has two stable states, it is termed a "bistable" multivibrator (also known as a flip-flop). If it can change from a stable state to another state for only a short period of time and then revert back to the initial stable state, it is referred to as a "monostable" multivibrator (also known as a one-shot). An "astable" multivibrator continuously changes states.

Exhibit 2-80 shows a simple bistable multivibrator and how it responds to a train of input pulses. The first pulse causes the bistable to go from a low output to a high output. The next pulse causes the bistable to go low again. This shows that each pulse causes the bistable to change to the opposite state, whatever its initial state is. As a result, there is one output pulse for every two input pulses.

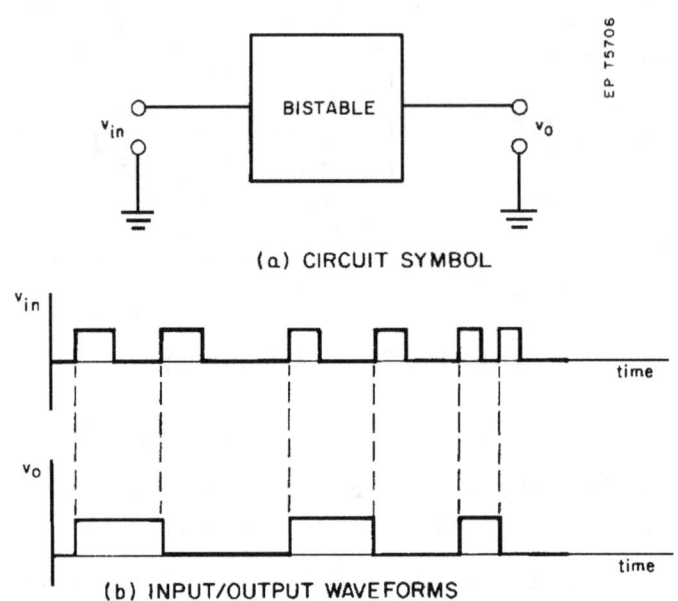

Exhibit 2-80 — Bistable Multivibrator

Exhibit 2-81 shows a simple monostable multivibrator. When the input pulse arrives, the output changes state for a predetermined period of time. It then reverts to its stable state until the next input pulse arrives.

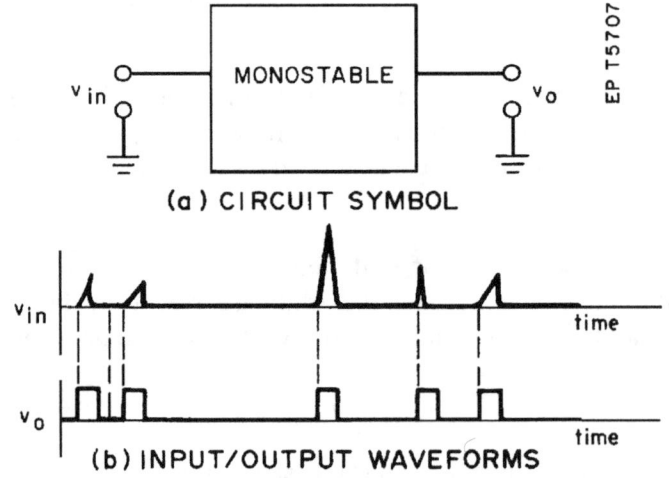

Exhibit 2-81 — Monostable Multivibrator

The third type of multivibrator, an astable multivibrator, is shown in Exhibit 2-82. This device has no input and simply changes state continuously, resulting in a train of pulses. The width of each pulse and the time between pulses can be precisely determined.

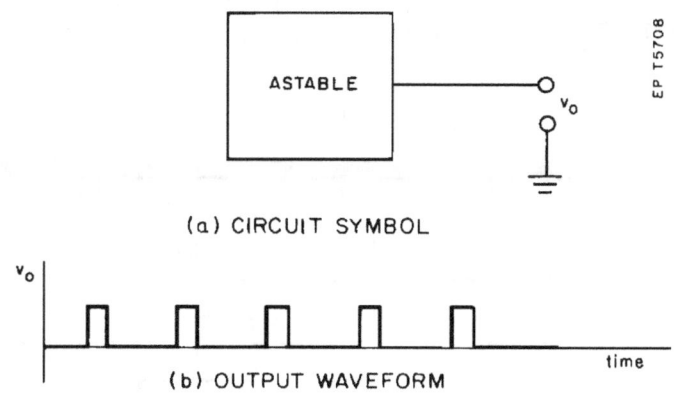

Exhibit 2-82 — Astable Multivibrator

Multivibrators are used in many places. The astable multivibrator is used as a source of continuous pulses or a clock generator for digital systems. A monostable multivibrator is used to take random width and amplitude input pulses and convert them to pulses of uniform amplitude and width. This is useful if the number of pulses that occur in a given time is to be determined. A bistable multivibrator is used as a flip-flop.

2.9.7 — Counters

Counters are sequential logic circuits used to count events and periods of time, or to put events into sequence. Counters are used as memory devices, addressing devices, and as frequency dividers.

The pulses or events to be counted are applied to the counter input, causing flip-flops in the counter to change state such that the binary numbers stored in the flip-flops are representative of the number of input pulses that have occurred. By observing the flip-flop outputs, the number of pulses that were applied to the input are determined. Counters find widespread use in digital circuits.

2.9.7.1 — The Ripple Counter

A 4-bit binary sequential counter is shown in Exhibit 2-83 along with a timing diagram. The counter is made up of J-K flip-flops. The circuit shown is triggering on the negative edge of the clock pulses. This is the meaning of the small circle on the clock pulse input.

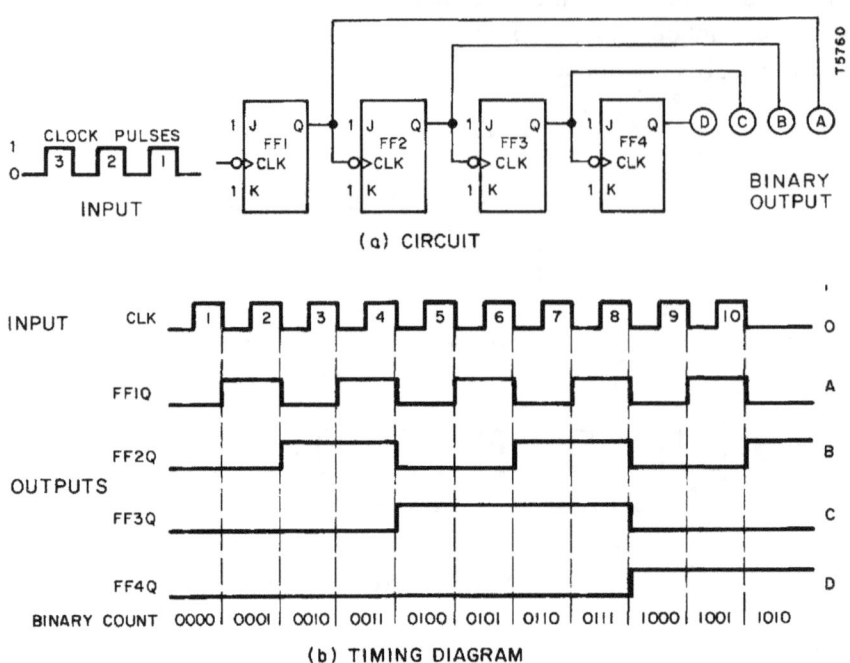

Exhibit 2-83 — Four Bit Binary Counter

The J and K inputs of all the flip-flops in Exhibit 2-83 each have a binary 1 applied to it continuously. The clock pulses toggle the first flip-flop (FF1). Note that the flip-flops trigger on the negative going edge of the clock pulse. As the clock pulse toggles FF1, the output of FF1 is sent to FF2 and causes it to toggle on the negative going edge of the output from FF1. The output of FF2 is sent to FF3 which also toggles. The counting continues with each flip-flop output triggering the next flip-flop on its negative going edge. Although not shown, most counters have a reset line to set the counter to zero. The changing of states is a chain reaction that ripples through the counter. This counter is also known as a modulo-16 counter or an asynchronous counter.

2.9.7.2 — Synchronous Counter

The counter shown above is an asynchronous counter, meaning each flip-flop did not change state with the clock pulse. All of the flip-flops in a synchronous counter change states at the same time. Therefore, a synchronous counter is capable of higher frequency operation. Exhibit 2-84 shows a 3-bit synchronous counter with the clock inputs all connected in parallel. If all the flip-flops are initially set to binary 0, the counter will count upwards in accordance with the truth table shown in Exhibit 2-84. Remember that a J-K flip-flop toggles if the J and K inputs have a binary 1 applied while the clock signal is applied.

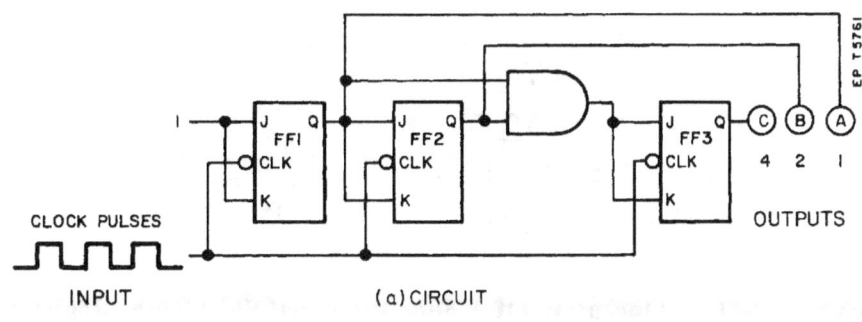

CLOCK PULSES

INPUT

(a) CIRCUIT

NUMBER OF CLOCK PULSES	BINARY COUNTING SEQUENCE			DECIMAL COUNT
	C	B	A	
0	0	0	0	0
1	0	0	1	1
2	0	1	0	2
3	0	1	1	3
4	1	0	0	4
5	1	0	1	5
6	1	1	0	6
7	1	1	1	7
8	0	0	0	0

(b) TRUTH TABLE

Exhibit 2-84 — Circuit For Three Bit Synchronous Counter

2.9.7.3 — Counters As Frequency Dividers

A common use for a counter is for frequency division. The circuit shown below divides the 60-Hz input by 60 to provide an output of 1 Hz. This circuit is actually composed of two counters, a divide by 10 and a divide by 6 counter connected in series (Exhibit 2-85). Frequency division is used in digital watches, frequency counters, oscilloscopes, and test equipment.

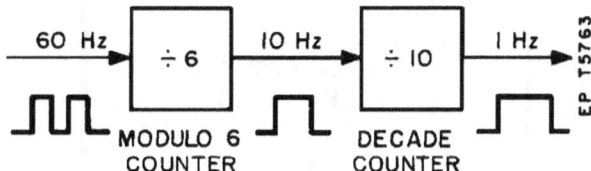

Exhibit 2-85 — Two Counter Frequency Divider

2.9.8 — Shift Registers

Another commonly used type of sequential logic circuit is the shift register. The storage elements in a shift register are cascaded such that the stored binary bits are moved or shifted from one element to another adjacent element. All of the storage registers are actuated at the same time by a single clock pulse. The basic concept of a shift register is depicted in Exhibit 2-86.

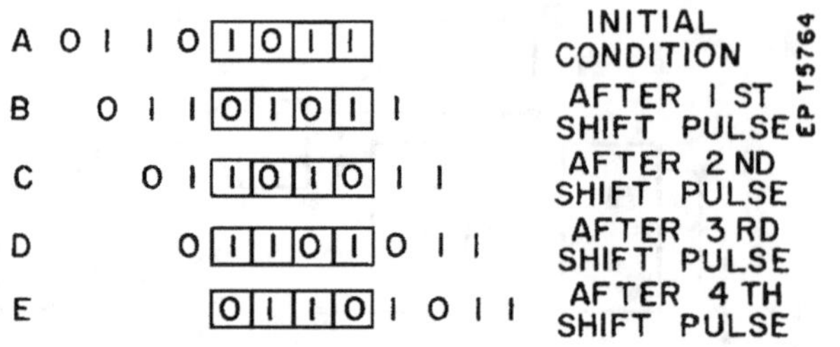

A 0 1 1 0 |1|0|1| | | INITIAL CONDITION

B   0 1 1 |0|1|0|1| | AFTER 1 ST SHIFT PULSE

C   0 1 |1|0|1|0| 1 | AFTER 2 ND SHIFT PULSE

D   0 |1|1|0|1|0 1 1   AFTER 3 RD SHIFT PULSE

E |0|1|1|0| 1 0 1 1   AFTER 4 TH SHIFT PULSE

EP T5764

Exhibit 2-86 — Operation Of A Four Bit Serial Shift Right Register

This shift register consists of four binary storage elements, such as flip-flops. The binary number 1011 is initially stored in the register. A series of input pulses representing another binary number, 0110, is generated externally and input to the register serially. As clock pulses are applied, the number stored in the register is shifted out to the right while the new number is shifted into the register. This occurs sequentially (or in series), one pulse at a time. This is called a serial shift right register.

Two basic types of shift registers commonly used are serial load and parallel load. A serial load shift register permits only 1 bit of data to be entered per clock pulse. A parallel load shift register permits all data bits to be entered at one time on the same clock pulse. Both types of shift registers are set up to shift data either to the right or to the left. In addition to the two basic types of inputs, combinations of inputs and outputs are also formed (e.g., parallel in-serial out or vice versa).

2.9.8.1 — Parallel Shift Registers

Exhibit 2-87 illustrates an 8-bit parallel in-parallel out shift left register. The shift register accepts the 8-bit binary number 01011101 and outputs the binary number 10111010. The input number was shifted to the left by one digit. Parallel shift registers produce either a left shifted or a right shifted number.

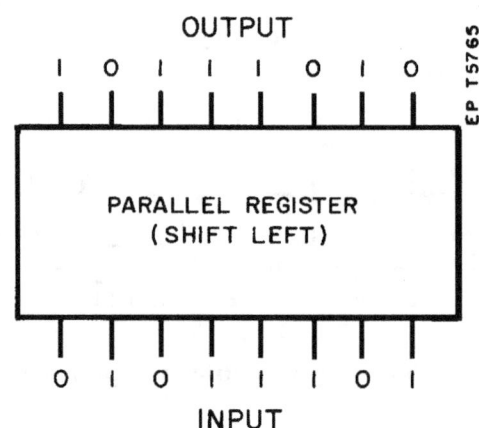

Exhibit 2-87 — Parallel In-Parallel Out Shift Left Register

2.9.8.2 — Serial/Parallel Shift Register

Exhibit 2-88 illustrates a serial in-parallel out shift register where the input arrives sequentially as a series of numbers separated in time from one another by the length of the clock pulse. After the eighth clock pulse, all eight input pulses have arrived and the output has taken the form shown in Exhibit 2-88.

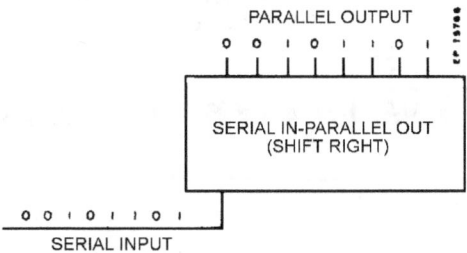

Exhibit 2-88 — Serial In-Parallel Out Shift Register

2.9.9 — Comparators

The purpose of a comparator is to compare two inputs and develop an output which identifies how they compare. The concept of using combinations of logic gates to make comparisons between binary numbers is illustrated in Exhibit 2-89. In this figure, one input signal (B) is inverted before it enters the AND gate.

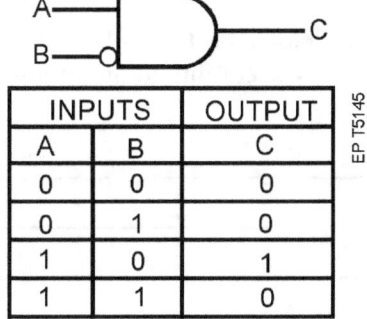

Exhibit 2-89 — Using An Inverter And An AND Gate To Develop An Output For A > B

Exhibit 2-90 illustrates a comparator. The particular device shown compares the 4-bit word A with the 4-bit word B and produces an output at terminal 1 if A < B, an output at terminal 2 if A = B, or an output at terminal 3 if A > B. The truth table shows that the output C = 1 only when A > B.

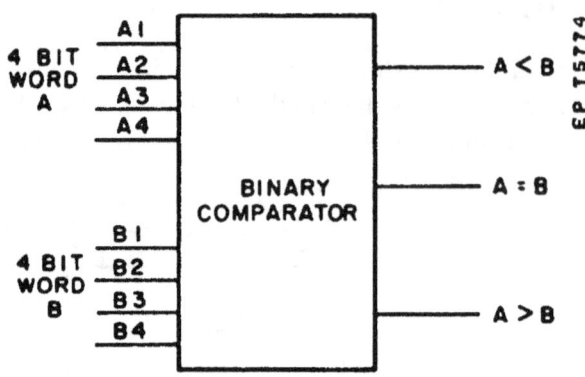

Exhibit 2-90 — Typical Four Bit Comparator

2.9.10 — Arithmetic Logic Units

The arithmetic-logic unit (ALU) is the section of a digital computer that, as the name implies, performs arithmetic and logic operations on data. Although modern ALUs perform many functions, the basic arithmetic operations (addition, subtraction, multiplication, and division) remain as their fundamental purpose. This section describes how the basic ALU operates.

2.9.10.1 — Half-Adder

A basic module used in binary arithmetic elements is the half-adder. The function of the half-adder is to add two binary digit, producing a sum and a carry according to the following binary addition rules:

Table 2-5 — Binary Addition Rules

Input Bits	Sum Bits
0 + 0	0
0 + 1	1
1 + 0	1
1 + 1	0 with a carry of 1

The logic symbol and diagram for a half-adder are shown in Exhibit 2-91.

INPUTS		OUTPUTS	
B	A	Σ	C_o
0	0	0	0
0	1	1	0
1	0	1	0
1	1	0	1
BINARY DIGITS TO BE ADDED		SUM	CARRY OUT
		XOR	AND

(a) TRUTH TABLE

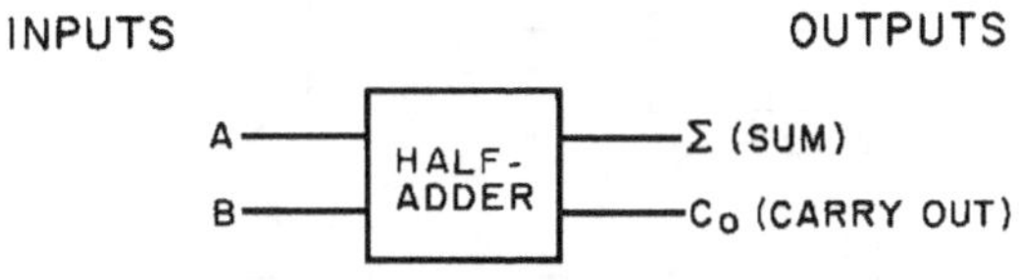

(b) CIRCUIT SYMBOLS

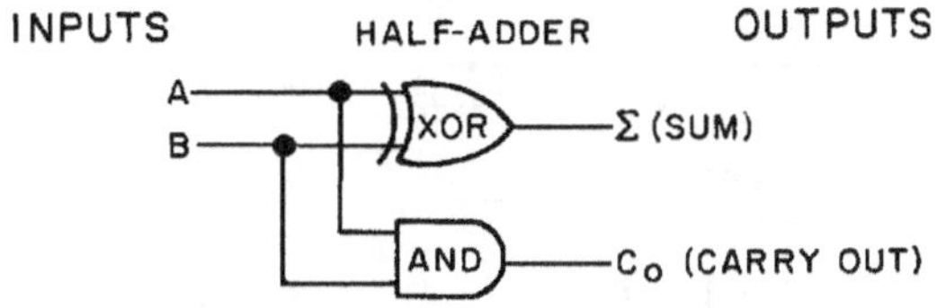

(c) LOGIC GATE DESIGN

Exhibit 2-91 — Half-Adder

2.9.10.2 — Full-Adder

When more than two binary digits are to be added, several half-adders are not adequate, since the half adder has no input to handle carries from other digits. The full-adder, shown in Exhibit 2-92, accepts the two bits to be added, plus a carry input for a carry generated by a previous stage. The two outputs are the same as for the half-adder, a sum bit and a carry bit.

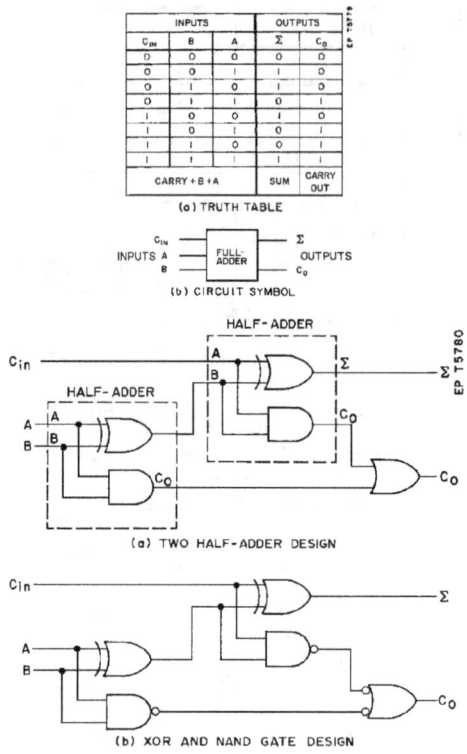

Exhibit 2-92 — Full-Adder

2.9.10.3 — Other Arithmetic Logic Units

Many types of ALUs are available as integrated circuit components. The comparator circuit discussed earlier is categorized as an ALU since it compares the two binary numbers, A and B, and determines which is larger. All ALUs are quite similar. For example, half-subtractors and full-subtractors are very similar to the half-adders and full-adders described above. Multipliers are based on the concept of repeated additions.

2.9.11 — Digital-To-Analog Conversion

Exhibit 2-93 illustrates a digital-to-analog converter (DAC), which is a device that takes a digital input signal and converts it to a single analog output signal. The four input lines, each of which will carry a voltage level representing a binary 1 or 0. The number of inputs is referred to as the resolution of the DAC. The output of this DAC ranges from 0 to 3 volts. A list of input-output relations for this DAC is shown in Table 2-6. By inspecting these relations, it can be seen that for each input value there is a corresponding analog output voltage.

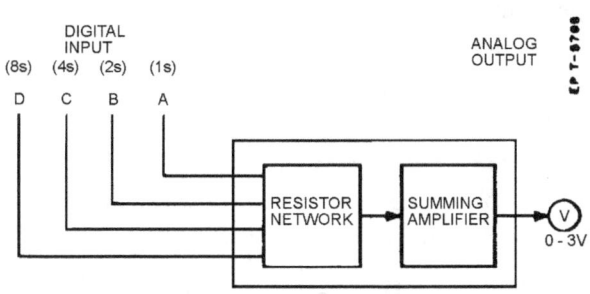

Exhibit 2-93 — Digital-To-Analog Converter

Table 2-6 — Truth Table For D/A And A/D Converters

Analog Input	Binary Output			
Volts	8s	4s	2s	1s
	D	C	B	A
0.0	0	0	0	0
0.2	0	0	0	1
0.4	0	0	1	0
0.6	0	0	1	1
0.8	0	1	0	0
1.0	0	1	0	1
1.2	0	1	1	0
1.4	0	1	1	1
1.6	1	0	0	0
1.8	1	0	0	1
2.0	1	0	1	0
2.2	1	0	1	1
2.4	1	1	0	0
2.6	1	1	0	1
2.8	1	1	1	0
3.0	1	1	1	1

Several types of D/A converters are available. The type illustrated in Exhibit 2-94 is a weighted resistor network followed by a summing amplifier. These circuits are called resistive ladder networks. The summing amplifier takes the output voltage from the resistor network and amplifies it to get the voltages shown in the right column of Table 2-6. The summing amplifier typically uses an IC type operational amplifier.

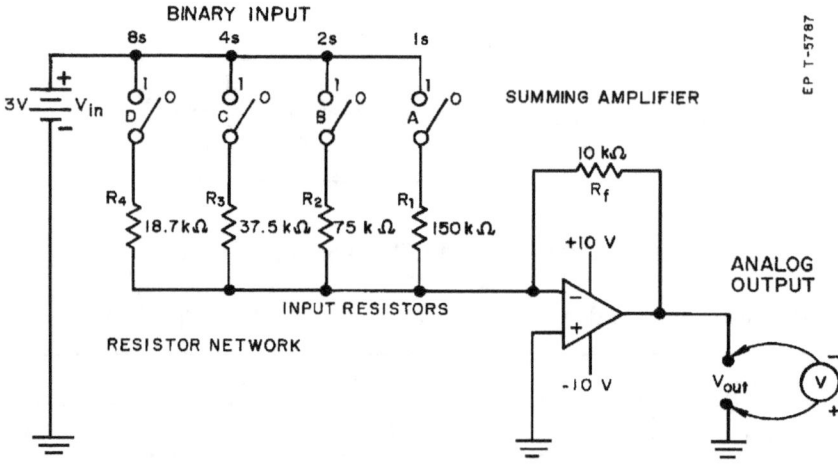

Exhibit 2-94 — Weighted Resistor D/A Converter

D/A converters are available with various voltage and current ratings. Troubleshooting requires a knowledge of the input voltage requirements on the various terminals and the output voltages expected for the various combinations of digital input.

2.9.12 — Analog-To-Digital Conversion

Exhibit 2-95 illustrates an analog-to-digital converter (A/D converter). The A/D converter is a device that takes an analog input signal and converts it to a digital output signal. The input is a single variable voltage. The voltage in this case varies from 0 to 3 volts. The output of the A/D converter is a 4-bit binary number.

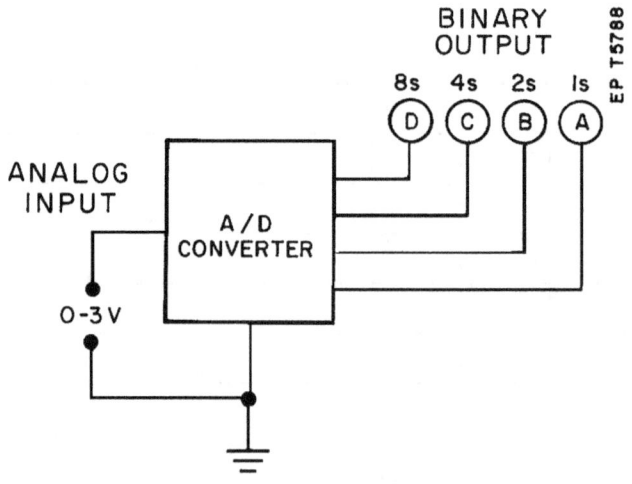

Exhibit 2-95 — A/D Converter

Zero volts applied to the input of the A/D converter outputs a binary 0000. A 0.2-volt input outputs a binary 0001. Note that each increase of 0.2 volt increases the binary count by 1. When the maximum of 3 volts is applied to the input, the output reads a binary 1111. The truth table for the analog-to-digital converter is just the reverse of the D/A converter truth table in Table 2-6; the inputs and outputs have just been reversed.

Section 2.10 — Microprocessor-Based Computer Systems

2.10.1 — Introduction

The purpose of this section is to explain the basic concepts of the design and operation of microprocessor-based computer systems. Today's computers are the result of a history of attempts made to create machines which could perform long sequences of calculations automatically. The most famous early attempt at creating such a machine was made by Charles Babbage, and English scientist and mathematician, who designed a mechanical computer which used cardboard cards with holes punched in them for introducing both the instructions and the raw data into the machine. Although he never successfully built this machine, he did succeed in establishing the fundamental principles on which today's electronic computers are based.

As time and technology progressed, computers that used relays, vacuum tubes, and eventually transistors were developed and used for a wide variety of uses. By the mid 1960s, computers were making use of integrated circuits, a new technology which allowed great numbers of transistors and other electronic components to be mounted in a single small package. Modern microprocessors are the culmination of integrated circuit developments which allow hundreds of thousands of active components to be packaged in volumes of a fraction of an inch, producing small, low cost, ultrafast, reliable computers with large amounts of memory. These features make computers ideal for use in solving business and scientific problems as well as in real-time control systems.

2.10.2 — Information Processing

This section presents the types of information processing that take place in a microprocessor. A theoretical microprocessor-based system is explained, and important concepts are introduced to demonstrate how a real system works. Prior to introducing the parts of a microprocessor system, an overview of a typical computer system is presented.

Exhibit 2-96 is a diagram of a simplified computer system consisting of three basic building blocks: a microprocessor, an external memory, and an input/output system. The **microprocessor** contains both internal memory and the components necessary to do basic math. Sometimes the microprocessor is referred to as the Central Processing Unit (CPU) of the system, because its function is similar to the function of large computer (or mainframe) CPUs. The **external memory** is the storage facility for information in the system. It consists of storage locations, each capable of storing a single bit of data in groups of n bits, where n is the number of bits in the data bus. The most commonly used data buses are those with 8, 16, or 32 bits. **An 8-bit data bus microprocessor-based system will be assumed throughout this section when specific reference to the data bus is made.** Finally, the **input/output** block contains the circuits that communicate with the world outside the computer system.

As an example of the operation of the computer in Exhibit 2-96, suppose that the input signal is the temperature of a tank of water. If that temperature reaches 150°F, an alarm is to sound. If the temperature reaches 160°F, the computer is to turn off the tank heaters. The computer would have these setpoints (150°F and 160°F) stored in external memory. The computer also would have stored in the external memory the instructions it needs or the sequence of steps it must take to make the desired comparison. When the computer is running, it samples and stores the input temperature in an internal memory location. It then fetches the lower setpoint (150°F) from the external memory and compares it with the input temperature previously stored. If the number previously stored is larger than the setpoint, the microprocessor "tells" the output equipment to initiate an alarm. Next, the microprocessor fetches the setpoint for heater turnoff (160°F) from the external memory and compares this with the input temperature previously stored. Again, if the number is larger than the setpoint, the microprocessor tells the output equipment to turn off the heaters. The computer now repeats the whole process over and over again.

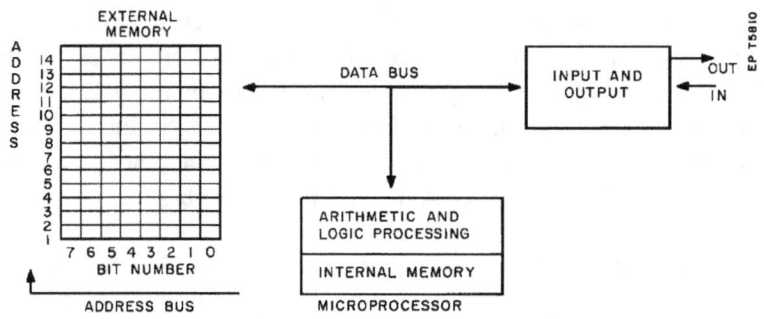

Exhibit 2-96 — Basic Computer System

In the material that follows, the contents of these three blocks are discussed in detail. It is helpful to refer back to Exhibit 2-96.

2.10.3 — Parts Of A Microprocessor-Based System

This section discusses the three basic components of any microprocessor-based system: the memory, the microprocessor, and the input-output system.

2.10.3.1 — Memory

Memory is the section of the microprocessor-based system used to store information. The memory is broken into a number of storage locations or addresses. Each address contains a group of bits, known as a word, which is handled by the computer as a unit. The size of the word, which depends on the computer system, can be 8, 16, 32 or even 64 bits for microprocessor-based computer systems. The eight bits of data are stored in eight flip-flops. Each location in the memory is identified by an address, so it can be referenced by the microprocessor. The rest of this section will discuss microprocessor-based systems with eight-bit word lengths .

2.10.3.2 — Microprocessor

An example of a microprocessor is shown in Exhibit 2-97. The microprocessor is a single integrated circuit usually contained in a 40-pin dual-in-line package about the size of a domino. This is the part of the system that controls the other parts of the system as well as performs all calculations. For example, it can move the information stored in the memory location from one place to another, it can add or subtract two numbers from memory locations, or it can compare two eight-bit quantities to determine which is larger. These simple operations make up the elements for the very complicated things that the computer is able to do.

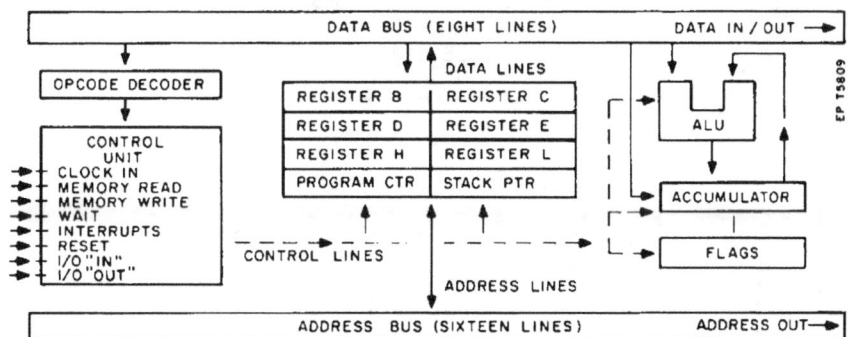

Exhibit 2-97 — Microprocessor Diagram

The microprocessor has internal memory locations called **registers** that are usually used by the system for short-term storage. The register where the microprocessor places the results of mathematical or logical operations is called the **accumulator**. For example, when numbers from other registers or from memory are added, the result is placed in the accumulator. The section of the microprocessor that does the actual add operation is the **arithmetic-logic unit** (ALU). This is the device designed into the microprocessor that provides its computing capability.

The microprocessor knows what to do by reading a specific section of the external memory called **program storage**. This area contains instructions to control the computer. For a microprocessor, these instructions are known as operation codes (opcodes). For example, an opcode may cause two numbers to be added. A specific register in the microprocessor, known as the program counter, keeps track of where in the program storage the computer will find the next instruction. A special register inside the microprocessor, called the opcode decoder, translates the opcode into the internal instructions to make the microprocessor do the required actions.

Normally, the microprocessor fetches the opcode located at the memory address pointed to by the number in the program counter, decodes the opcode, performs the indicated procedure, adds one to the program counter, and starts the process over again. The result is that the program is executed sequentially, one opcode at a time.

The microprocessor also has a register of **"flags"** to keep track of events that have happened inside the microprocessor while the program is running. Flags are individual bits that are tested for a set (1) or not set (0) condition. They are set when a specific condition is met in the processor and are subsequently tested by an opcode supplied by the programmer. For example, most microprocessors have a flag that tests to determine if the accumulator has a value of zero. If a zero exists in the accumulator, the zero flag is set and it remains so until the program causes it to be changed. This flag is useful if the programmer wants to know that a zero resulted from an operation. A special routine can then be performed.

A microprocessor-based system needs to have a clock to keep things going. The clock is actually a special circuit that turns on and off several million times per second. A typical clock signal of two megahertz is shown in Exhibit 2-98. It pulses the microprocessor every half of a millionth of a second. The microprocessor uses this signal to time its actions. The opcodes tell the processor what to do, and the clock tells it when to do it.

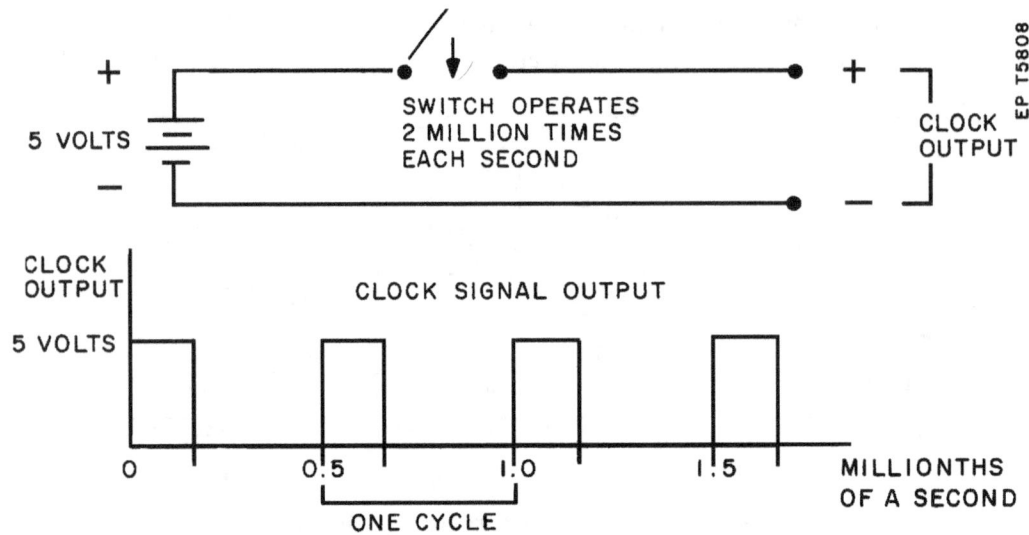

Exhibit 2-98 — Clock Pulses For A Microprocessor

2.10.3.3 — Input/Output (I/O)

The methods of input and output are entirely dependent on the specific application of the processor. If the system controls a microwave oven, the input would consist of the buttons on the front of the oven, and the output would consist of the device used to turn the microwaves on and off or control their intensity. For a personal computer, input devices would be such equipment as a keyboard, a mouse, or a scanner and output devices would be a monitor, printer, plotter, etc.

One way that a microprocessor handles input and output is called memory mapped I/O. Using this method, a memory address is used as an input location and another is used for output. The microprocessor accesses this kind of I/O in the same manner as it accesses memory. The only difference is that when the microprocessor "stores" information at that specific location, the information is sent to an output device. The result could be that an alarm sounds or a printer prints. Likewise, if a computer uses memory-mapped input when it accesses a specific input location, it really gets the last thing that the input device "had to say."

How does the microprocessor know that some device outside the system is calling for servicing? A technique known as programmed I/O is used. The microprocessor routinely checks (polls) each device for new information, similar to checking a mailbox for letters that have arrived. The program stored in the system external memory contains routines that periodically cause the microprocessor to leave what it is doing (jump) and check for new information at the input devices. These jumps are known as **subroutine calls**. If no new information is presented at the input, the microprocessor picks up where it left off (return).

The purpose of the discussion presented above is to provide an overview of the operation of a microprocessor-based system. The basic concepts needed to understand these systems have been briefly described. The discussion that follows covers a simple example of a microprocessor program. After that, some specific types of components used in these systems and troubleshooting are discussed.

2.10.4 — Programming A Microprocessor-Based System

This section briefly discusses how a microprocessor-based control system is programmed, introducing some basic programming concepts.

Exhibit 2-99 is a simple version of a microprocessor-based control system. This system uses memory-mapped I/O located at address locations 20 and 21. The program storage is located in the lower section of the memory as shown. The microprocessor is similar to the one described in the previous section. It contains an ALU, a flag register, internal registers labeled B, C, D, E, H, L, an opcode decoder, a stack pointer, and an accumulator. The accumulator is referred to as register A in this example program. If an instruction tells the microprocessor to put "5" in A, a "5" is loaded (or stored) in the accumulator.

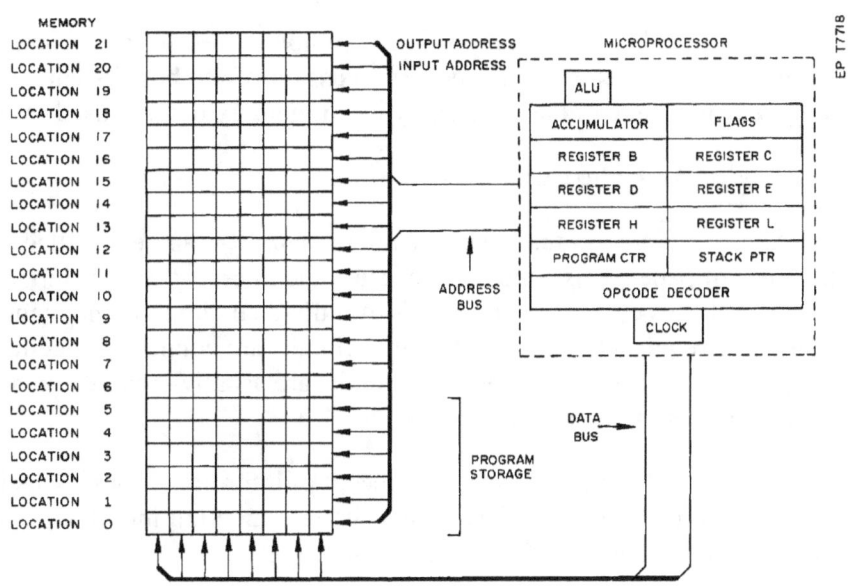

Exhibit 2-99 — Example Of A Microprocessor-Based System

Normally, the microprocessor executes program instructions sequentially in memory, but it is possible to cause the microprocessor to go back in memory and do part of the program over again. Opcodes called jump statements do this. They place the desired opcode memory address in the program counter register, making the microprocessor return to that location. When a program is designed with a jump statement as described here, the program is said to loop. These loops are useful in performing a task repeatedly. The loop would go on forever unless some other type of opcode, called a **conditional instruction**, is installed inside the loop. A conditional instruction makes the microprocessor jump out of the loop to the rest of the program when some specified condition is obtained.

A program is written using coded words called **mnemonics**. The microprocessor does not understand these words. The mnemonic program must be converted (or assembled) into the binary eight-bit numbers that are understood by the microprocessor. When the program is assembled, the location of each instruction in memory is determined. Program assembly is a complicated process and is, itself, very often performed by a computer.

Computer programs, as written by the programmer, normally contain three parts: labels, statements, and comments. Labels are short words listed down the left-hand column of a program that provide a destination for jump statements during program execution. The label, when assembled, becomes the address of the opcode.

Statements are the actual instructions performed by the microprocessor as it executes the program. The instructions include moving numbers from one register to another, moving numbers to specific memory locations, adding and subtracting numbers to the contents of the accumulator, and many other possibilities.

Comments are remarks that help explain the program. Comments are often placed in parenthesis. When the program is assembled, the contents of the braces are ignored. Comments have no effect on the program.

2.10.5 — Microprocessor Hardware

This section discusses some of the hardware components commonly used in microcomputers and microprocessor-based control systems. It includes a discussion of the bus system used in a microprocessor-based system and some of the different types of memory devices used for information storage.

2.10.5.1 — The Circuit Board And The Bus System

A major advantage of integrated circuits is that an entire microcomputer can be installed on a single circuit board smaller than this sheet of paper. This is possible because the microprocessor chip and other required circuits are integrated into packages containing thousands of individual circuit components on a single piece of silicon. The microcomputer requires that these integrated circuits are connected together in a fashion that allows them to communicate with each other. Integrated circuits are usually mounted on a printed circuit board with one or more laminated layers of conductors.

Microprocessor-based systems are designed around the concept of information flow. This information flows as electronic pulses that are carried on groups of wires called busses which fall into three categories: the data bus, the address bus, and the control bus.

The **data bus** is a set of eight wires that connects all the information processing components of the system. Information is passed as pulses representing binary numbers. As an example, if the microprocessor intends to place a number in a memory location, the number is placed on the data bus in binary form and sent to the memory. (If a 16-bit microprocessor is used, the data bus has 16 lines rather than 8.)

The address bus is a set of 16 lines on the printed circuit board that accesses specific storage locations in the memory. Each eight-bit memory location has an address corresponding to a particular combination of "on" signals on the 16 address lines. As an example, if the address bus were only one wire, only two locations ("1" or "0") could be addressed. If there were two wires, four locations could be addressed, ("00", "01", "10", and "11"). With 16 lines, a total of 65,536 separate locations can be addressed. This number is the limit on the amount of memory that a 16-bit address bus can access.

The **control bus** is a set of several lines and associated logic circuits that coordinate the operations of the various chips in the microcomputer. The microprocessor provides this information to the memory through separate lines. If one of the lines is active, the memory stores the information from the data bus. If the other is active, the memory provides the information to the data bus.

Many other control lines are used in a typical microprocessor-based system. Some examples of these are:

Reset	This line is connected to most of the major components of the system to reset them to their power-up state. This causes them to be in a known state when the initial transients from turning the system on are over. The microprocessor then begins executing instructions at the start of the program storage locations
Wait	This line is used by circuitry outside the microprocessor to stop the operation of the microprocessor while something else takes place.
Interrupt	This is a line used by components outside the system to cause the microprocessor to jump to a special type of program known as an interrupt service routine. An interrupt service routine is a special program or subroutine written by the programmer to perform some function in response to the incoming interrupt signal.

An actual system might have many other control lines. The specifics of the particular components used in a circuit would be found in the component technical manual.

2.10.5.2 — Memory

Up to this point in the discussion of microprocessor-based systems, little attention has been given to the manner in which the computer memory stores information. During the 1950s, a 625-K memory (65,536 bits of storage) would be contained in three cabinets, each approximately the size of a large vending machine. Today, this amount of information is stored on a single integrated circuit chip. The miniaturization of memory devices is as important to the development of microprocessor-based systems as is the invention of the microprocessor itself.

Computer memory falls into two broad categories: **read-write memory (RAM)**, and **read only memory (ROM)**. A RAM is used to store information temporarily during program execution. A ROM only stores the computer program or other data in use, and it cannot be changed.

The abbreviation, RAM, for read-write memory actually stands for random access memory. Both RAM and ROM are random access in the sense that by sending an address to the memory, information is retrieved out of any memory location desired without having to search through previous locations to find the desired information. In spite of this, RAM and ROM are abbreviated as such to distinguish one from the other.

The most commonly used RAM in control applications is called **static RAM**. This type of memory consists of flip-flops. One set of pins on the memory is for inputting the address desired; one set of pins allows data flow (information) in or out of the memory; and separate pins tell the memory whether an operation is a memory read or a memory write. Static memory is highly reliable. Once information is stored in a flip-flop, it stays unchanged until either the location undergoes a write operation or the memory loses power.

ROM is computer memory that is read by the computer or control system, but it cannot be used for "on line" storage. This type of memory is commonly used for program storage in microprocessor-based systems because it does not lose the stored information when the system loses power. An example of ROM is the cartridge that is plugged into a video game. This cartridge stores the entire program in a few ROMs, and when the cartridge is changed, the computer in the video game terminal has actually been reprogrammed.

There are three types of commonly used ROM: **programmable ROM** (PROM), **erasable programmable ROM** (EPROM), and **electrically erasable programmable ROM** (EEROM). PROMs consist of single bit memory locations that are constructed of tiny circuits containing semiconductor "fuses." With these fuses unblown, the locations indicate a "0" when read. If a high voltage (30 volts) is applied to the location, the fuses are blown and the locations indicate a logic "1". This type of ROM cannot be erased once it is programmed. If the program needs to be changed, a new PROM must be programmed and installed in the system. Actually, this process is performed quickly using a device called a PROM programmer. The new PROM is plugged into a socket on the programmer; a button is pushed; and the new PROM is automatically programmed in a few seconds.

EPROMs are similar to PROMs except that they are erased by shining ultraviolet light through a small window on the top of the chip. This type of memory is programmed by using the same device as is used for PROMs. EPROMs save the cost of buying the new PROM when reprogramming. No way exists to change only a section of an EPROM. Once an EPROM is exposed to ultraviolet light, the whole memory is erased.

EEROMs have the advantage of being programmable in the same circuit in which they normally serve. Once programmed, they can have specific locations reprogrammed by applying a high voltage at the desired address location. This type of memory still retains its information when power is lost. The reason that this type of memory is not called a RAM is that the procedure to write (store the information) is more complicated than the normal static RAM. The voltage required for reprogramming an EEROM is usually much higher than the normal 5-volt logic signal in the circuit, and the time required to perform a store operation is much longer than the nanoseconds required for normal static RAM.

2.10.5.3 — The Microprocessor

The microprocessor acts as the center of the bus system. It has some specific needs that the technician must understand. The first and simplest need is to have a steady and properly adjusted power supply. In most systems, the microprocessor requires only one supply at 5 volts, but this supply is required to be within 5% (or 0.25 volts) of that value. If the voltage drifts low, the system may appear to work fine, but it may make errors on a sporadic basis. As a result, the program running may fault while the system appears to be working. The first step in finding the cause of any problem with a microprocessor circuit is to check the operating voltages of all power supplies in the system.

Another requirement of the microprocessor is a clock. In some cases this clock is external to the microprocessor. The clock output is a pulse (generally a square wave) with a frequency between 2 and 10 megahertz. The specific frequency required depends on the type of microprocessor. If the system is not working, this clock frequency should be checked. The system fails if the frequency drifts too high or too low. Some microprocessors have the clock circuitry inside their own integrated circuit. These devices usually require that an external crystal be attached, similar to the crystal oscillator of a radio transmitter. In these cases when the oscillator is not functioning properly, the crystal should be checked for proper electrical connection. Otherwise, the microprocessor or the crystal needs to be replaced.

2.10.5.4 — Bus Interfacing With External Devices

The techniques used to connect the microprocessor-based system to circuit components outside the microprocessor and memory are varied and complicated. Just a few years ago, several circuit boards packed with components were required to interface a system to a television-type display, a gas discharge display, or to a printer. Today, these functions and many others are done using single chips. As a result, hundreds of integrated circuits are available for doing specific tasks that are themselves as complicated as the microprocessor. One of these devices, a **universal asynchronous receiver-transmitter** (UART), is common in industrial controls circuitry using microprocessor-based systems and is discussed in a later paragraph. (It is important to understand that the UART is only an example of an interfacing device, and the device used in a specific application probably has different characteristics.)

Two common ways that a microprocessor-based system communicates with its environment are **parallel communication** and **serial communication**. Using parallel communication, signals are sent out of the system as they appear on the data bus. In this type of communication, the information is passed as a parallel set of eight lines carrying pulses, and the receiver at the other end of the communication needs to interpret the data and make use of it in some particular fashion. This type of communication is limited to short simple runs, because it requires several wires for the data alone and is highly susceptible to radio frequency interference. The second form of data transmission is serial communications. In this case, data on the data bus is received as eight parallel lines and converted to a string of pulses on a single line. Although other lines are required for control, the data are all carried on the single line. Serial communication, for example, is suitable for a telephone line.

The UART is designed to convert parallel data inside the microprocessor system to serial data for transmission. The same type of device, installed in the receiving machine, converts received serial information to parallel data for the data bus of the receiving machine. The basic principle of how parallel-to-serial conversion is done is simple. The data are received from the data bus in parallel form and are stored in a memory register inside the UART. The UART then shifts the data out onto the serial output one bit at a time starting with the lowest bit. Refer to Exhibit 2-100 for the following example.

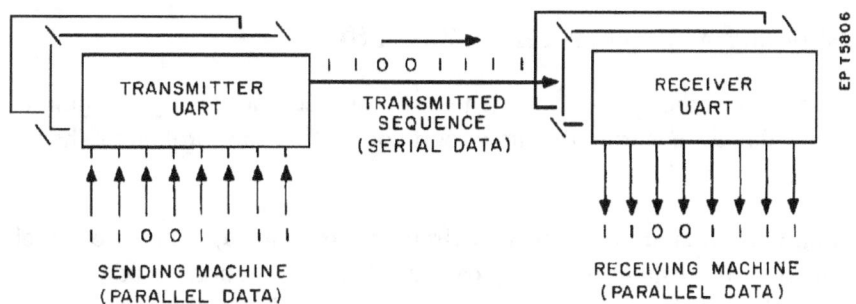

Exhibit 2-100 — Serial Data Transmission

If the binary number 11001111 was placed on the parallel port (which is connected to the data bus) of the UART, and the UART received a command to "send", the binary number would be sent out on the serial port as the sequence: 1,1,1,1,0,0,1,1. The UART at the receiving end would receive this data and load an internal register with the number 11001111. Finally, this number would be placed on the data bus of the receiving machine.

The complex part of the operation of a UART is the coordination of the device with the rest of the system. The UART receives input and output commands from the microprocessor on separate control bus lines. It has special flags that the microprocessor checks to determine if data have been completely transmitted (if it is functioning as a sender) or received (if it is functioning as a receiver). Other flags indicate that the UART has data ready to receive from the serial port or has just been sent over the serial port. Troubleshooting the UART is limited to checking the power supply voltages and using the signature analyzer. Like the microprocessor, the UART is a single chip, usually packaged in a 40-pin dual in-line package.

A second device commonly used in microprocessor-based control systems is the **parallel input/output interface** (PIO). The PIO is designed to transmit data in parallel or one byte at a time instead of one bit at a time as the UART does. This represents a faster data transmission for a system with a given speed. Unlike serial communication, parallel communication has not been well standardized industry-wide. The term PIO is adopted in this manual for convenience, but this is not an industry-wide standard. Other terms such as PIA, PPI, and PDC are also commonly used to name a parallel interface integrated circuit.

2.10.5.5 — The Power Supply

The power supply provides the regulated d-c power for the microprocessor and its associated logic circuits. Most microprocessors have been developed to operate with few different voltages. The best chips need only one, 5 volts. This voltage must be extremely stable or the whole system can fault; the clock will not operate properly; and the logic internal to the microprocessor will have a difficult time distinguishing logic high states from logic low states.

Power supplies in computer systems use voltage regulators to control the system supply voltages. These are integrated circuits the size of a quarter or smaller containing many circuit components. During troubleshooting efforts, one of the first things to check is the voltage of all power supplies in the system. Power supply voltage drift is the most common cause of faulty computer operation. Sometimes the whole system appears to work fine, but it will make mistakes due to a low or high operating voltage. This kind of fault is particularly bad, because it may go undetected if operating voltages are not routinely checked.

2.10.6 — Troubleshooting A Microprocessor-Based System

The electronic repair aspects of the system outside the microprocessor usually are simple. They consist of checking power supply voltages, ensuring that no circuit connections are broken or shorted, and determining if integrated circuit chips have failed.

The most common fault encountered with a printed circuit board is a bad connection, such as a broken solder joint. These bad connections are usually found by careful visual inspection. In addition, when searching for faults on a circuit board of this type, careful inspection should also be made for cracks in the board material that could have resulted in breaking a circuit line.

As discussed earlier, normal signal tracing and other conventional troubleshooting techniques are not productive when attempting to find faults in a microprocessor-based system. The microprocessor communicates with the other components in the circuit using pulses placed on the system's busses. The timing of these pulses is dependent upon the specific program running on the machine. It became clear early in the design of these systems that a new type of troubleshooting and testing needed to be developed.

One troubleshooting technique used in these systems is called signature analysis. It works on the principle of running a specific predetermined test program on the system and examining the way that voltage fluctuates on various lines in the system. If the waveforms in these lines were checked using an oscilloscope, the pattern generated would look like random square pulses, and it would not be possible to determine if things were operating properly. What is needed is a device that observes the pulses over a specific time period and provides information on whether or not the pulses seen were the expected sequence. Such a device is known as a signature analyzer. The analyzer connects to the system clock to provide timing, and a probe is touched to a bus line in the system. If the microprocessor is executing a test program, a four-digit combination of letters and numbers is displayed on the face of the analyzer. The technician needs only to check if these numbers are the same as those given for that line in the technical manual to determine if a problem exists. If the numbers and letters are different than expected, the manual gives further specific test points to test to isolate the fault. This technique is easy to do and has proven to be reliable. Before signature analysis is performed on a system, it is advisable to check two other parameters in the system. Both of these were discussed earlier in this chapter. One is the system power supply voltages, and the other is the system clock operation. If either of these is out of specification, the system will not work properly.

2.10.7 — Digital Data Transmission

Most data in digital systems are transmitted directly through wires between components or through the conductors in printed circuit boards. In some cases, fiber optic light guides are used. Many bits of data must be transmitted from one place to another. If all the data were sent at one time over parallel wires or conductors, a large number of wires would be needed. Instead, the data are generally sent over a single pair of wires in serial form and reassembled into parallel data at the receiving end.

The two common ways of digital data transmission are known as parallel communication and serial communication. Using parallel communication, an eight-bit digital number is sent over eight parallel wires to the receiver. Each wire has the voltage pulse that corresponds to a binary 1 or a binary 0, and all the data are sent at the same time. Parallel communication is usually limited to short runs, because it requires several wires for the data alone.

Most data transmission in digital systems and in microprocessors is done using serial communication. In this case, the data are sent as a string of pulses on a single line. Paragraph 2.10.5.4 discussed the use of a universal asynchronous receiver-transmitter (UART) to convert parallel data into serial data for transmission purposes. Paragraph 2.9.8 discussed several types of shift registers (e.g., parallel in-serial out and serial in-parallel out) which are used for this purpose. The data are converted into serial data and sent over a single pair of wires connecting the sender and the receiver. The serial data are transmitted in accordance with a strict protocol that specifies the voltage levels and allows the receiver to check the data and ensure that there were no errors in transmission.

There are several different protocols used to control serial data transmission. The following is a list of the more common protocols: The serial data transmission channel between two microprocessor-based systems is often called a **serial data link** (SDL), and it has two basic formats:
1. synchronous data transmission
2. asynchronous data transmission.

Asynchronous data transmission is characterized by using special bit patterns to separate groups of data bits. In synchronous data transmission, the groups of data bits can be transmitted back-to-back, but special characters must be included at the beginning and end of the transmission.

2.10.7.1 — Asynchronous Serial Data Transmission

Asynchronous serial data transmission is generally used for data exchange between two microprocessor-based instruments or between a microprocessor-based instrument and its associated digital meters. Exhibit 2-101 shows a diagram of a typical asynchronous serial data transmission system. Data bits are sent to the transmitter (UART) in parallel where they are converted to serial data and formatted for asynchronous serial transmission. Such transmission is characterized by transmitting data in groups, also called characters, of a predetermined number of bits. Each character includes reserved bits used to control the data flow and to check the validity of the data being transmitted.

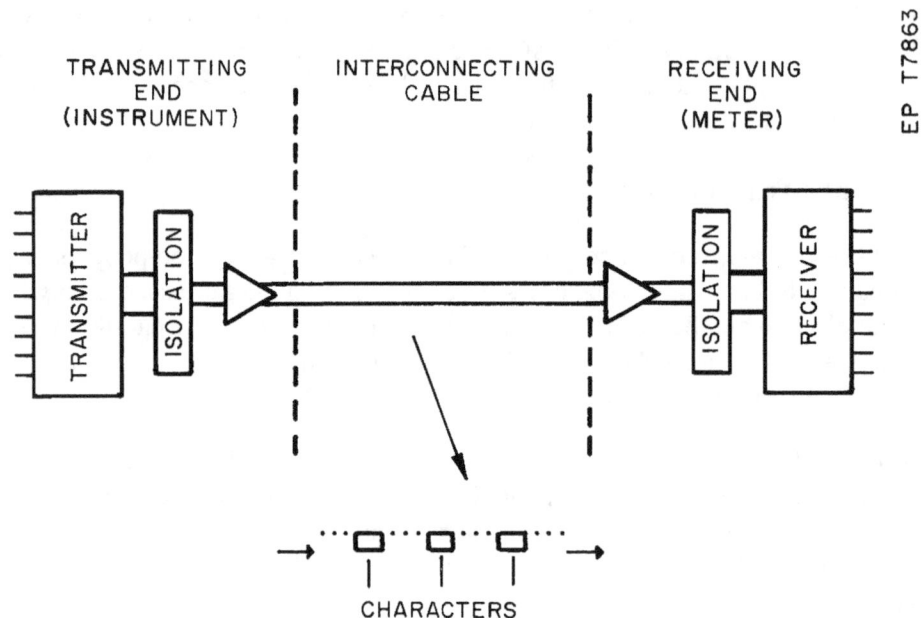

Exhibit 2-101 — Typical Asynchronous Serial Data Transmission Scheme

2.10.7.2 — Synchronous Serial Data Transmission

Synchronous serial data transmission is generally used for high-speed data transmission between two microprocessor-based instruments. The synchronous serial data transmission format is similar to the asynchronous format, but instead of transmitting only one set of data bits at a time, any number of such sets of data can be sent. The synchronous transmission format is characterized by "frames" of information rather than characters. A frame consists of start, stop, control, and several sets of data. The transmission is called synchronous, since the data characters are transmitted sequentially with no idle space or time between them requiring both ends (transmitting and receiving) of the system to be synchronized. In general, the same clock signal controls both ends of the transmission channel. However, this is not necessary if a device on the receiving end can sample and lock into the frequency of the incoming stream of data. Synchronous data links are also called **high-speed data links** (HSDL). A typical HSDL scheme is shown in Exhibit 2-102.

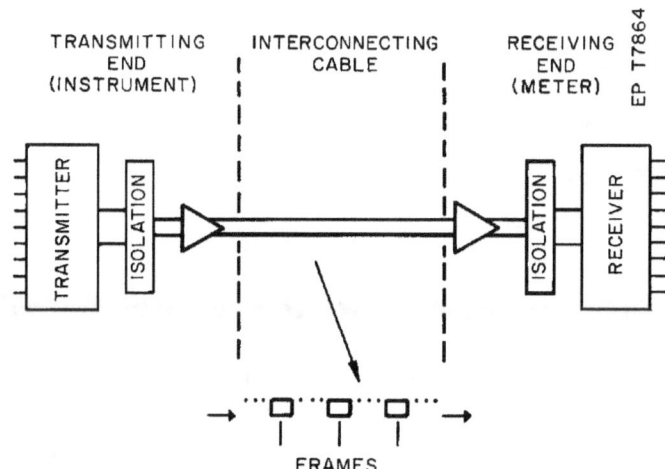

Exhibit 2-102 — Typical High-Speed Data Link Scheme

2.10.7.3 — Troubleshooting

If a problem is suspected with the transmission link, the signature analyzer and oscilloscope are used to check the transmitting and receiving UARTs. If they are not at fault, the next check is the cable and its associated connectors. If an outright short or open circuit is suspected, an ohmmeter is used to check for continuity. The cable connectors should also be checked. For less obvious faults, a capacitance bridge, a megohmmeter, and the time domain reflectometer are used to evaluate the performance of the cable.

(Intentionally Blank)

Chapter 3 — REACTOR THEORY REVIEW

Section 3.1 — Basics

3.1.1 — The Atom

Atoms are the individual structures that constitute the basic units of the chemical elements. As shown in Exhibit 3-1, atoms consist of two basic parts:
- the electron cloud
- the nucleus

.

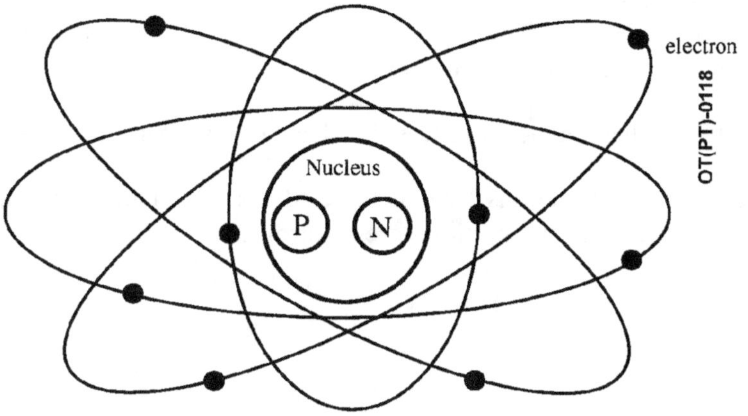

Exhibit 3-1 — Schematic Of The Atom

The electron cloud consists of negatively charged electrons in orbits surrounding the nucleus. The electron cloud contains very little of the atom's mass, but occupies most of its volume. Dimensions of the electron cloud, on the order of 10^{-8} cm, set the atomic dimensions and vary little from one element to the next.

The nucleus contains the majority of the atom's mass and lies at the center of the atom. Its dimensions are on the order of 10^{-12} cm. The size of the nucleus varies not only from element to element but from isotope to isotope. The general expression used to determine the radius of a nucleus is:

Equation 3-1

$$R = R_0 A^{\blacklozenge}$$

where:

R = the nuclear radius

R_0 = a constant in the range, 1.1×10^{-13} to 1.5×10^{-13}

A = the atomic mass number.

Neutrons and protons are generically known as **nucleons**. Neutrons and protons have approximately the same mass, which is roughly 2000 times the mass of a single electron. Neutrons and protons can transform into one another through a process called beta decay (Section 3.2).

All nucleons interact thorough the very short range, attractive, strong nuclear force. Due to its short range, this force saturates so that each nucleon can only interact with a limited number of other nucleons.

All protons are positively charged and interact, as well, through the long range, repulsive Coulomb force. This force does not saturate and a proton interacts with all other protons in the nucleus.

3.1.2 — Particles

In the physics world, the term particle generally refers to any, very small part of matter, such as a molecule, atom, or electron. These particles were discussed in Paragraph 3.1.1. A few other particles are worth mentioning since they will appear later in this chapter.

The position, e^+, is a positively charged electron. It is the electron's antiparticle or its antimatter.

The beta, β, particle refers to those "electrons" that are ejected from the nucleus during beta decay (Section 3.2). β^- particles are common electrons. β^+ particles are positrons.

Two other particle types appear in beta decay:
- the neutrino, ν
- the antineutrino, $\overline{\nu}$

The particles discussed so far: electrons, neutrinos, and nucleons are **fundamental** particles and can be used to construct other **"particles."**. Macro-particles are constructed from fundamental particles, primarily protons and neutrons.

One important macro-particle in the reactor business is the alpha particle, α. An alpha particle consists of two neutrons and two protons. Basically, it is a helium nucleus.

Two other macro-particles that occasionally appear are the deuteron and triton. The deueron consists of one proton and one neutron and is basically the nucleus of ^2H, deuterium. The triton consists of one proton and two neutrons and is basically the nucleus of ^3H, tritium.

3.1.3 — Energy Range

For core physics, the energy range over which neutrons live extends from 0 to 10 MeV and is divided into three segments as shown in Exhibit 3-2. The energy cut points should be viewed as approximate and adjusted when needed.

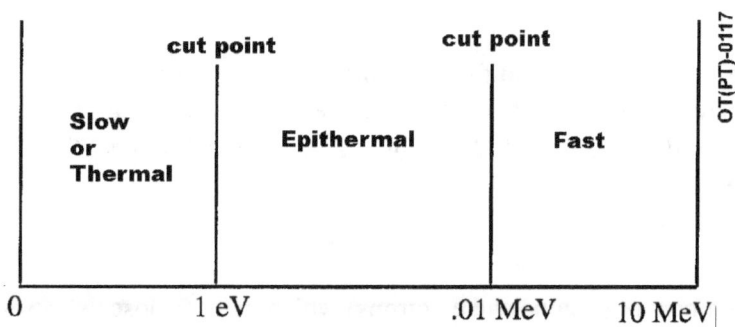

Exhibit 3-2 — Energy Classification For Core Neutrons

Each of the three energy regions embraces a physical phenomenon that is significant to the reactor business. The neutrons born in the fission process of ^{235}U all possess energies that lie in the fast energy range. Less than one percent of these fission neutrons have energies between 10 MeV and 21 MeV and are only important in special shielding situations. Neutrons that slow down (see Section 3.6) from the fast range to the thermal range pass through the epithermal range where resonances (see Section 3.5) in the cross sections exist. Neutrons in the slow or thermal energy range are in thermal equilibrium with their surroundings where they may lose or gain energy.

Reactors can be classified as thermal, epithermal, or fast, depending upon the energy of the **majority** of the neutrons that induce fission.

3.1.4 — Nuclear And Atomic Terminology

3.1.4.1 — Nuclide

The nuclear species characterized by the proton number Z and the mass number A is the nuclide $^{A}_{Z}X$.

3.1.4.2 — Isotopes

Atoms with the same number of protons, Z, (and when electrically neutral, the same number of electrons), are isotopes. Since an atom's chemical behavior depends on the **number** and **arrangement** of electrons, isotopes behave the same chemically.

3.1.4.3 — Isobars

Atoms with the same mass number A are isobars.

3.1.4.4 — Isotones

Atoms with the same number of neutrons (A - Z) are isotopes.

3.1.4.5 — Isomers

Atoms with the same atomic number Z and the same mass number A, but different amounts of internal energy are isomers. If there are two isomers for a given A and Z, one is the ground state configuration and the other exists in the excited state. Isomers differ in other nuclear properties (besides internal energy) such as rate of radioactive decay.

3.1.4.6 — Ion

A neutral atom (equal numbers of protons and electrons) which gains or loses electrons is an ion. The common method used to designate an ionized atom is to add an indication of the ion's net charge as a superscript on the right of the chemical symbol. For example, $_Z^A X^{-1}$ has one more electron and $_Z^A X^{+1}$ has one less electron than the electrically neutral atom $_Z^A X$.

3.1.4.7 — Element

A substance consisting of atoms all having the same atomic number, Z, is an element.

3.1.4.8 — Compound

A substance composed of two or more elements chemically combined in a fixed proportion is a compound. The formula for a compound is a list of the atoms present in the smallest group of atoms which possesses the chemical characteristics of that compound.

3.1.4.9 — Fission Fragment

A fission fragment is one of two lighter nuclei produced directly from fission. Fission fragments are highly ionized and lose their kinetic energy very quickly, over a small distance as the result of electrostatic interactions with surrounding atoms.

3.1.4.10 — Fission Product

A fission product is a fission fragment that has lost its kinetic energy and has attracted the orbital electrons necessary to neutralize its charge.

3.1.4.11 — Prompt Neutron

A neutron that appears 10^{-14} seconds after fission from the decay of a fission fragment is considered a prompt neutron. The most probably prompt neutron energy is 2.0 MeV.

3.1.4.12 — Delayed Neutron

A delayed neutron comes from a delayed neutron emitter. The delayed neutron emitter is the result of a fission fragment, known as a delayed neutron precursor nuclide, decaying to the delayed neutron emitter and leaving the emitter in an excited state that has an energy greater than the binding energy of the last neutron. It is the beta decay of the precursor nuclide that controls the appearance of the delayed neutron. The half-lives of delayed neutron precursors range from hundredths of a second to 1 minute. The most probable delayed neutron energy is 0.5 MeV.

3.1.4.13 — Effective Full Power Hour

The Effective Full Power Hour (EFPH) is the unit often used to determine how long the core will last. It is natural to seek an answer in units of time. However, time alone is meaningless because during any time period, the core may be operated at any power from zero to full-rated power. The EFPH combines two important quantities, power output and time, that are important in measuring the endurance of the core.

An EFPH is the amount of energy produced by the reactor while operating at full power for one hour. If operating at half-power (50%), two hours of operation would be required to expend 1 EFPH.

3.1.5 — Formulae, Constants, And Conversion Factors

3.1.5.1 — Absolute Temperature

The absolute temperature, T($^\circ$K), given in degrees Kelvin, is related to the Fahrenheit temperature, T($^\circ$F), by

Equation 3-2

$$T(K) = \frac{5}{9}\left[T(F) - 32\right] + 273.16$$

Using this equation, the temperature 68°F corresponds to 293.16 $^\circ$K, or in terms of the most probable energy, 0.0253 eV.

3.1.5.2 — Atomic Concentration

In a sample which contains **m** grams of a single element and occupies a volume V, the number of atoms N of the element per unit volume is given by

Equation 3-3

$$N = \frac{N_A\left[\dfrac{m}{GAW}\right]}{V} = \frac{\rho N_A}{GAW}$$

where:

$\mathbf{N_A}$ = Avogadro's Number, 6.02×10^{23} atoms/GAW

GAW   = Gram Atomic Weight

ρ = the elemental density, **m**/V.

If the isotopic abundance of a given isotope of an element is known, then the corresponding atomic concentration of that isotope can be found by multiplying the atomic concentration given by Equation 3-3 by

$$\frac{a/o}{100\%}$$

to obtain

Equation 3-4

$$N_i = \frac{\rho N_A}{GAW} \times \frac{(a/o)_I}{100\%}$$

where:

$\quad\quad\quad$ **N_i** $\quad\quad$ = concentration of isotope i

$\quad\quad$ **$(a/o)_I$** $\quad\quad$ = abundance of isotope i.

3.1.5.3 — Atomic And Isotopic Mass

One gram is a very large mass compared to the mass of a single atom. A smaller unit of mass, called the atomic mass unit (amu) and defined as 1/12 the mass of one neutral ^{12}C atom, is often used for convenience. Because the nucleus of a ^{12}C atom contains 12 nucleons, one amu is very close to the mass of one nucleon.

Equation 3-5

$$1\,amu = \frac{mass\ of\ ^{12}C\ atom}{12} = \frac{1.9925 \times 10^{-23}\,gram}{12} = 1.66604 \times 10^{-24}$$

$$proton\ rest\ mass = 1.007276\,amu$$

$$neutron\ rest\ mass = 1.008665\,amu$$

For historical reasons, the mass of one atom of a nuclide is referred to as its atomic weight (AW).

The isotopic mass of an atom is its mass relative to the mass of a standard ^{12}C atom.

Equation 3-6

$$Isotopic\ mass\left(^A_Z X\right) = \frac{mass\ of\ ^A_Z X}{mass\ of\ ^{12}_6 C}(12\,amu)$$

Except for ^{12}C, the isotopic mass of an atom is approximately, but not exactly, equal to its mass number A in amu. Except for very precise calculations, isotopic masses can be approximated by the atomic mass numbers in amu.

3.1.5.4 — Electrical Potential Energy

Every electrically charged body possesses an energy field that interacts with other charged bodies. The energy of interaction between the fields of a charged body and any other charge may be expressed in terms of electrostatic (Coulomb) potential energy. This potential energy PE for a charge +**e** in the field of a nucleus of charge **Ze** is given by

Equation 3-7

$$PE = \frac{KZe^2}{r}$$

where:

e	= the charge of a single proton
r	= the distance from the center of the nucleus
K	= a constant.

Outside the nuclear surface the potential energy decreases as 1/**r**. Inside the nucleus, the behavior of the electrostatic potential is more complex.

3.1.5.5 — Electron Volt

When an electron is accelerated through a potential difference of one volt, it gains a kinetic energy (as potential energy is converted to kinetic) equal to 1.6×10^{-19} joule. This unit of energy is very useful and convenient when dealing with subatomic particles and is given the name **electron volt** (eV). Other appropriate units for subatomic particles are the keV and MeV.

$$1\,eV = 1.6 \times 10^{-19} \text{ joule}$$

$$1\,keV = 1.6 \times 10^{-16} \text{ joule}$$

$$1\,MeV = 1.6 \times 10^{-13} \text{ joule}$$

3.1.5.6 — Photons Properties

The amount of energy carried by a single photon is

Equation 3-8

$$E = h\nu$$

where:

h	= Planck's constant, 6.63×10^{-34} joule-sec

$$\nu \qquad = \text{frequency of the photon.}$$

The momentum carried by a photon is

Equation 3-9

$$P = h/\lambda$$

where λ = the photon's wavelength.

The frequency and wavelength of the photon are related by the equation

Equation 3-10

$$c = \nu\lambda$$

where **c** = the speed of light, 3×10^{10} cm/sec.

The photon's momentum given by Equation 3-8 can be expressed as

Equation 3-11

$$p = \frac{h}{\nu} = \frac{h\nu}{c}$$

3.1.5.7 — Relativistic Principles

Albert Einstein's Special Theory of Relativity states that mass **m** and Energy **E** (coulombs) are equivalent and are related through the speed of light, **c** (mass).

Equation 3-12

$$E = mc^2$$

This equation shows that an enormous amount of energy is associated with a very small amount of mass. One of the most frequently used conversion factors is the energy equivalent of one atomic mass unit (1 amu) in MeV.

Equation 3-13

$$E = (1\,\text{amu})(c^2) = 931.5\,\text{MeV}$$

Therefore,

Equation 3-14

$$c^2 = 931.5\,\text{MeV/amu}$$

An object's mass is observed to increase with velocity according to the equation:

Equation 3-15

$$m = \frac{m_0}{\sqrt{1 - v^2/c^2}}$$

where:

m = mass of the object when moving with velocity v

m₀ = mass of the object at rest.

It is customary to call **m** and **m₀** the **relativistic mass** and the **rest mass** respectively.

The **rest mass energy E₀** is given by

Equation 3-16

$$E_0 = m_0 c^2$$

and the **total relativistic energy E** is

Equation 3-17

$$E = mc^2$$

Therefore, the kinetic energy **T** becomes the difference between the total relativistic energy and the rest mass energy.

Equation 3-18

$$T = E - E_0 = (m - m_0)c^2$$

Section 3.2 — Stability Of The Nucleus

3.2.1 — Neutron-to-Proton Ratio

The strong nuclear force and the electromagnetic (Coulomb) force are the two fundamental forces responsible for stability in the nucleus. The strong nuclear force is a very short range attractive force exerted by all nucleons. Due to its short range, the strong force saturates so that one nucleon can interact only with a limited number of nearest-neighbor nucleons. The Coulomb force is a long range repulsive force exerted by each proton on the other protons in the nucleus. Since its range is essentially infinite, the Coulomb force does not saturate.

For the nucleus to be stable and its nucleons remain together, the nuclear force must be greater than the repulsive Coulomb force. For this reason and the nature of the two forces, neutrons are needed as the "extra glue" to hold the protons together in the nucleus. As the atomic number Z increases, the neutron-to-proton ratio increases to exert the additional attractive force needed to overcome the repulsive Coulomb force between protons, Exhibit 3-3.

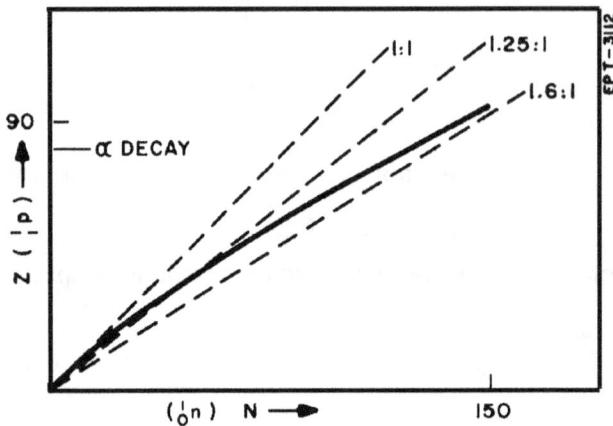

Exhibit 3-3 — Neutron-to-Proton Ratio

3.2.2 — Line Of Stability

When all combinations of N and Z are considered, many nuclides are possible with the stable ones forming a small subset of these. Shown as a solid black line on the Z versus N plot of Exhibit 3-3, the stable nuclides lie in a region known as the **Line of Stability** or **Peninsula of Stability**. The Line of Stability has a finite thickness and lies in that region where the neutron-to-proton ration is between 1:1 and 1:6. The Line of

Stability ends with $_{83}^{209}\text{Bi}$. Those nuclides not on the Line of Stability are unstable and will eventually decay to some nuclide that lies on the Line of Stability.

3.2.3 — Mass Defect

A survey of the atomic masses shows that the atomic mass of any nucleus is less than the sum of its constituent particles in the free state. Each specific nucleon that is added to a nucleus gives up a different amount of mass in order to belong to that nucleus. The total amount of mass given up by the constituents, ΔM, can be calculated from

Equation 3-19

$$\Delta M = (A - Z)M_n + ZM\left(_1^1H\right) - M\left(_Z^A X\right)$$

where:

$\quad\quad\quad\quad \mathbf{\Delta M} \quad\quad = \text{mass defect}$

$\quad\quad\quad\quad \mathbf{M_n} \quad\quad = \text{mass of the neutron}$

$M(_1^1H)$ = mass of ordinary hydrogen, ^1H

$M(_Z^AX)$ = mass of nuclide with Z protons and A nucleons (A–Z) neutrons.

When ΔM is given up by the constituents, mass is converted into energy that is liberated from the nucleus (often in the form of gamma rays). A mass of 1 amu is equivalent to 931.5 MeV of energy.

3.2.4 — Binding Energy

The energy equivalent of the mass defect is known as the **binding energy**. This is the amount of energy released when the constituents come together to form the nucleus. In order to break up a nucleus into its component neutrons and protons, an amount of energy equivalent to the binding energy must be supplied to the nucleus.

3.2.5 — Binding Energy Per Nucleon (BE/A)

Since each nucleon gives up a different mass amount to join the nucleus, contributions to the nuclear binding energy also vary. The <u>average</u> binding energy per nucleon is obtained by dividing the total binding energy by the mass number A. When BE/A is plotted versus A, the curve in Exhibit 3-4 results with a peak near A = 60.

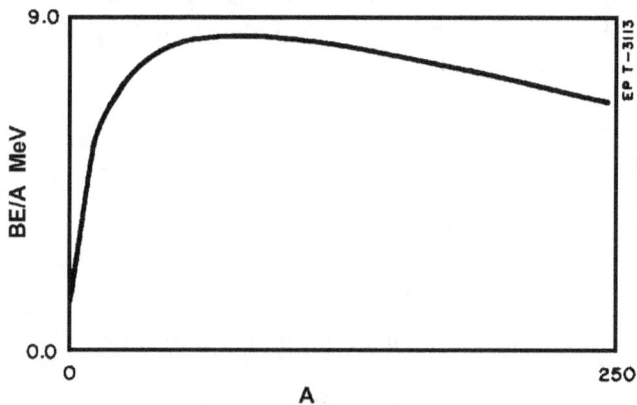

Exhibit 3-4 — Binding Energy Per Nucleon

BE/A is a measure of stability for the nucleus. Larger values for BE/A imply a more stable nucleus since more mass was given up on average by its constituents. The largest values of BE/A occur for the iron isotopes. Nuclides lying to the left or right of iron in Exhibit 3-4 can become more stable and gain a larger BE/A by giving up additional mass through the fusion and fission processes, respectively.

3.2.6 — Radioactive Decay

Radioactive decay or radioactivity is the phenomenon of the unstable atom or nucleus becoming more stable through the spontaneous emission of radiation (either as particles or electromagnetic as in gamma or x-rays).

The fundamental nature of radioactivity is that it is a statistical process that can be described as a probability that the atoms or nuclei of a substance will undergo a spontaneous transmutation. Further, it is a statistical one-shot process; when a particular nucleus has decayed, it cannot repeat the identical event again. Radioactive decay is governed by the following:

- The probability of decay per unit time is independent of the age and history of the nucleus. As long as a nucleus has not decayed, the probability for it doing so during the next second remains a constant.
- The probability of decay per unit time is the same for all nuclei of the same type.
- The probability of decay per unit time is independent of external influences such as temperature, pressure, volume, or chemical form.

There is no way of predicting the time when any one nucleus will decay. However, during an infinitesimally small time interval, dt, the probability that an unstable nucleus will decay is directly proportional to this time interval.

Equation 3-20

$$\lambda dt = \text{probability that a nucleus decays in time dt}$$

where λ = the decay or disintegration constant.

All decays, regardless of the differences in the emitted particle or the rates at which the disintegrations take place, follow a single law: The Law of Radioactive Decay. Given a sample of unstable nuclei, the average number that survive after some time is given by:

Equation 3-21

$$N(t) = N_0 e^{-\lambda t}$$

where:

$N(t)$ = number surviving after a period of time **t**

N_0 = initial number of unstable nuclei

λ = total decay constant in units of reciprocal time.

The time during which one-half of any collection of radioactive nuclei decays ($T_{\frac{1}{2}}$) is obtained by setting $N = N_0/2$ and $t = T_{\frac{1}{2}}$ in Equation 3-21 to yield:

Equation 3-22

$$T_{\bullet} = \frac{\ln 2}{\lambda} = \frac{0.693}{\lambda}$$

Exhibit 3-5 a plot of N/N_0 versus $t/T_{\frac{1}{2}}$, illustrates an exponentially decreasing concentration of radioactive nuclei as a function of time. For example, after 2 $T_{\frac{1}{2}}$'s, $\frac{1}{4} = \frac{1}{2} \cdot \frac{1}{2}$ of the initial concentration remains.

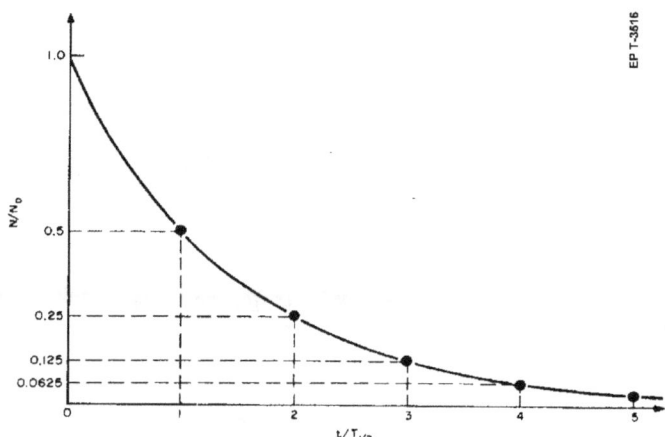

Exhibit 3-5 — N/N_0 Versus $t/T\frac{1}{2}$: Exponential Decrease In Radioactive Nuclei Concentration With Time

3.2.7 — Activity

A useful quantity describing radioactive decay is the **activity** when it is defined as the number of disintegrations (or decays) per second.

Equation 3-23

$$A(t) = \lambda N(t)$$

The activity **A(t)** has the same exponential time dependence as the number of radioactive nuclei **N(t)** present at time **t**.

Equation 3-24

$$A(t) = A_0 e^{-\lambda t}$$

where:

$\qquad$ **A(t)** $\qquad$ = activity after a period of time **t**

$\qquad$ **A$_0$** $\qquad$ = initial activity

It is the activity **A(t)** which is usually measured rather than **N(t)** since it is easier to monitor the radiation emitted during the decay process rather than the nuclei that are undergoing the transformation.

A convenient unit of activity is the **curie** which is defined to be:

Equation 3-25

$$1\,\text{curie} = 1\,\text{Ci} = 3.7 \times 10^{10} \ \text{decays/second}.$$

The commonly used subunits of the curie are the millicurie (mCi), the microcurie (µCi), and the micromicrocurie (µµCi):

Equation 3-26

$$1\,\text{mCi} = 10^{-3}\ \text{curie} = 3.7 \times 10^{7}\ \text{decays/second}.$$

Equation 3-27

$$1\,\mu\text{Ci} = 10^{-6}\ \text{curie} = 3.7 \times 10^{4}\ \text{decays/second}.$$

Equation 3-28

$$1\,\mu\mu\text{Ci} = 10^{-12}\ \text{curie} = 3.7 \times 10^{-2}\ \text{decays/second}.$$

The **specific activity** is defined as the activity per gram and units of Ci/gm, mCi/gm, etc. that is appropriate.

3.2.8 — Alpha Decay

Alpha decay is a decay process that primarily occurs among heavy mass nuclides beyond lead when it is necessary for the nucleus to relieve Coulomb repulsion. In this process, the parent nucleus emits an α particle ($^{4}_{2}\text{He}$ nucleus) with a kinetic energy that is typically in the range of 4 to 9 MeV. The daughter nucleus may be left in an excited state from which it decays to the ground state by emitting one or more gamma rays. Alpha decay reduces the mass and charge of the parent nucleus as shown by the following equations:

Equation 3-29

$$^{A}_{Z}\text{X} = ^{A-4}_{Z-2}\text{Y} + ^{4}_{2}\alpha$$

or

Equation 3-30

$$^{A}_{Z}\text{X} = ^{A-4}_{Z-2}\text{Y} + ^{4}_{2}\alpha + \gamma$$

3.2.9 — Beta Decay

The disintegration of nuclei with the emission of a beta particle is a process whereby the charge of a nucleus is changed without changing the total number of nucleons. This decay process moves the nucleus closer to the line of stability (or to where the line would be if it were extended into the heavy mass region beyond bismuth) with a mass that is slightly smaller than before the process took place. There are three modes of beta decay; all of which may leave the daughter nucleus in an excited state from which it decays through gamma emission.

3.2.10 — Beta-Minus (β^-) Decay

Beta-minus decay is the process by which a nucleus with too many neutrons converts an excess neutron into a proton while simultaneously ejecting an electron (β_-) and an antineutrino $\bar{\nu}$. Beta-minus decay changes the identity and charge of the parent nucleus as shown by the following equation:

Equation 3-31

$$\ _{Z}^{A}X \rightarrow \ _{Z+1}^{A}Y + \ _{-1}^{0}\beta^- + \bar{\nu}$$

Beta-minus decay is the primary decay among fission products trying to reach the line of stability.

3.2.11 — Beta-Plus (β^+) Decay

Beta-plus decay is the process by which a nucleus with too many protons converts an excess proton into a neutron while simultaneously ejecting a positron (β^+) and a neutrino (ν). Beta-plus decay changes the identity and charge of the parent nucleus as shown by the following equation:

Equation 3-32

$$\ _{Z}^{A}X \rightarrow \ _{Z-1}^{A}Y + \ _{1}^{0}\beta^+ + \nu$$

Beta-plus decay can be found in reactors when delayed neutron emission or (n, 2n) reactions produce a nucleus that is deficient in its number of neutrons.

3.2.12 — Electron Capture (EC) Decay

Electron capture decay is also a process by which a nucleus with too many protons converts a proton into a neutron, but through capturing an orbital electron (usually an inner electron from the K or L shells) instead of ejecting a positron. Like beta-plus decay, a neutrino is also emitted from the nucleus. Electron capture decay changes the identity and charge of the parent nucleus as shown by the following equation:

Equation 3-33

$$\ _{Z}^{A}X + \ _{-1}^{0}e \rightarrow \ _{Z-1}^{A}Y + \nu$$

Fluorescent x-rays are emitted from the daughter nucleus as the outer shell electrons move to fill the vacancies created by the capture of the inner orbital electrons.

Electron capture decay competes with beta-plus decay and occurs in some instances when it is physically impossible for the nucleus to decay by beta-plus decay.

3.2.13 — Gamma Decay

Gamma decay is the process by which an excited nucleus shed its excess energy through the emission of one or more gamma rays as it reaches its ground state configuration. This decay process does not alter the number of nucleons A or number of protons Z in the nucleus as shown by the following equation:

Equation 3-34

$$\ _{Z}^{A}X^{*} \rightarrow \ _{Z}^{A}X + \gamma$$

Gamma decay is possible after any induced reaction or decay process provided the nucleus is left in an excited state.

Section 3.3 — Reactions

The definition of a "nuclear reaction" differs slightly depending upon the source. Some sources consider radioactive decay to be a nuclear reaction where a number of unstable nuclei undergo a transmutation to another species. In this type of reaction, the isolated population of the decaying nuclei decreases with time.

Other sources consider a nuclear reaction to be a process in which a nucleus reacts with another nucleus, an elementary particle (neutron, proton, etc.), or a photon to produce one or more other nuclei and possibly other particles in a very short period of time. Reactions in which the target nucleus is struck with an incoming projectile are called induced nuclear reactions. Some consider induced reactions to be the opposite of a radioactive decay since induced reactions can lead to an increase in a population of unstable nuclei whereas radioactive decay leads to a decrease in the population of unstable nuclei (with the nucleus moving towards increased stability).

Strictly speaking, it is probably best to consider a "nuclear reaction" to be any nuclear process where the final configuration differs from the initial configuration. Therefore, both decay and induced reactions meet the requirements. It may be more proper to refer to induced reactions, as described above, as "collisions" instead of "reactions" since the projectile is colliding with the nucleus in some manner and is therefore more indicative of the type of process.

3.3.1 — The General Reaction Equation

The entire collision process may be represented by the general collision equation:

Equation 3-35

$$A + B \rightarrow (C)^{*} \rightarrow C + D$$

where:

$\quad\quad\quad$ **A** $\quad\quad$ = incident particle

$\quad\quad\quad$ **B** $\quad\quad$ = the target nucleus

(C)*   = the compound nucleus

C   = the product nucleus

D   = the outgoing particle

In addition to the asterisk symbolizing its excited state, the compound nucleus is enclosed in parentheses to distinguish it from an ordinary excited state.

An example of an induced nuclear reaction written in this format is:

Equation 3-36

$$\ _0^1n + \ _1^1H \rightarrow (\ _1^2H)^* \rightarrow \ _1^2H + \gamma$$

The collision process may be written in a more compact format as:

Equation 3-37

$$A\,(B, C)\,D$$

In this format, the compound nucleus does not appear. In this shorter format the previous example appears as:

Equation 3-38

$$\ _1^1H\,(n, \gamma)\ _1^2H$$

A collision reaction involving these incident and emitted particles is designated an (n, γ) reaction.

3.3.2 — The Compound Nucleus

The compound nucleus theory is part of Bohr's liquid drop model of the nucleus and is valid for nuclear reactions when the incident particle has an energy of 10 MeV or less. Implementation of this theory was necessary to account for the time delay between the incident particle entering the nucleus and the outgoing particle exiting the nucleus.

There are two independent stages to the existence of a compound nucleus. The first stage is **formation** where the incident particle enters the nucleus with its excitation energy. This excitation energy is shared in a random fashion among all of the nucleons until the nucleus forgets how it obtained this energy.

The second stage is called **breakup**. It occurs after the excitation energy reconcentrates on one or a group of nucleons from which it either ejects those nucleons or sheds the excess energy in the form of a gamma ray. For the nucleus, the end result is that it returns to its ground state. The way the compound nucleus breaks up or disintegrates depends only on its properties, such as its energy and angular momentum, but not on the specific way it was formed.

The formation and breakup can occur in more than one way, subject to the conservation laws:

Equation 3-39

$$\left.\begin{array}{c} {}^{0}_{1}p + {}^{13}_{6}C \\[2mm] {}^{2}_{1}H + {}^{12}_{6}C \\[2mm] {}^{4}_{2}\alpha + {}^{10}_{5}B \end{array}\right\} \rightarrow \left({}^{14}_{7}N\right)^{*} \rightarrow \left\{\begin{array}{l} {}^{14}_{7}N + \gamma \\[2mm] {}^{13}_{7}N + {}^{1}_{0}n \\[2mm] {}^{13}_{6}C + {}^{1}_{1}p \\[2mm] {}^{12}_{6}C + {}^{2}_{1}H \end{array}\right.$$

3.3.3 — Classification Of Induced Nuclear Reactions

Induced nuclear reactions may be broadly classified as either **scattering** or **absorption**. Each of these categories consist of various types.

A nuclear reaction in which the incident and exiting particle are the same type and the overall result is the transfer of energy from the particle to the nucleus is called **scattering**. There are two types of neutron scattering that are important for nuclear reactors: **elastic** and **inelastic**. The difference between the two types depend upon whether kinetic energy is conserved in the system. The scattering of neutrons by nuclei plays a highly important role in the operation of nuclear reactors.

An induced nuclear reaction in which the incident and exiting particles are different types and in which the target and product nuclei are different is called **absorption**. The types of neutron-induced absorption reactions are more numerous than for scattering. The ones important to the reactor business include radiative capture, fission, and particle emission.

3.3.4 — Elastic Scattering

Elastic scattering is the simplest consequence of a neutron colliding with a nucleus. It occurs with all nuclei at all energies. In this type of process, both kinetic energy and linear momentum are conserved. Because linear momentum must be conserved, some of the incident particles kinetic energy is transferred to the target nucleus. However, the sum of the kinetic energies after the collision equals sum of the kinetic energies before the collision. If the target is initially at rest, the kinetic energy before is just the kinetic energy of the incident particle. After the collision, the nucleus is left in the same internal nuclear state, usually the ground state, as it had before the reaction. A collision between two billiard balls, in which neither is damaged or otherwise changed, is the standard example of an elastic collision.

An example of elastic scattering in nuclear reactors that is important to the moderating process is neutrons elastically scattering from hydrogen in water:

Equation 3-40

$$ {}^{1}_{0}n + {}^{1}_{1}H \rightarrow {}^{1}_{1}H + {}^{1}_{0}n$$

In the equation above, it is noteworthy to observe that no compound nucleus, $\left({}^{2}_{1}H\right)^{*}$, was shown. Typically, compound nucleus formation for elastic scattering takes place when a resonance condition exists between the neutron's incident energy and the energy of the excited state for the compound nucleus. This resonance scattering occurs frequently in the epithermal energy region. In the thermal and fast energy regions, the compound nucleus is not formed for elastic scattering since the neutron does not enter the nucleus. Rather, the neutron interacts with the nuclear potential at the nuclear surface. This process is generally referred to a **potential scattering**.

3.3.5 — Inelastic Scattering

Inelastic scattering is a more complicated scattering reaction when compared to elastic scattering. In this type of process, linear momentum is conserved, but kinetic energy is not. In this scattering process, the nucleus captures a neutron and forms a compound nucleus. Some of the incident neutron's kinetic energy is transferred to the target nucleus to satisfy conservation of linear momentum. However, the neutron departs the compound nucleus after giving the nucleus much more energy than necessary to satisfy conservation of linear momentum. The extra energy left behind by the neutron leaves the nucleus in an excited state. Shortly after the neutron leaves, the nucleus shed this excess energy by emitting a gamma ray and returning to its ground state. Gamma rays emitted in this process are referred to as inelastic (scattering) gamma rays. Therefore, kinetic energy is converted to radiative energy in the form of a gamma ray. The sum of the neutron's kinetic energy after the collision, the target's recoil kinetic energy, and the inelastic gamma ray must equal the neutron's kinetic energy before the collision.

For this process to take place, the incident neutron must have a kinetic energy that is greater than the first excited state of the target nucleus since the nucleus must be in at least its first excited state if it is to retain any extra kinetic energy from the neutron. In light mass nuclei the first levels are approximately 1 MeV above ground state while in heavy mass nuclei the first excited state is on the order of 0.1 MeV above the ground state. Therefore, inelastic scattering is a threshold process and will not occur for neutrons of all energies.

When the incident neutron's kinetic energy is large enough that it is possible to leave the nucleus in an excited state above the first excited state, gamma emission often proceeds to the ground state by a series or cascade of gamma emissions where the gammas are given off one after another.

Cross section magnitudes for inelastic scattering are typically on the order of a few barns. These values are generally less than those for elastic scattering except in the fast energy region where the magnitudes are comparable.

Inelastic scattering is only a minor player in the slowing down process in water moderated thermal reactors, Paragraph 3.5.1. This is primarily due to hydrogen possessing no inelastic scattering cross section and the fact that the threshold for most inelastic scattering is of the order of 1 MeV and >0.1 MeV for light and heavy nuclei respectively.

3.3.6 — Radiative Capture

In this type of induced reaction, the target nucleus captures an incident particle, namely a neutron, emits a gamma ray, and takes on the identity of the compound nucleus's nuclide in its ground state. These radiative capture reactions are often simply termed captures. Radiative capture is one of the simplest consequences of a neutron colliding with a nucleus and like elastic scattering, it occurs with all nuclides at all neutron energies.

The gamma rays are called **capture gammas** since their existence is due to the capture of the incident particle. Their emission coincides with the break up of the compound nucleus. The energy of the capture gamma or the total energy of capture gammas emitted in the cascade corresponds to the binding energy of the incident particle to the product nucleus. These binding energies can be quite large. Their instantaneous appearance and larger energies cause capture gammas to be a concern as a radiation source and must be taken into consideration when calculating core power and operational doses.

Equation 3-41

$$
{}_{0}^{1}\text{n} + {}_{1}^{1}\text{H} \rightarrow \left({}_{1}^{2}\text{H}\right)^{*} \rightarrow {}_{1}^{2}\text{H} + \gamma\,(\approx 2.25\,\text{MeV})
$$

The addition of another neutron to the nucleus in neutron induced reactions alters the n/p ratio and the product nucleus may or may not be stable. In Equation 3-41, the product nucleus was stable. Equation 3-42 is one where the product nucleus is unstable.

Equation 3-42

$$
{}_{0}^{1}\text{n} + {}_{27}^{59}\text{Co} \rightarrow \left({}_{27}^{60}\text{Co}\right)^{*} \rightarrow {}_{27}^{60}\text{Co} + \gamma\,(\approx 7.5\,\text{MeV})
$$

For neutron induced reactions, product nuclei that are unstable move toward stability by beta decay. Often in the beta decay process, the product nucleus is transformed to possess one less neutron and one more proton (β^{-} decay) and is left in an excited, bound state. From the excited state, the nucleus quickly gives off a gamma ray in order to arrive at its ground state.

Equation 3-43

$$
{}_{27}^{60}\text{Co} \rightarrow {}_{28}^{60}\text{Ni}^{*} + \beta^{-} + \bar{\nu} \rightarrow {}_{27}^{60}\text{Ni} + 2\gamma
$$

The gamma rays that are emitted under the conditions discussed above are called **activation gammas**. The time at which they appear after the capture even is controlled by the beta decay. These times are typical of those encountered in beta decay, but since they are much longer than it takes for the capture process to conclude, their appearance is delayed. Since activation gammas are controlled by the beta decay process, their appearance can take place after the reactor is shutdown. For this reason, they are a radiation source during shutdown and must be taken into consideration when calculating shutdown dose rates.

Activation gamma energies are typical of the energy level differences in the nucleus. Total energy of these gammas is much less than the energy associated with capture gammas.

3.3.7 — Fission

Induced fission is that nuclear reaction which causes the nucleus to split into two lighter fragments of comparable mass after absorbing an incident particle. This process is discussed in depth in Section 3.4.

3.3.8 — Particle Emission

A particle emission reaction results when a nucleus absorbs one type of particle to form a compound nucleus and the compound nucleus decays by emitting a different type of particle. If the product nucleus retains any excitation energy, it may decay by gamma emission. Radiative capture and fission are not considered particle emission reactions.

Particle emission reactions sometimes involve charged particles, either as the incoming or exiting particle. These charged particles are influenced by Coulomb force arising from the protons in the nucleus in addition to the strong nuclear force. The Coulomb force creates a potential barrier that hinders the motion of the charge particle. Charged particle emission reactions are often threshold reactions since additional energy is needed to deal with the Coulomb barrier.

Equation 3-44

$$\ _{0}^{1}n + \ _{5}^{10}B \rightarrow \left(_{5}^{11}B \right)^{*} \rightarrow \ _{3}^{7}Li + \ _{2}^{4}\alpha$$

Equation 3-45

$$\ _{0}^{1}n + \ _{8}^{16}O \rightarrow \left(_{8}^{17}O \right)^{*} \rightarrow \ _{7}^{16}N + \ _{1}^{1}p$$

Equation 3-46

$$\ _{2}^{4}\alpha + \ _{8}^{18}O \rightarrow \left(_{10}^{22}Ne \right)^{*} \rightarrow \ _{10}^{21}Ne + \ _{0}^{1}n$$

Equation 3-47

$$\ _{0}^{1}n + \ _{92}^{235}U \rightarrow \left(_{92}^{236}U \right)^{*} \rightarrow \ _{92}^{234}U + 2\ _{0}^{1}n$$

Equation 3-44 and Equation 3-45 are neutron induced reactions that produce charged particles. Equation 3-44 is one of several exceptions to charged particle reactions possessing thresholds. In this case, the Coulomb barrier is low compared to the α's energy so that it escapes the nucleus quite easily. Equation 3-45 represents a charged particle emission reaction that has a threshold; in this case, it is on the order of 10.2 MeV. This reaction is responsible for the formation of radioactive ^{16}N, a source of high energy gammas.

The reaction in Equation 3-46 is endothermic (endoergic), requiring the incident α to possess a kinetic energy of at least 0.7 MeV. The Coulomb barrier for the α and ^{18}O has a height of ~3.6 MeV. The 0.7 MeV mass deficiency and the 3.6 MeV barrier are no problem for alpha particles from the decay of heavy mass nuclides since these alphas have energies in the range of 4 to 9 MeV.

Equation 3-46 and Equation 3-47 are particle emission reactions where the exiting particle is one or more neutrons. These serve as examples of non-fission neutron sources. All are threshold reactions.

In the (n, 2n) reaction the kinetic energy of the incident neutron must be equal or greater than the sum of the binding energies for the last two neutrons in the compound nucleus. If the kinetic energy of the incident neutron just exceeds the threshold energy for the (n, 2n), the two neutrons that are emitted from the compound nucleus possess energies in the thermal energy range. The product nucleus will generally be neutron deficient for stability and hence be a positron emitter.

Section 3.4 — Fission

3.4.1 — Introduction

Nuclear fission is the process of an atomic nucleus splitting into two lighter fragments of comparable mass accompanied by a release of energy. This process proceeds with relative ease among heavy mass nuclides as a means of relieving Coulomb repulsion. Viewed as a decay mode for a nucleus not undergoing neutron or photon bombardment, spontaneous fission is second to alpha decay in relieving Coulomb repulsion for heavy mass nuclides whose Z is less than 106. For those nuclides with Z greater than 106, spontaneous fission dominates over alpha decay.

In terms of nuclear reactions, fission may be sped up by inducing it with neutrons or gamma rays. These particles can supply energy to the heavy mass nucleus so it overcomes the Coulomb barrier that delays the nucleus from breaking apart instantaneously. However, with the additional energy throughout the nucleus, there is no guarantee that fission takes place. When the nucleus absorbs a neutron or gamma, a compound nucleus if formed from which it may break apart into fission fragments or it may rid itself of excess energy by emitting a gamma ray. For the ^{236}U compound nucleus (formed by ^{235}U absorbing a neutron), fission results ~85% of the time while ~15% of the time a gamma is emitted and ^{236}U is left in its ground state.

Equation 3-48

$$_{0}^{1}\text{n} + {}_{92}^{235}\text{U} \rightarrow \left({}_{92}^{236}\text{U} \right)^{*} \rightarrow 2\text{fp} + \text{n(s)} + \gamma\text{(s)}$$

Equation 3-49

$$_{0}^{1}\text{n} + {}_{92}^{235}\text{U} \rightarrow \left({}_{92}^{236}\text{U} \right)^{*} \rightarrow {}_{92}^{236}\text{U} + \gamma$$

From a physics standpoint, fission is noteworthy because of the following features in each fission event:
- an enormous amount of energy released
- the number of neutrons released
- radioactive by-products.

More energy is released in a single fission event (~200 MeV) than in typical chemical combustion reactions (~ 4eV). With ~2.43 neutrons released on average (^{235}U fission) in a single fission event, self-sustaining chain reactions are possible. The unstable fission fragments are themselves radioactive by-products of the fission process.

3.4.2 — Liquid Drop Characteristics

The liquid drop model is the model used most frequently to explain the energetics of the fission process. The nucleus is visualized as an oscillating drop of incompressible liquid. If the oscillations are violent enough such that the surface tension cannot hold the nucleus together, the coulomb repulsion causes the nucleus to break apart. With spontaneous fission, millions upon millions of oscillations occur before the nucleus splits. Since extra energy is available in induced reactions, the nucleus undergoes far fewer oscillations before breaking apart.

Before splitting, the nucleus has a neutron-to-proton ratio that is appropriate for heavy mass nuclei (~1.57). Initially, the fission fragments have this same ratio since the neutrons and protons are homogeneously in the liquid drop representation. However, this ratio is inappropriate for nuclei containing a Z that is roughly one-half of that belonging to the heavy nucleus. Some of these fission fragments decay to a somewhat more stable nucleus by β^- decay as they move back to the line of stability. A competitor to β^- in the fission products is delayed neutron emission which allows an unstable fission product to simultaneously become more stable and change fission product decay chains by emitting a loosely bound neutron.

3.4.3 — Fission Fragment Yield Curve

Each fission event, for heavy mass nuclei in the region around ^{235}U, results in the formation of two, unequal fission fragments with widely varying mass numbers. Studies of ^{235}U fission have found 53 possible ways of splitting the nucleus resulting in 106 fission products results from the breakup of any single nucleus. However, an experimentally determined quantity known as the **fission product yield**, **fission yield**, or just simply **yield** is useful in predicting what fission products appear. Fission product yield is defined as:

Equation 3-50

$$\text{Fission Yield} = Y_i\% = \frac{N}{F}(100)$$

where:

Y_i = the yield

N = the number of fission products of a particular type i

F = the number of fission events.

When a large sample of fission events are studied, a double humped yield curve results that is nearly symmetric about mass 118 ($118 = 0.5 \cdot 236$). The fission yield peaks about a pair of fragments of approximate mass 95 and 140, corresponding to the light and heavy fragments formed from fission. Since the observed values of fission yield range from something as small as 10^{-5} percent to something as large as 7 percent, the ordinate scale in Exhibit 3-6 is logarithmic.

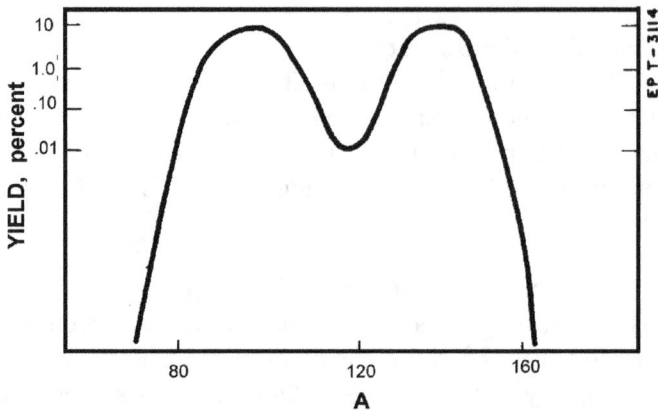

Exhibit 3-6 — Fission Fragment Yield Curve

The double humped character of the fission yield distribution is not consistent with the simple liquid drop model of fission which predicts that symmetric fission is the most probably way the nucleus splits. This would result in a fission yield distribution with a single hum centered around A/2 where A is the atomic mass number of the fissioning nucleus. Symmetric fission occurs for lighter mass nuclei like bismuth. To understand the double humped nature of the fission yield curve for a nucleus like ^{235}U, the shell model and magic numbers must be used with the liquid drop model, see Reference 3, Section 1.4.2.1.

3.4.4 — Critical Energy

The Q value for spontaneous fission at the moment of fission before neutrons are emitted is given by:

Equation 3-51

$$Q = \left[M(A,Z)) - M(A_1,Z_1) - M(A_2,Z_2) \right] c^2$$

where:

Equation 3-52

$$M(A,Z) = \text{the mass of the fissioning nucleus}$$
$$M(A_1,Z_1) \text{ and } M(A_2,Z_2) = \text{fission product masses}$$
$$c = \text{velocity of light}$$

M(A, Z) may represent the ground state of a nucleus undergoing spontaneous fission or the positive ground state of the compound nucleus formed by a target nucleus acquiring a neutron. When Q is positive, fission is possible. Since there are 53 ways of splitting the ^{235}U nucleus, there are 53 different Q values, these average out to ~200 MeV for nucleus in the vicinity of ^{235}U.

The potential energy of the liquid drop in different stages of fission has been determined as a function of the degree of liquid drop deformation. The form of the potential energy is shown in Exhibit 3-7 where potential energy E is plotted against the parameter r which is a measure of the degree of separation. When r = 0, the two fission products overlap to form the fissioning nucleus where the amount of the potential energy is nothing more than the energy equivalent of the fissioning nucleus's rest mass in the ground state, M(A, Z).

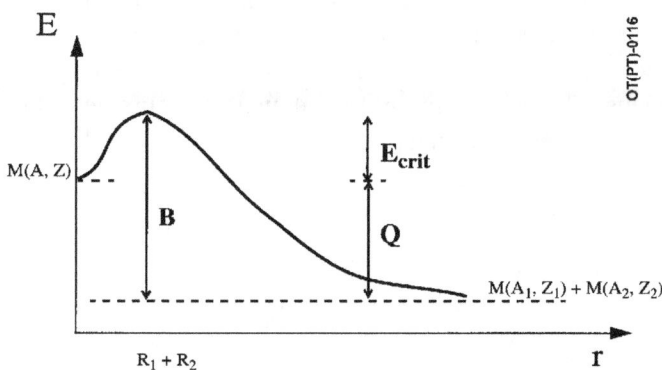

Exhibit 3-7 — Variation Of Fission Fragment Potential Energy With Fragment Separation

The meaning of the graph can be seen more easily if the fission process is viewed in reverse. The smallest value of the potential energy occurs when the two fission products are separated by an infinite distance and is just the energy equivalent for the sum of the fission product rest masses, $M(A_1, Z_1) + M(A_2, Z_2)$. If R_1 and R_2 are the radii of the two fission products. $r = R_1 + R_2$ the potential energy is just the sum of the rest mass energy at infinity and the electrostatic (Coulomb) energy resulting from the mutual repulsion of the two positively charged fission products. The value of the electrostatic energy when $r > R_1 + R_2$ is:

Equation 3-53

$$E_{coul} = \frac{kZ_1Z_2e^2}{r}$$

where:

E_{coul}	= the electrostatic Coulomb energy
Z_1 and Z_2	= the atomic numbers of the fission products
e	= the electron charge unit
k	= the electrostatic constant.

When the two fragments just touch the electrostatic energy is given by:

Equation 3-54

$$\left(E_{coul}\right)_{r=R_1+R_2} = \frac{kZ_1Z_2e^2}{R_1+R_2}$$

When $r < R_1 + R_2$ both the nuclear and electrostatic forces come into play and the electrostatic energy ceases to increase as l/r. Due to the attractive nature of the nuclear force, the electrostatic energy contribution decreases in the manner indicated in the graph so that the total potential energy at $r = 0$ is just the rest mass energy of the fissioning nucleus. The peak in the potential energy occurs near the point where $r = R_1 + R_2$ and is referred to as the fission barrier, B so that:

Equation 3-55

$$B = \frac{kZ_1Z_2e^2}{R_1+R_2}$$

The difference between Q and B is the **critical energy** of fission and is the amount of energy that must be supplied to the ground state of the fissioning nucleus to overcome the barrier and split apart.

Equation 3-56

$$E_{crit} = B - Q$$

where:

E_{crit} = the critical energy of fission.

Spontaneous fission, which requires the fission fragments to penetrate the Coulomb barrier, remains an unlikely event when B > Q.

For the ^{236}U compound nucleus, E_{crit} is ~ 5.3 MeV. For photon induced fission, this excitation energy must be supplied by the energy of the gamma ray for fission to take place. For neutron induced fission, this excitation energy comes from the neutron bringing energy <u>and</u> kinetic energy. Odd mass nuclei ^{233}U, ^{235}U, and ^{239}Pu are fissionable by thermal neutrons since the binding energy brought to the compound nucleus is larger than E_{crit} and no kinetic energy is required for the fission process. These nuclides are called **fissile** since they fission with neutrons of any energy.

Target nuclei with even mass numbers like ^{234}U, ^{232}Th, and ^{238}U are called fertile since they fission only with neutrons in the fast energy range, but produce fissile nuclei through neutron absorption and beta decay. When a fertile nucleus captures a neutron to form a compound nucleus, the Q value of the compound nucleus is smaller than E_{crit}. For fission to take place the energy difference must be supplied by the neutron's kinetic energy.

Heavy mass nuclei with an odd A (odd N) are fissile while even A nuclei are fertile. This distinction is due to the difference in neutrons bringing energy for even N nuclides relative to odd N nuclides. When the target is odd A (odd N), the addition of a neutron creates an even N compound nucleus where extra binding energy is available due to the complete pairing of all neutrons. When the target is even A, the additions of a neutron creates an odd N compound nucleus where the unpaired neutron cannot gain extra binding energy.

3.4.5 — Neutrons From Fission

A discreet number of neutrons is released in each fission event. This number varies depending upon the fragments into which the compound nucleus divides. Some fragments release none while some may emit as many as 5 or 6. The measured average number of neutrons per fission, ν, emitted by both fragments of ^{235}U in thermal neutrons is 2.43. Most of the neutrons from fission are emitted within 10^{-14} seconds of the break up and are referred to as **prompt** neutrons.

Approximately 0.65% of all fission neutrons from ^{235}U are **delayed** neutrons and arise due to the decay of delayed neutron precursors in the fission product decay chains. These delayed neutrons are particularly important in reactor control because they dramatically increase the effective time between successive neutron generations, thus limiting the rate at which power can increase or decrease. The number of delayed neutrons relative to all fission neutrons is so important to controlling the reactor and used so frequently in kinetics calculations that the symbol β has come to take on a special meaning when working with delayed neutrons.

Equation 3-57

$$\beta = delayed\,neutron\,fraction = \frac{number\,of\,delayed\,neutrons}{all\,fission\,neutrons}$$

Delayed neutrons require fewer collisions to thermalize since their average energy is ~0.4 MeV (compared to ~ 2 MeV for prompt neutrons). Delayed neutrons travel shorter distances while slowing down since the mean free path at lower energies is smaller than that from prompt neutrons at higher energies. Because they travel shorter distances, delayed neutrons are less likely to leak from the reactor and more likely to cause thermal fission. Primarily because of these effects, the fraction of all fissions induced by delayed neutrons to the fission process, and **effective delayed neutron fraction**, $\bar{\beta}$, is defined where:

Equation 3-58

$$\bar{\beta} = effective\,\beta = \gamma\beta$$

and γ is calculated or experimentally determined constant usually greater than 1.

Although β is a constant material property, $\bar{\beta}$ is a function of material and core geometry where the geometry determines the leakage. Typically, for a large reactor, γ might be 1.02 while γ of ~ 1.1 might be appropriate for a small reactor.

3.4.6 — Energy From Fission

One of the most distinctive features of the fission process is the amount of energy released in a single event. Most heavy mass nuclei in the neighborhood of ^{235}U release ~ 200 MeV on average in a single fission event.

Most of the energy, ~90%, appears instantaneously with the breakup of the nucleus while the remainder appears delayed as a result of the fission product decay. Since the half-lives of the fission products range anywhere from a fraction of a second to thousands of years, the delayed energy is a concern not only for reactor operation, but shutdown as well. However, the delayed energy reaches an equilibrium days after reactor operation.

All of the energy released through the fission process is not recoverable for use. Appearing in the beta decay of the fission products, antineutrinos have a tiny cross section for interaction with matter and remove energy from the reactor. This energy loss is partially compensated by non-fission, neutron capture events. The most prominent of these events are the many (n, γ) reactions along with ^{10}B (n, α) and ^{16}O (n, p) events taking place in the core.

Table 3-1 — Energy From ^{235}U Fission

Form	Emitted Energy MeV	Recoverable Energy MeV
Instantaneous Energy		
Kinetic energy of fission fragments	165.6	165.6
Kinetic energy of fission neutrons	4.8	4.8
Fission gamma rays	7.7	7.7
Neutron capture gamma rays	7.5	7.5
Total instantaneous energy	185.6	185.6
Delayed Energy		
Kinetic energy of beta particles	7.2	7.2
Delayed neutrons	~ 0	~ 0
Fission product decay gamma rays	7.2	7.2
Antineutrinos	10.2	0
Total delayed energy	24.6	14.4
Total energy released per fission	210.2	200.0

Section 3.5 — Nuclear Reaction Cross Sections

3.5.1 — Microscopic Cross Section

The probability of a particular reaction occurring between an incident particle and a target nucleus is proportional to the effective "target area" presented by the nucleus to the incident particle. This "target area" is often larger than the geometrical cross sectional area presented by the nucleus to the incident particle.

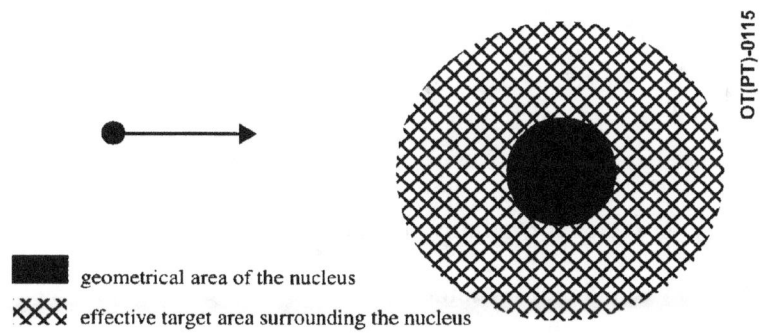

OT(PT)-0115

■ geometrical area of the nucleus

XXX effective target area surrounding the nucleus

Exhibit 3-8 — Nucleus With Surrounding Effective Target Area

The effective "target area" is called the microscopic cross section, σ, of the nucleus. The unit for the microscopic cross section is the barn, b; 1 b = 10^{-24} cm^2.

Each reaction type has its own cross section. These are termed "partial" cross sections since they are part of the total target area. The total microscopic cross section, σ_t, is the total effective target area presented to an incident particle by a nucleus for all reactions. The partial cross sections can be summed to find the total:

Equation 3-59

$$\sigma_t = \sum_i \sigma_i$$

where "i" is the summation index over all possible partial cross sections.

Since all induced nuclear reactions are either scattering or absorption reactions, σ_t may be written as the sum of the absorption part, σ_a, and the scattering part, σ_s.

Equation 3-60

$$\sigma_t = \sigma_a + \sigma_s$$

For neutron induced reactions, the scattering cross section is the sum of its elastic scattering, σ_{el}, and inelastic parts, σ_{in}.

Equation 3-61

$$\sigma_s = \sigma_{el} + \sigma_{in}$$

For neutron induced reactions, the absorption cross section is the sum of many more parts

Equation 3-62

$$\sigma_a = \sigma_c + \sigma_f + \sigma_{n,\alpha} + \sigma_{n,p} + \sigma_{n,2n}$$

where:

        c   stands for radiative capture

        f   stands for fission.

For non-fissile nuclides in the thermal energy range, radiative capture is the dominate reaction process and

$$\sigma_a \approx \sigma_c .$$

3.5.2 — Energy Dependence Of Microscopic Cross Sections

The microscopic cross section for any induced nuclear reaction varies with
1.   the identity of the target nucleus
2.   the identify of the incident particle
3.   the kinetic energy of the incident particle.

This section briefly reviews some of the major features of the energy dependence of neutron induced cross sections. Exhibit 3-9 graphically summarizes the general trend for absorption and scattering cross sections.

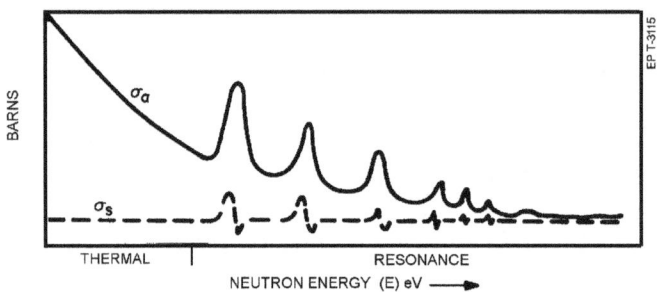

Exhibit 3-9 — General Shape Of Microscopic Cross Sections

3.5.2.1 — Thermal Neutrons

The absorption cross section, σ_a, varies inversely with the neutron's velocity or square root of its kinetic energy:

Equation 3-63

$$\sigma_a = \frac{C1}{V} = \frac{C2}{\sqrt{KE_n}}$$

where:

$$\mathbf{V} \qquad = \text{neutron's velocity}$$

$$\mathbf{KE_n} \qquad = \text{neutron's kinetic energy}$$

$$\mathbf{C1, C2} \qquad = \text{constants.}$$

Equation 3-63, called the "1/V law", is usually written as:

Equation 3-64

$$\sigma_a = \sigma_{ao}\left(\frac{v_o}{v}\right)$$

where v_0 is the reference speed and σ_{ao}, called the reference cross section, is the value of σ_a at $v = v_o$. The reference speed is usually chosen as $v_o = 2200$ meter/second which corresponds to a neutron kinetic energy of 0.0253 eV at 68°F.

The most notable example of 1/V absorption is the $^{10}B(n, \alpha)$ reaction that extends well into the epithermal energy range. The 1/V law for slow neutrons applies not only to (n, α) reactions, but (n, γ) and (n, f) reactions as well.

The scattering cross section, σ_s, is constant over the thermal range with magnitudes varying from ~ 1 barn for light nuclei to ~ 100 barns for heavy nuclei.

3.5.2.2 — Epithermal Neutrons

The most distinctive feature of neutron cross sections in the epithermal energy range is the presence of pronounced maxima in the cross sections at discrete energies. These maxima, which are called **resonances**, correspond to the presence of discrete energy levels in the compound nucleus and are superimposed on the smooth cross section, for either absorption or scattering, extending from the thermal energy range. The resonance condition exists whenever the kinetic energy of an incoming neutron is such that the excitation energy $(Q + KE_n)$ matches one of allowed energy levels. When this condition is met, the compound nucleus is more likely to form.

Since the nuclear energy level spacing decreases with increasing excitation energy and as the number of nucleons in the nucleus increases, the neutron cross section resonances for medium and heavy nuclei lie very close together at higher neutron energies. Eventually, when these close lying energy levels overlap each other, the neutron cross section again exhibits a smooth variation with energy. The region where resonance peaks are separated, generally $KE_n < 5$ keV, is called the **resolved resonance energy range**. The region where resonance peaks overlap, generally for $KE_n > 5$ keV, is called the **unresolved resonance energy range**.

Resonance scattering, for the most part, does not alter the slowing down of neutrons as discussed in Section 3.6. However, resonance absorption can be important particularly in fissionable nuclides such as ^{235}U and ^{238}U, since capture resonances can keep the neutron from causing fission. For example, ^{238}U is so good at resonance capture with no fission taking place that it acts as a poison in this energy region. By contrast, ^{235}U also absorbs well in the resonance region but can fission. Specifically, about 1/3 of ^{235}U resonance absorptions produce fission.

3.5.2.3 — Fast Neutrons

The absorption cross section, σ_a, decreases faster than 1/V so that it is generally smaller than the scattering cross section in the fast energy region. The reason for this is primarily due to increased competition form the particle emission reactions.

The variation of σ_s with KE_n is generally described as "smooth and rolling" and tends to approach a value equal to the physical cross section of the target nucleus, πR^2, where R is the radius of the target nucleus.

Threshold reactions start to appear in this energy region. An endothermic reaction exhibit a zero cross section below its threshold energy value and a characteristic steep rise in magnitude as KE_n increases above the threshold.

3.5.3 — Macroscopic Cross Section

The product of the number density, N, and the microscopic cross section, σ, occurs so frequently that it is given the symbol Σ and is called the **macroscopic cross section**. For each microscopic cross section there exists a corresponding macroscopic one:

Equation 3-65

$$\Sigma = N\sigma$$

With N the number of atoms or nuclei per unit volume and σ the area per atom or nuclei, the unit for the macroscopic cross section is "per unit distance" or most commonly cm^{-1}. The macroscopic cross section is the probability that a particle interacts in traveling a unit distance through a macroscopic sample of material. This is a very useful quantity since it links the probability of interaction with the material density. Another way of thinking about the macroscopic cross section is as the total nuclear target area of a material

Like microscopic cross sections, macroscopic cross sections are additive.

Equation 3-66

$$\Sigma_t = \Sigma_a + \Sigma_s$$

Equation 3-67

$$\Sigma_t = \Sigma_{el} + \Sigma_{in} + \Sigma_c + \Sigma_f + ...$$

Various regions are often composed of a mixture of materials. Each material may be either a pure element or a chemical compound. The macroscopic cross sections used for that material must include the contribution of each nuclide present. Thus, the macroscopic cross section for a mixture or compound is obtained by summing the individual macroscopic cross sections of its components, according to the general equation,

Equation 3-68

$$\Sigma = \Sigma^A + \Sigma^B + \Sigma^C + ...$$

where A, B, C refer to each component of the mixture.

3.5.4 — Mean Free Path

The average distance a particle travels before undergoing a collision is called its **mean free path** and is inversely related to the macroscopic cross section of the material through which it is passing.

Equation 3-69

$$\text{mean free path} = \lambda = 1/\Sigma$$

The total mean free path is the average distance that a particle travels in a material before undergoing a collision of any type. This distance is just the inverse of the total macroscopic cross section.

Equation 3-70

$$\lambda_t = 1/\Sigma_t$$

The absorption mean free path is the average distance that a particle travels in a material before undergoing any type of absorptive collision. This distance is just the inverse of the total macroscopic absorption cross section.

Equation 3-71

$$\lambda_a = 1/\Sigma_a$$

The scattering mean free path is the average distance that a particle travels in a material before undergoing any type of a scattering collision. This distance is just the inverse of the macroscopic scattering cross section.

Equation 3-72

$$\lambda_s = 1/\Sigma_s$$

While cross sections (probabilities) are additive, mean free paths are not.

Equation 3-73

$$\lambda_t \neq \lambda_a + \lambda_s$$

Rather,

Equation 3-74

$$\frac{1}{\lambda_t} = \frac{1}{\lambda_a} + \frac{1}{\lambda_s}$$

Since the mean free path depends upon the macroscopic cross section, and the cross section is a function of the incident particle's energy, the mean free path is also a function of energy.

3.5.5 — Attenuation

If a beam of particles at a given energy strikes a slab of material of thickness X and total cross section Σ_t, only a fraction of these particles pass through the slab without undergoing any reaction. If ϕ_0 is the number of particles per second incident perpendicular to a unit area of slab surface, then the number exiting the slab per unit area per second, ϕ_f, is given by

Equation 3-75

$$\phi_f = \phi_0 e^{-\Sigma_t X}$$

This equation defines the attenuation of a particle beam by a given thickness of material and is called the **Attenuation Law**. More generally, the Attenuation Law may be written as:

Equation 3-76

$$\phi(x) = \phi_0 e^{-\Sigma_t x}$$

Form this form it can be realized that the number of particles decreases exponentially with position in the slab. Because of the exponential nature of attenuation, the number of particles never becomes identically zero, not even at arbitrarily large distances for very large slabs.

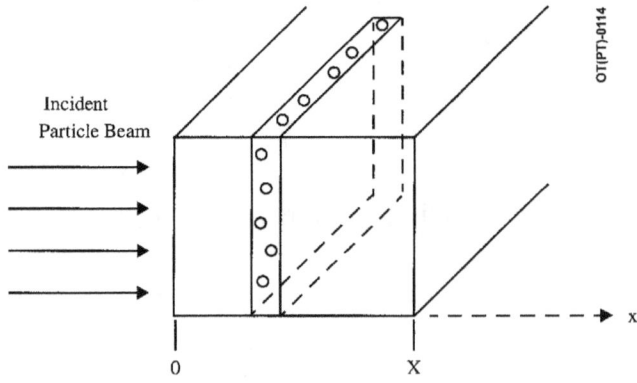

Incident
Particle Beam

Exhibit 3-10 — Beam Attenuation

Section 3.6 — Neutron Slowing Down

Prompt neutrons born in the fission process have energies greater than 0.1 MeV with an average energy of 2 MeV. Delayed neutron energies are lower than those for the prompt with the delayed neutron average being approximately 0.4 MeV. The highest probability of these neutrons inducing other fission events and contributing to the chain reaction occurs in the thermal energy range where the fission cross section is highest. These higher energy neutrons are thermalized through scattering events where they lose some of their kinetic energy to the surrounding bulk material. However, any time a neutron collides with a nucleus there is also the probability for the interaction event to be an absorptive one; this removes the neutron from the cycles. Likewise, neutrons that leak out of the core are removed from the cycle and cannot contribute to the chain reaction.

3.6.1 — Mechanisms

Elastic and inelastic scattering reactions are the only types of interactions that cause the neutron to lose energy without removing them from the cycle. Of the two types of scattering reactions, elastic scattering collisions are the most important since they occur at all neutron energies and their cross sections are generally larger than the cross sections for inelastic scattering.

Inelastic scattering is only a minor player in the slowing down process. This is due to
1. the process being a threshold one where the threshold is on the order of several keV and
2. the magnitude of the cross section, typically on the order of a few barns, being smaller than the cross section for elastic scattering. However, for some nuclides, the cross section magnitudes for elastic and inelastic scattering become comparable in the fast energy range.

3.6.2 — Materials

The effectiveness of any material as a moderator is dependent upon three things:
1. the amount of energy lost in a single collision
2. magnitude of the scattering cross section
3. magnitude of the absorption cross section.

Good moderators are those that have a large energy loss per collision, large scattering cross section, and low absorption cross section.

The maximum energy lost in any single collision increases as the mass of the target nucleus decreases. This means that light nuclei contribute more effectively to the slowing down process and are considered good moderators. In reactors, the most commonly used moderators are:
- ordinary water (H_2O)
- heavy water (D_2O)
- beryllium
- graphite (carbon)

Boron and lithium are both light mass nuclei, but are not good moderators due to their high absorption cross sections. Likewise, helium is a light mass nuclei, but is not a good moderator despite its low absorption cross section since its material density causes its macroscopic scattering cross section to be quite low.

3.6.3 — The Probability Density Function For The Energy Of Scattered Neutrons

Given a neutron's initial energy E_0, the minimum energy that it retains after a single elastic scatter is αE_0 with:

Equation 3-77

$$\alpha = \frac{(m_T - m_n)^2}{(m_T + m_n)}$$

where:

m_T = the target's mass

m_n = the neutron's mass

The maximum possible neutron energy loss in a single elastic collision is:

Equation 3-78

$$\Delta E_{max} = E_0(1 - \alpha)$$

The possible final energies of the scattered neutron are:

Equation 3-79

$$\alpha E_0 \leq E \leq E_0$$

A neutron possessing an energy very nearly equal to E_0 after the collision corresponds to a case in which it merely grazes the target without transferring any energy while an energy of αE_0 corresponds to a head-on collision with the nucleus with the maximum permitted exchange of energy. After a single elastic collision, a neutron has the same probability of possessing a specific energy between E_0 and αE_0 as any other specific energy within the allowable range. In other words, the energy distribution of scattered neutrons is independent of the energy after scattering. This permits development of a probability density function for the energy of scattered neutrons, P(E). As shown in Exhibit 3-11, P(E) is uniform and non zero between E_0 and αE_0, and zero outside that range reflecting the fact that neutrons scatter with equal probability into the allowed range.

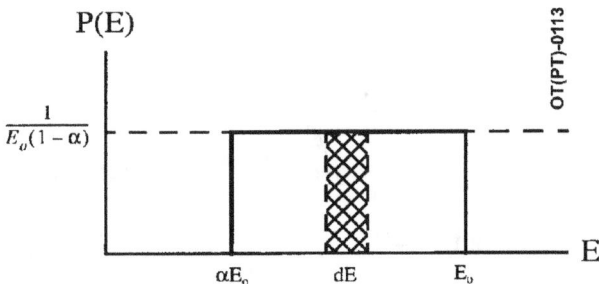

Exhibit 3-11 — Probability Density Function P(E) Versus Energy E Of Elastically Scattered Neutrons

The average value of neutron energy after a single elastic scattering $\bar{E}_1$, is

Equation 3-80

$$\bar{E}_1 = E_0 \left(\frac{1+\alpha}{2} \right)$$

The ratio $\bar{E}_1/E_o = (1+\alpha)/2$ is independent of the initial energy. This means that on the average each elastic scattering event reduces neutron energy by the faction $(1 + \alpha)/2$ which depends only on the mass of the target nucleus.

3.6.4 — Average Logarithmic Energy Decrement

The average change in the logarithm of neutron energy for one collision is the **average logarithmic energy decrement** and is denoted by the symbol ξ.

Equation 3-81

$$\xi = 1 + \frac{\alpha \ln \alpha}{1-\alpha}$$

where ξ, like α, is independent of initial neutron energy so long as elastic scattering is isotropic in the center of the mass system. Given A as the ratio of m_T/m_n, ξ may be expressed as:

Equation 3-82

$$\xi = 1 + \frac{(A-1)^2}{2A} \ln \left(\frac{A-1}{A+1} \right)$$

For large values of A, ξ is closely approximated by:

Equation 3-83

$$\xi \approx \frac{2}{A + \frac{2}{3}}$$

For A = 2, the error of Equation 3-83 is only 3.4%.

The value of ξ when A approaches 1, hydrogen is:

Equation 3-84

$$\lim_{A \to 1} \xi = 1$$

When a moderating medium contains several nuclides, the overall moderating ability of that medium is characterized by a weighted average of the ξ values for each nuclide, where the weighting factor for each nuclide is its macroscopic scattering cross section. For a medium containing two nuclides,

Equation 3-85

$$\xi = \frac{\Sigma_{S1}\xi_1 + \Sigma_{S2}\xi_2}{\Sigma_{S1} + \Sigma_{S2}}$$

where:

ξ_1, ξ_2 = the ξ values for the two nuclides

Σ_1, Σ_2 = the macroscopic scattering cross sections for the two nuclides.

3.6.5 — Slowing Down Power And Moderating Ratio

As a means to quantify materials as moderators, two parameters have been defined: Slowing Down Power and Moderating Ratio.

Slowing Down Power, $\xi\Sigma_s$, is defined as the average logarithm of neutron energy per unit path length:

Equation 3-86

$$\xi\Sigma_s = \xi/\lambda_s$$

Although the slowing down power is a satisfactory measure of a material's ability to reduce neutron energy, it does not account for the possibility that an absorption in that material may remove the neutron before it is thermalized. To account for absorption, the slowing down power is divided by the macroscopic absorption cross section to yield the moderating ratio:

Equation 3-87

$$\xi\Sigma_s/\Sigma_a$$

For either of these parameters, a larger value indicates a more effective moderator. Some values for typical moderators are presented in Table 3-2. In terms of the slowing down power, ordinary water is the best. In terms of the moderating ratio, heavy water far surpasses the others. Unfortunately, heavy water is expensive due to its scarcity. Hence, most thermal reactor designs choose to use less efficient but cheaper and more readily available moderators.

Table 3-2 — Slowing Down Parameters Of Typical Moderators*

Moderator	ξ	$\rho(gm/cm^3)$	$\xi\Sigma_S$	$\xi\Sigma_S/\Sigma_a$
H_2O	0.920	1.0	1.35	71
D_2O	0.509	1.1	0.176	5670
Be	0.209	1.85	1.58	143
C	0.158	1.60	0.060	192
* Duderstadt, James J. and Louis J. Hamilton, "Nuclear Reactor Analysis," John Wiley and Sons, New York, 1976.				

Slowing down power is affected by temperature Σ_s depends upon the material's number density. As temperature increase, the material's density decreases due to expansion. Hence, the number density and Σ_s decrease as well which causes the slowing down power to decrease.

Section 3.7 — Thermal Equilibrium

3.7.1 — Neutron Energy Equilibrium

In water-moderated thermal reactors, fast neutrons (born in the fission process) primarily lose kinetic energy through collisions with hydrogen in the water and enter thermal energy range. Once in the thermal range the neutrons possess kinetic energies that are comparable in magnitude to the kinetic energies of the atoms and molecules of the surrounding reactor material. (Since the reactor material is at some temperature greater than absolute zero, its atoms and molecules have velocities and kinetic energies.) In this situation, the neutrons can lose energy to or gain energy from the surrounding so that there is a distribution of energies (velocities) in the thermal neutron population. The kinetic energies (velocities) are distributed statistically according to the Maxwell-Boltzmann Distribution Law. Considering a large sample of thermal neutrons, there is no net energy change for all the neutrons when equilibrium is established.

3.7.2 — Maxwell-Boltzmann Distribution

In the ideal environment of an infinite, non-absorbing media, a burst of fast neutrons eventually establishes a Maxwell-Boltzmann distribution (in energy or velocity) with its surroundings. The distribution of neutron energies forms a curve that peaks at the most probable neutron energy and drops rather sharply to zero neutrons at zero energy. The drop in the curve is less rapid for energies greater than the peak energy with more neutrons possessing energies greater than the most probable energy than there are with energies less than the most probable energy.

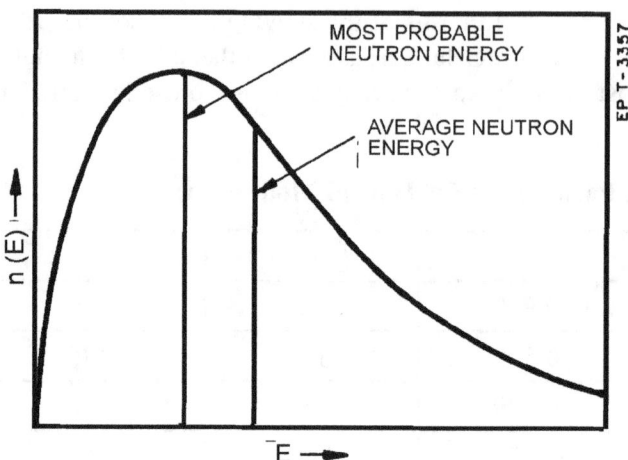

Exhibit 3-12 — Ideal Maxwell-Boltzmann Neutron Distribution With Energy

In the absence of absorption (or in the presence of very small amounts of absorption), the most probable neutron energy is given by:

Equation 3-88

$$E(eV) = 8.61 \times 10^{-5} \times T(K)$$

An increase in the temperature of the media increases the amount of energy available for exchange with the thermal neutrons and produces three changes in the Maxwell-Boltzmann distribution:
- the peak height is lowered
- the peak energy is shifted to the right
- the distributions broadened (so that more neutrons are at higher energies).

Thus, an increase in temperature **hardens** the thermal neutron distribution by shifting more neutrons to higher energies. The total area under the curve (or number of neutrons) remains the same.

3.7.3 — Deviation From Maxwell-Boltzmann

Reactors are not the ideal environment in which to establish a Maxwell-Boltzmann distribution of thermal neutrons since they are <u>finite</u>, contain <u>absorbing</u> materials, and possess a <u>continuous source</u> of fast neutrons.

In the thermal range, 1/V absorption cross sections increase as $1/\sqrt{E}$ when the energy decreases. As a result, more neutrons with low energy are removed from the thermal neutron distribution that neutrons with higher energy. Consequently, absorption lowers the peak of the distribution and shifts the distribution to higher energy. This process is referred to as **absorption hardening**.

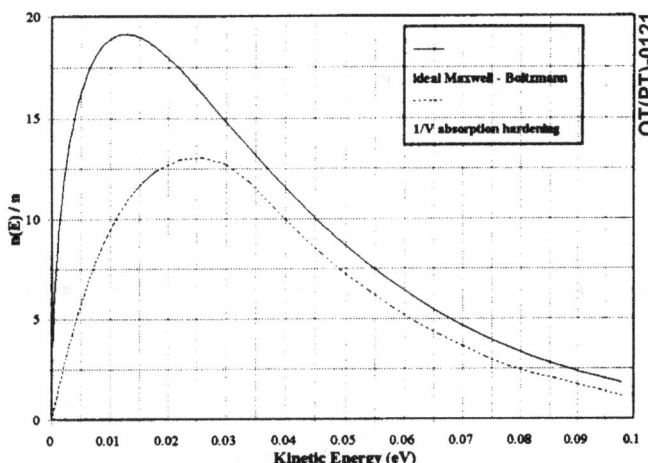

Exhibit 3-13 — The Effect Of 1/V Absorption Hardening On An Ideal Maxwell-Boltzmann Distribution

The continuous production of fast neutrons in the fission process gives rise to a continuous source of lower energy neutrons entering the thermal region from the epithermal region. This source is known as the **slowing down source**. As a result, the neutron source is not turned off (non-burst) and the number of neutrons at the upper end of the thermal range is much larger than what is expected from an ideal Maxwell-Boltzmann distribution.

The finite reactor size has an effect on the thermal neutron distribution that is much smaller than those due to the absorption and continuous source. Since the mean free path increases as the neutron's energy increases, those neutrons at the upper end of the thermal range travel larger distances before colliding than those at the lower end of the energy range. Consequently, more of the neutrons at the upper end of the thermal range leak, escape, from the finite core than those from the lower end of the thermal range. This phenomenon is known as **diffusion cooling**. As a result, the final distribution of neutrons in the upper half of the thermal range is slightly less than what would be present from the slowing down source.

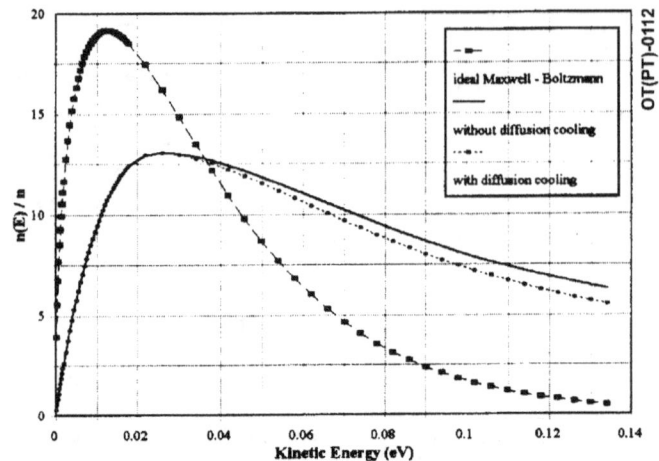

Exhibit 3-14 — The Effect Of Diffusion Cooling On A 1/V Absorption-Hardened Distribution Of Thermal Neutrons

Section 3.8 — Neutron Density, Flux, Reaction Rates, And Power

3.8.1 — Neutron Density

The number of neutrons in a unity volume is known as the neutron density, n. Typical units for neutron density are neutrons/cm^3.

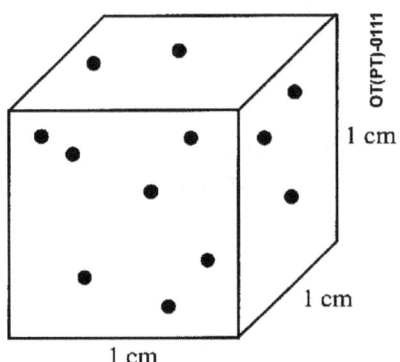

Exhibit 3-15 — Neutron Density

In a reactor core, n is not constant, but varies with position. For this reason, n can be expressed as $n(\bar{r})$ to identify it with a location in some coordinate system.

The thermal neutron density n_{th} is the number of **neutrons/cm^3** that have energies within the thermal energy region.

3.8.2 — Flux

The chance that a nucleus will react with a neutron (from a group of neutrons) depends upon the number of neutrons approaching the nucleus (neutron density) and the velocity with which the neutrons are passing. Higher neutron velocities, v, imply more neutrons approaching the nucleus per unit time.

n and v occur together so often that a special quantity known as the (neutron) flux ϕ is defined as the scalar product of these variables.

Equation 3-89

$$\phi = nv$$

With **n** possessing **neutrons/cm^3** and v possessing units of **cm/sec**, the units for ϕ are **neutrons/cm^2–sec**. Note that Equation 3-89 is completely general and **neutrons** can be replaced with any type of particle such as **gammas, alphas**, etc.

Often it is necessary to work with the flux for the thermal energy region only. Therefore, the thermal flux ϕ_{th} can be defined as:

Equation 3-90

$$\phi_{th} = n_{th}v$$

where $\bar{v}$ is the average speed of the thermal neutrons.

3.8.2.1 — Physical Interpretation

Given a unit volume (1 cm^3) centered about some location, the flux at that location is equal to the total path length traveled in one second by all the neutrons initially occupying that unit volume. This interpretation is illustrated in Exhibit 3-16. The individual neutron paths show abrupt changes in direction due to scattering from nuclei (not shown), and are of varying lengths because they have a distribution of speeds.

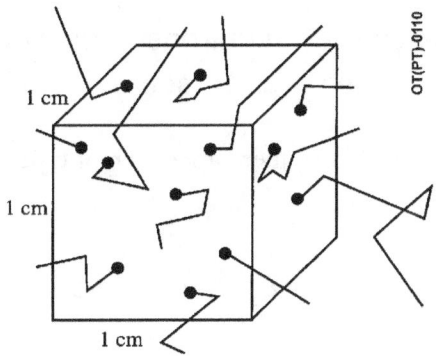

Exhibit 3-16 — Neutron Path Lengths In A Unit Volume

If the path lengths for each neutron over a duration of one second are summed, the result is a total path length quantity with dimensions of **neutron-cm**. (Another way of arriving at this is to consider the average path length, $\overline{p}$ for the group of neutrons $n \bullet \overline{p}$ gives the same result with the appropriate units of neutron-cm.) The total path length divided by the volume over which the paths are summed yields the average flux over that volume with units given by:

Equation 3-91

$$neutron - cm \bullet \frac{1}{cm^3} = \frac{neutron}{cm^2}$$

Since this process has taken place over a period of 1 second, the units expressed by Equation 3-91 may be modified to yield

Equation 3-92

$$\frac{neutron}{cm^2 - sec}$$

3.8.3 — Reaction Rates

A reaction rate is the number of nuclear reactions of a particular type that take place in unit time. Generally, in a material or region of interest, a reaction rate R is the product of the flux ϕ and the macroscopic reaction cross section Σ .

Equation 3-93

$$R = \phi\Sigma$$

Typical units for reaction rates are:

Equation 3-94

$$\frac{reactions}{cm^3 - sec} = \frac{neutrons}{cm^2 - sec} \times \frac{reactions}{cm}$$

Just as there are different types of cross sections, there are different types of reaction rates:

Equation 3-95

$$R_a = \phi\Sigma_a, R_f = \phi\Sigma_f, etc.$$

3.8.4 — Power

3.8.4.1 — Power Density

The energy release per fission event is constant (200 MeV for thermal fission of ^{235}U). The power density (power/cm^3) is the product of the fission reaction rate (fissions/cm^3–second) and the energy released per fission.

Equation 3-96

$$PD = kR_f = k\phi_{th}\Sigma_f$$

where:

$\quad$ **PD** $\quad$ = the power density (typically in units of watts/cm^3)

$\quad$ **R_f** $\quad$ = the fission reaction rate

and k is the energy per fission defined as follows:

$$k = \varepsilon k'$$

where:

$\quad$ **k'** $\quad$ = a constant that contains the core volume, etc.

$\quad$ **ε** $\quad$ = the fast fission factor.

ε accounts for the fissions that occur while neutrons are slowing down to thermal energies.

The total reactor power is obtained by summing all the power densities in each cm^3 of the reactor core.

3.8.4.2 — Power Operations

During reactor power operations, ^{235}U depletes causing N^{235} and therefore Σ_f^{235} to decrease. The decrease in N^{235} is linear due to the <u>same</u> number of ^{235}U nuclei fissions needed to maintain <u>constant</u> power. The number of ^{235}U surviving at any time, N^{235}(t), is given by

Equation 3-97

$$N^{235}(t) = N^{235}_o - (K_1 \bullet EFPH)$$

where:

$\quad$ **N^{235}_o** $\quad$ = the initial concentration of ^{235}U at BOL

$\quad$ **K_1** $\quad$ = a constant that includes power density.

To achieve a specified power throughout core life, the product $\Sigma_f\phi_{th}$ must be constant at any time. When Equation 3-96 is substituted into Equation 3-97 and solved for ϕ_{th}, the result is

Equation 3-98

$$\phi_{th} = \frac{K_2}{N^{235}_o - (K_1 \bullet EFPH)}$$

where:

$\mathbf{K_2}$ = a constant that includes power.

Equation 3-98 shows a hyperbolic increase in ϕ_{th} as the core accumulates EFPH. The ϕ_{th} and $N^{235}(t)$ behavior as a function of time or EFPH is shown schematically in Exhibit 3-17.

Since the fission rate is constant for constant power operation, the number of fast neutrons born by fission per second and the number of neutrons thermalized per second are also constant. Because thermal capture and fission macroscopic sections in the reactor decrease with life due to depletion of fuel and burnable poisons, the thermal neutrons scatter for a longer time before being absorbed. This causes the ratio of thermal neutrons to fast neutrons to increase with core lifetime.

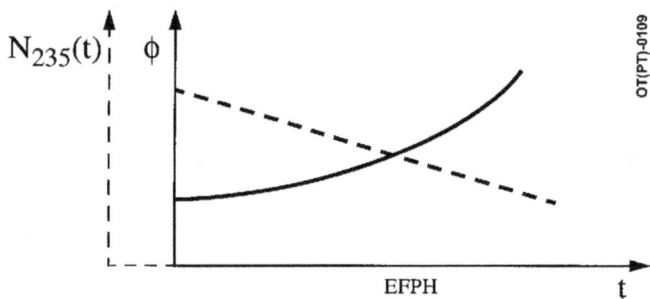

Exhibit 3-17 — Schematic of ϕ_{th} And $N^{235}(t)$ Trends With Core Life

Section 3.9 — Slowing Down, Diffusion, And Migration Lengths

3.9.1 — Slowing Down Length

Neutrons travel in straight lines between collisions. While an absorption terminates the neutron's progress, each scattering collision may cause the direction of a neutron's flight to change. From birth to thermalization, a neutron might typically follow a path such as the solid line in Exhibit 3-18.

The "crow-flight" distance, r_f, from the neutron's point of birth to its point of thermalization is much less than the neutron's path length. The **slowing down length, L_s,** can be shown to be related to r_f in the following manner:

Equation 3-99

$$L_s^2 = \frac{1}{6}\overline{r_f^2}$$

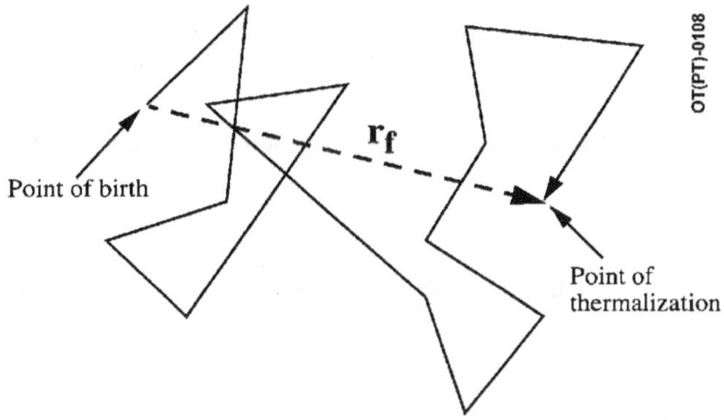

OT(PT)-0108

Exhibit 3-18 — Schematic Representation Of Typical Neutron Slowing Down Path

3.9.1.1 — In Terms Of Slowing Down Power

The average path length between collisions is equal to the scattering mean free path, λ_s. If the number of collisions to thermalize a neutron is C, then the total path length is $\lambda_s C$. This path length is much longer than the dashed line in Exhibit 3-18. For a random process such as the one shown in Exhibit 3-18, there is a mathematical rule which states that the average straight-line (crow-flight) distance from start to finish is proportional to the square root of the number of path segments times the average segment length. Thus,

Equation 3-100

$$L_s = \text{constant}\,(\lambda_s \sqrt{C}) = \text{constant}\,\frac{\sqrt{C}}{\Sigma_s}$$

Or, upon squaring both sides of Equation 3-100:

Equation 3-101

$$L_s^2 = \text{constant}\,\frac{C}{\Sigma_s^2}$$

Equation 3-101 is limited in its application because it does not account for resonance absorption. The presence of resonance absorption means that a fast neutron may not complete its thermalization prior to absorption. Therefore, it may undergo a few collisions, travel a shorter distance from its birthplace, and be less likely to leak out of the core. It can be shown that the presence of resonance absorption reduces the average number of collisions per fast neutron from C to pC, where p is the resonance escape probability (Paragraph 3.10.2.2).

Equation 3-102

$$L_s^2 = \text{constant} \frac{pC}{\Sigma_s^2}$$

The physical meaning of Equation 3-102 becomes clearer with two simple manipulations. Knowing that $C\xi$ is a constant, C can be eliminated. With ξ present in the equation Σ_s can be grouped with it to form the slowing down power. Therefore,

Equation 3-103

$$L_s^2 = \text{constant} \times \left(\frac{p}{\xi\Sigma_s} \right) \times \frac{1}{\Sigma_s}$$

3.9.1.2 — Relation To The Diffusion Equation

The simplified diffusion equation for the fast energy range is

Equation 3-104

$$D_{fast}B_{fast}^2\phi_{fast} + \left(\Sigma_\alpha^{fast} + \Sigma_{rem}^{fast} \right)\phi_{fast} = S_{fast}$$

Leakage + Absorption + Slowing Down = Fast Neutron Production Source

where:

D_{fast} = the fast diffusion coefficient

B_{fast}^2 = the fast buckling

ϕ_{fast} = the fast flux

Σ_α^{fast} = macroscopic absorption cross section in the fast energy range

Σ_{rem}^{fast} = removal cross section from fast energy range to thermal

S_{fast} = source of fast neutrons.

In the absence of absorption, the slowing down length can be expressed in terms of the diffusion coefficient and the removal cross section:

Equation 3-105

$$L_S^2 = \frac{D_{fast}}{\Sigma_{rem}^{fast}}$$

It can be shown that

Equation 3-106

$$D_{fast} \cong \frac{1}{3\Sigma_s(1-\overline{\mu})}$$

and

Equation 3-107

$$\Sigma_{rem}^{fast} \cong \frac{\xi\Sigma_s}{\ln(E_0/E_{th})}$$

where:

$\overline{\mu}$ = the average cosine of the scattering angle

E_0 = the neutron's birth energy

E_{th} = thermal energy.

When Equation 3-106 and Equation 3-107 are substituted into Equation 3-105, L_S^2 is expressed as:

Equation 3-108

$$L_S^2 = \frac{\ln(E_0/E_{th})}{3\xi\Sigma_s^2(1-\overline{\mu})}$$

When scattering is isotropic such that $\overline{\mu} = 0$, Equation 3-108 simplifies to an expression that is similar to Equation 3-103.

3.9.1.3 — Importance Of L_S

Equation 3-108 shows the dependence of the slowing-down length upon Σ_S and ξ. A large L_S is associated with a large mean free path (small Σ_S) and a large nuclear mass (small ξ).

For a given energy loss per collision, a neutron in a medium of large λ_S needs to travel a greater distance (an average) to slow through a given energy interval than it must in a medium with a small λ_S. Consequently, the neutrons slowing down in the former system tend to spread out more than those in the latter.

If the scattering cross section is fixed, then a neutron slowing down in a medium with a large nuclear mass (small ξ) makes many more collisions to slow to a given energy than it would in a medium of small nuclear mass. Thus, an effect that tends to lengthen the slowing down process (more collisions) yields a larger L_S.

These elementary conclusions have considerable significance in selection of suitable moderators for a given reactor type. If a small thermal reactor is desired, the choice of moderators is limited to those materials which have the smaller values of L_S. The goal is to slow down the fission neutrons in as small a volume as practical. This is accomplished only if the characteristic dimension of the moderator is much larger than the slowing-down length.

Equation 3-109

$$L_s \ll \text{moderator size}$$

For any reactor of finite size, neutron leakage occurs. Larger values of L_S imply greater amounts of leakage.

3.9.2 — Diffusion Length

Like fast neutrons, thermal neutrons travel in zig-zag paths between collisions in a given region, thermal neutrons terminate by absorption.

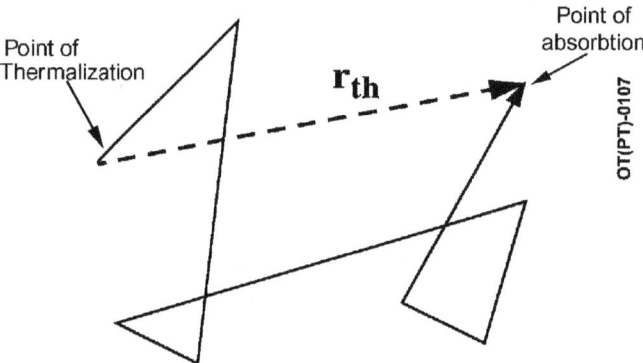

Exhibit 3-19 — Schematic Representation Of Typical Thermal Diffusion Path

The "crow-flight" distance r_{th} from the neutron's point of thermalization to its point of absorption is much less than the neutron's path length. The (thermal) diffusion length L can be shown to be related to r_{th} in the following manner:

Equation 3-110

$$L^2 = \frac{1}{6}\overline{r_{th}^2}$$

3.9.2.1 — Relation To The Diffusion Equation

The simplified diffusion equation for the thermal energy range is:

Equation 3-111

$$D_{th}B_{th}^2\phi_{th} + \Sigma_a^{th}\phi_{th} = \Sigma_{rem}^{fast}\phi_{fast}$$

where:

D_{th} = the thermal diffusion coefficient

$L_S^2 = (L_S^2)_{ref}(\rho_{ref}/\rho)$ = the thermal buckling

ϕ_{th} = the thermal flux

Σ_a^{th} = macroscopic absorption cross section in the thermal group.

The diffusion length is related to the thermal group constants in the following manner:

Equation 3-112

$$L^2 = \frac{D_{th}}{\Sigma_a^{th}}$$

The thermal diffusion coefficient is given by:

Equation 3-113

$$D_{th} = \frac{1}{3\Sigma_s^{th}(1-\overline{\mu})}$$

When scattering is isotropic such that $\overline{\mu} = 0$, Equation 3-113 simplifies so that substitution into Equation 3-112 it yields:

Equation 3-114

$$L^2 = \frac{1}{3\Sigma_a^{th}\Sigma_s^{th}}$$

3.9.3 — Density Changes

Since scattering in the moderator dominates all scattering in the reactor, L_S^2 behavior strongly correlates with the moderator's behavior. The core average scattering cross section Σ_S is primarily the scattering cross section of the water. The scattering cross section for the water is directly proportional to the density of the water, ρ, which is inversely proportional to the temperature. The resonance escape probability p is dependent upon the inverse of the slowing down power:

$$p \propto 1/\xi\Sigma_s$$

Now, L_S^2 can be determined approximately at any temperature if a reference length and density are known, $(L_S^2)_{ref}$ and ρ_{ref}.

Equation 3-115

$$L_S^2 \propto \frac{p}{\xi\Sigma_s^2} \propto \frac{(\rho/\rho_{ref})}{(\rho/\rho_{ref})^2} \propto (\rho_{ref}/\rho)$$

So,

Equation 3-116

$$L_S^2 = (L_S^2)_{ref}(\rho_{ref}/\rho)$$

Equation 3-114 shows the L^2 dependence upon Σ_a and Σ_s. Σ_a is strongly dependent upon the fuel and poison loadings while Σ_s is primarily due to moderator choice. Fuel and poison material densities change little with temperature and have little affect upon L^2 in this regard. For water-moderated cores, Σ_s is highly influenced by the water's macroscopic cross section. As stated for L_S^2.

Equation 3-117

$$\Sigma_S = (\Sigma_S)_{ref}(\rho/\rho_{ref})$$

which implies

Equation 3-118

$$L^2 \propto \frac{\rho_{ref}}{\rho}$$

so that

Equation 3-119

$$L^2 = (L^2)_{ref}\left(\frac{\rho_{ref}}{\rho}\right)$$

3.9.4 — Migration Length

The **migration length** M is a measure of the straight-line distance traveled by a neutron from its birth in the fast energy region to its absorption in the thermal energy region, and therefore depends upon both the slowing-down length, L_S, and the thermal diffusion length L. The migration length is defined by the relation:

Equation 3-120

$$M = \sqrt{L_S^2 + L^2}$$

When Equation 3-108 and Equation 3-114 are substituted into Equation 3-120 for L_S^2 and L^2, respectively, it is easy to see that M depends solely upon the material properties.

The **migration area** M^2 is the mean square distance from the point of birth as a fast neutron to death in thermal capture.

Section 3.10 — Neutron Life Cycle And The Six-Factor Formula

3.10.1 — Neutron Life Cycle

The power generated by a reactor is proportional to the thermal neutron density, n_{th}

. The process by which n_{th} changes with time is called neutron multiplication. Neutron multiplication can be understood by following an average group of neutrons as they slow down and diffuse. A schematic overview showing possible neutrons fates is presented in Exhibit 3-20. Some of the (original) fission neutrons induce fission reactions and liberate a new group of neutrons establishing a cycle. This second group is viewed as a new "generation" of fission neutrons. The neutron multiplication process continues in this way from generation to generation. The ratio of the number of fission neutrons produced in any two successive generations determines whether n_{th} and reactor power are constant or changing.

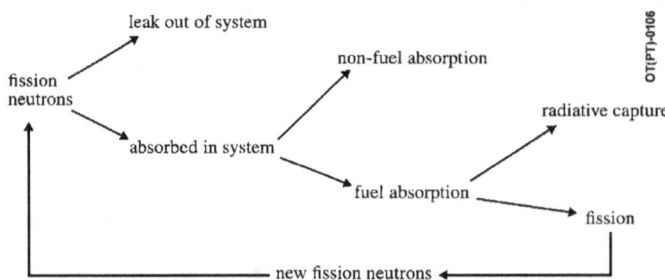

Exhibit 3-20 — Schematic Overview Of Neutron Life Cycle Showing Possible Fates

It is possible to expand the simple cycle depicted in Exhibit 3-20 and show more details and considerations when analyzing such a system. What results is the schematic of Exhibit 3-21 where significant steps in the life cycle are area blocks.

Each block represents a process which acts upon neutrons entering the process so that a gain or loss of neutrons results. The ratio of neutron output to neutron input for a process is called its multiplication factor. The ratio of the process' neutron loss to its neutron input is called its fractional neutron loss. Likewise, a fractional neutron gain is defined. The multiplication factor of a process possessing a net neutron loss is less than one, while a process having a net neutron gain has a multiplication factor greater than one.

The block with rounded corners represents the start of the life cycle for the N_i neutrons, marking the beginning of generation i. The history of the generation is followed by moving clockwise around the cycle, multiplying the number of neutrons entering each block in turn by its multiplication factor. The order in which blocks follow one another represents the most likely order of events (in a thermal reactor).

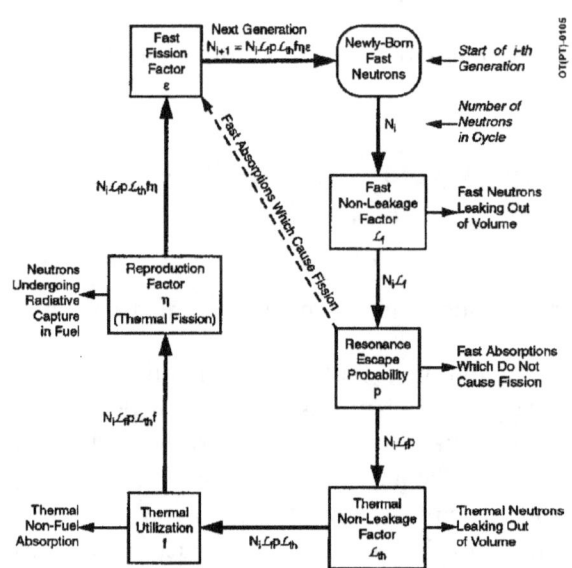

Exhibit 3-21 — Neutron Life Cycle In An Arbitrary Volume Of The Core

3.10.2 — Six-Factor Formula

When the six multiplication factors of Exhibit 3-21 are applied to the neutrons of generation i, the number of neutrons starting generation i + 1 result.

Equation 3-121

$$N_{i+1} = N_i \cdot \ell_f \cdot \ell_{th} \cdot p \cdot f \cdot \eta \cdot \varepsilon$$

The effective multiplication factor k_{eff} is defined as the ratio of the number of neutrons in one generation to the number in the preceding generation.

Equation 3-122

$$k_{eff} = \frac{N_{i+1}}{N_i}$$

Combining Equation 3-121 and Equation 3-122, the most common form of the six factor formula results:

Equation 3-123

$$k_{eff} = \ell_f \cdot \ell_{th} \cdot p \cdot f \cdot \eta \cdot \varepsilon$$

where the six factors are detailed in the following sections.

3.10.2.1 — Fast Non-Leakage Factor, ℓ_f

The **fast non-leakage factor** or probability is defined as the fraction of neutrons beginning each generation that do not leak out while slowing down. It may be expressed as the ratio:

Equation 3-124

$$\ell_f = \frac{\text{Number of Fast Neutrons which Do Not Leak Out while Slowing Down}}{\text{Number of Neutrons that Start to Slow Down}}$$

This is equivalent to:

Equation 3-125

$$f = \frac{\begin{array}{l}\text{Neutron Absorption Rate} + \\ \text{Neutron Thermalization Rate}\end{array}}{\begin{array}{l}\text{Neutron Absorption Rate} + \\ \text{Neutron Thermalization Rate} + \\ \text{Neutron Leakage Rate}\end{array}}$$

which yields the following when simple diffusion theory constants are used:

Equation 3-126

$$f = \frac{\Sigma_{rem}^{fast} + \Sigma_{a}^{fast}}{\Sigma_{rem}^{fast} + \Sigma_{a}^{fast} + D_{fast}B^2}$$

When fast absorption is considered, Equation 3-105 L_S^2 becomes

Equation 3-127

$$L_S^2 = \frac{D_{fast}}{\Sigma_{rem}^{fast} + \Sigma_{a}^{fast}}$$

Substituting Equation 3-126 and Equation 3-127 yields

Equation 3-128

$$f = \frac{1}{1 + L_S^2 B^2}$$

3.10.2.2 — Resonance Escape Probability, p

Of the slowing down neutrons which do not leak out, the resonance escape probability is the fraction that become thermalized. Expressed as a ratio, p becomes:

Equation 3-129

$$p = \frac{\text{Number of Neutrons that Reach Thermal Energy}}{\text{Number of Fast Neutrons which Do NOT Leak Out while Slowing Down}}$$

Using this definition and simple diffusion theory, p becomes

Equation 3-130

$$p = \frac{\Sigma_{rem}^{fast}}{\Sigma_{rem}^{fast} + \Sigma_a^{fast}}$$

3.10.2.3 — Thermal Non-Leakage Factor, $_{th}$

The **thermal non-leakage factor** is defined as the fraction of thermal neutrons that do not leak out of the reactor. Because thermal neutrons must either leak out of the core or be absorbed, an equivalent but more useful definition is the fraction of thermal neutrons that are absorbed in the reactor. Expressed as a ratio, $_{th}$ becomes:

Equation 3-131

$$_{th} = \frac{\text{Number of Thermal Neutrons Absorbed in the Core}}{\text{Number of Neutrons that Reach Thermal Energy in the Core}}$$

Expressed in terms of simple diffusion theory constants, $_{th}$ becomes:

Equation 3-132

$$_{th} = \frac{\Sigma_a^{th}}{\Sigma_a^{th} + D_{th}B^2}$$

Using Equation 3-112

Equation 3-133

$$_{th} = \frac{1}{1 + B^2 L^2}$$

3.10.2.4 — Thermal Utilization, f

Of all the thermal neutrons absorbed in the reactor, the fraction of these that are absorbed in the fuel is known as the **thermal utilization factor**.

Equation 3-134

$$f = \frac{\text{Number of Thermal Neutrons absorbed in the Fuel}}{\text{Number of Thermal Neutrons Absorbed in the Core}}$$

Expressed in terms of macroscopic cross sections, f becomes:

Equation 3-135

$$f = \frac{_{fuel}\Sigma_a^{th}}{_{fuel}\Sigma_a^{th} + _{nonfuel}\Sigma_a^{th}}$$

3.10.2.5 — Reproduction Factor, η

The **reproduction factor** is defined as the number of fission neutrons produced per thermal neutron absorbed in the fuel. Expressed as a ration, η becomes:

Equation 3-136

$$\eta = \frac{\text{Number of Neutrons Produced by Thermal Fission}}{\text{Number of Thermal Neutrons Absorbed by the Fuel}}$$

η is calculated by:

$$\eta = \frac{\nu \cdot {_{fuel}\Sigma_f^{th}}}{_{fuel}\Sigma_a^{th}}$$

where:

ν = the average number of fast neutrons produced per thermal fission.

3.10.2.6 — Fast Fission Factor, ε

The **fast fission factor**, ε, is the ratio of the total fission rate (fast plus thermal) to the thermal fission rate. If it is assumed that the number of neutrons released per fission is the same for fast and thermal fission (an assumption that is very good), then many ε may also be defined as the number of neutrons produced by all fissions per neutron produced by thermal fission. Expressed as a ratio:

Equation 3-137

$$\varepsilon = \frac{\text{Number of Fission Neutrons Produced by All Fission Events}}{\text{Number of Fission Neutrons Produced by Thermal Fission}}$$

As equivalent expression uses production rates:

Equation 3-138

$$\varepsilon = \frac{\text{Rate of Neutron Production by All Fission Events}}{\text{Rate of Neutron Production by Thermal Fission}}$$

Expressed in terms of simple diffusion theory constants and fluxes, ε becomes:

Equation 3-139

$$\varepsilon = \frac{v_{fast}\Sigma_f^{fast}\phi_{fast} + v_{th}\Sigma_f^{th}\phi_{th}}{v_{th}\Sigma_f^{th}\phi_{th}}$$

As previously stated,

$$v_{fast} \cong v_{th}$$

and Equation 3-139 simplifies:

Equation 3-140

$$\varepsilon = 1 + \frac{\Sigma_f^{th}\phi_{fast}}{\Sigma_f^{th}\phi_{th}}$$

The fast fission factor accounts for the fact that some of the fast neutrons absorbed while slowing down to thermal energies may result in fission.

Section 3.11 — Buckling, Leakage, And Flux Shapes

3.11.1 — Buckling

In mechanical design, the bending of a column or other such support is called **buckling**. Reactor designers borrowed this term as a measure of the overall curvature of the flux. Equation 3-141 shows the functional relationship between the flux and the buckling:

Equation 3-141

$$B^2 = -\frac{\vec{\nabla}^2\phi}{\phi}$$

where:

B^2 (not B) = the buckling

ϕ = the flux.

Mathematically, the ratio of the second derivative of a function to the function is a measure of the curvature. In terms of diffusion theory where fast and thermal fluxes are used, it is possible to define a fast buckling B_{fast}^2 and thermal buckling B_{th}^2. It is also proper to apply buckling to the core as a whole entity or to a region within a core or a region outside the core such as a reflector.

An infinite reactor system has $B^2 = 0$ and a corresponding uncurved or flat flux. Positive values of B^2 associated with the finite dimensions of a reactor core are viewed as the flat flux bending or "buckling" to match zero-flux boundary conditions. The larger the magnitude of B^2, the greater the curvature required for criticality and the smaller the geometry occupied by the flux.

If the geometry (shape and size) of a reactor is defined, the value of the buckling is fixed. For a finite, cylindrical core of height H and radius R, the buckling is:

Equation 3-142

$$B^2 = \left(\frac{\pi}{H}\right)^2 + \left(\frac{2.405}{R}\right)^2$$

As either H or R increases and the surface-to-volume ration decreases, B^2 decreases.

Buckling changes with rod motion. Consider a cylindrical core controlled with a bank of rods. The "effective" core is essentially the unrodded segment of the cylinder since the flux is greater here than in the rodded segment. As the rod bank is withdrawn, the effective height of the core increases and the buckling decreases.

3.11.2 — Leakage

In addition to absorption and scattering, a neutron may cross a region boundary (exterior boundary of a core) and escape from it. The escape of fast neutrons is called fast leakage and, like resonance absorption, reduces the chance of a neutron reaching thermal energies and continuing the chain reaction. Likewise, the escape of thermal neutrons is called thermal leakage and reduces the thermal neutron population and the chance of fissioning to continue the chain reaction.

The further a neutron travels in the process of slowing down or in thermal diffusion, the greater is its probability of encountering the region's (core's) surface and leaking out. The probability of a fast or thermal neutron leaking out is proportional to the slowing down length, L_S and diffusion length, L, respectively.

Using Equation 3-128, an expression for the fast non-leakage probability, the probability of leaking a fast neutron is:

Equation 3-143

$$1 - {}_f = 1 - \frac{1}{1 + L_S^2 B^2} = \frac{L_S^2 B^2}{1 + L_S^2 B^2}$$

Using Equation 3-128, an expression for the thermal non-leakage probability, the probability of thermal neutrons leaking is:

Equation 3-144

$$1- \text{\small th} = 1-\frac{1}{1+L^2B^2} = \frac{L^2B^2}{1+L^2B^2}$$

The thermal non-leakage probability in a reflected core is normally greater than unity since thermal neutrons "leak" into the core from the reflector. Therefore, since $\text{\small th}$ can be greater than 1.0, it is proper to call $\text{\small th}$ the thermal non-leakage <u>factor</u> rather than non-leakage <u>probability</u>.

3.11.3 — Flux Shapes

From the diffusion equation (see Equation 3-104 or Equation 3-111), $DB^2\phi$ represents the neutron leakage out of or into a region. Therefore, $DB^2\phi$ assumes the role of "macroscopic leakage cross section" with units of inverse distance (cm^{-1}). As the surface-to-volume ratio increases, B^2 increases and the leakage from the core increases.

In any material, $D > 0$. Since a negative flux has no physical significance, $\phi > 0$. Since the leakage rate in any region may be either positive, zero, or negative depending on the balance between neutron production and loss rates, it follows that B^2 can be positive, zero, or negative and that B^2 has the same sign as the leakage rate.

The sign of $\nabla^2\phi$ has a simple relationship to the shape of $\phi(x)$: in a plot of $\phi(x)$ versus x, the sign of $\nabla^2\phi$ is directly related to the direction of curvature. When $\nabla^2\phi$ is negative and B^2 is positive, $\phi(x)$ versus x has a negative curvature downward as shown in Exhibit 3-22. Conversely, when $\nabla^2\phi$ is positive and B^2 is negative, $\phi(x)$ versus x has a positive curvature and is concave upward.

In a region where B^2 is positive, positive leakage exists due to neutron production exceeding neutron loss. In a region where B^2 is negative, negative leakage exists due to neutron loss exceeding neutron production. In general, a reactor contains both regions where $B^2 > 0$ and regions where $B^2 < 0$. In order to accomplish the transfer of neutrons with $B^2 > 0$ to a region with $B^2 < 0$, a neutron current must be present at the boundary between the two regions; this current must flow from the region with $B^2 > 0$ into the region with $B^2 < 0$. As illustrated in Exhibit 3-22, the larger the curvature, hence the buckling, the more neutrons flow between regions. For the reactor as a whole, B^2 values for individual regions are summed to produce an overall value for the core.

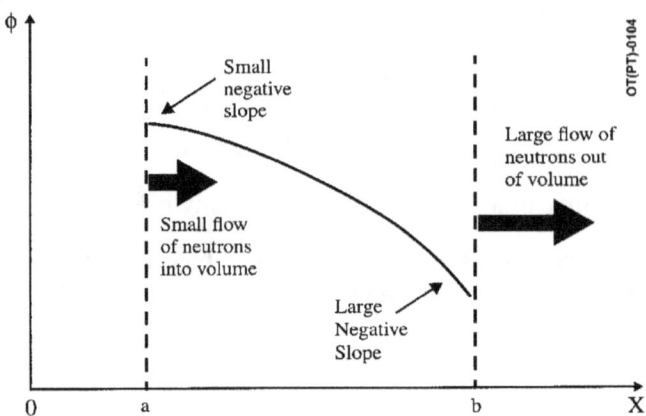

Exhibit 3-22 — Variation Of Flux Showing Degree Of Leakage

Because the distribution of the neutron flux in a three-dimensional, heterogeneous reactor is complex, it is helpful to first consider the flux distribution of much simpler cases.

3.11.3.1 — Bare Homogeneous Reactor

A homogeneous reactor contains no discrete regions; instead, all core materials are "homogenized" together to form a single region comprising the entire core.

Because a bare homogeneous reactor has no material surrounding its external surfaces, neutrons crossing this surface have no chance of scattering back into the reactor and are lost. The resulting fast and thermal leakage causes the neutron flux at the reactor boundaries to be very low. Neutrons near the center of a reactor, on the other hand, have relatively little chance of leaking out and are more likely to be thermalized before reaching the reactor's exterior surface. Therefore, the neutron flux tends to be high near the center of the reactor. For a bare homogeneous cylindrical reactor, the neutron flux is greatest at the reactor center and decreases smoothly radially outward from the centerline. Exhibit 3-23 illustrates this radial neutron flux shape. The flux just outside the reactor surface is not equal to zero due to the neutrons leaking from the reactor.

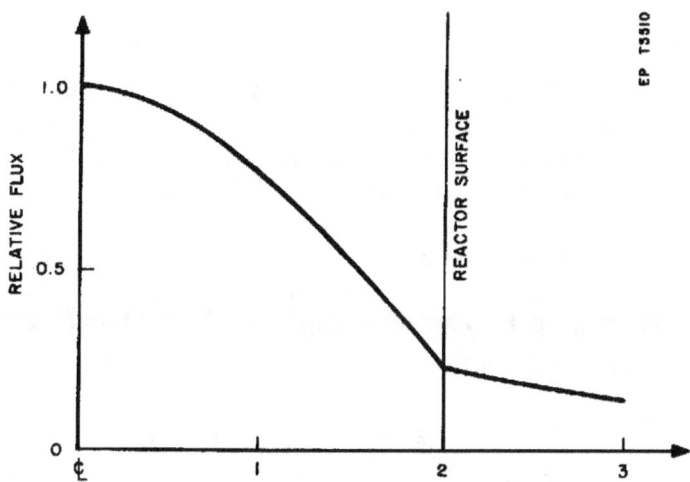

Exhibit 3-23 — Relative Neutron Flux Versus Radial Distance From A Bare Homogeneous Cylindrical Reactor's Centerline

Exhibit 3-24 shows a similar plot for the axial neutron flux shape, the flux at each point along the vertical axis of the cylindrical reactor. Like the radial flux, the axial flux is highest at the reactor center and decreases to lower values at the top and bottom boundaries of the reactor.

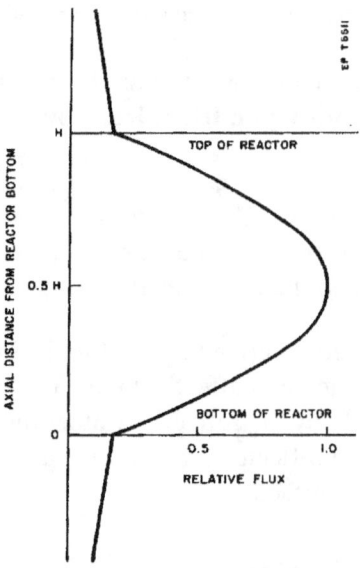

Exhibit 3-24 — Relative Neutron Flux Versus Axial Position For A Bare Homogeneous Cylindrical Reactor Of Height H

In a bare homogeneous reactor, the fast and thermal fluxes have very nearly the same shape, since both are subject to the processes of diffusion, leakage, etc. Therefore, Exhibit 3-23 and Exhibit 3-24 are representative of both fast and thermal fluxes.

3.11.3.2 — Reflected Reactor

Actual reactors are surrounded by an unfueled region called a reflector. Because it is mostly composed of moderator, it has a large scattering but small absorption cross section. As a result, fast neutrons leaking from the reactor are scattered repeatedly in the reflector where a large fraction are slowed down and returned to the reactor as thermal neutrons with a chance to contribute to the chain reaction.

The simplest reflected reactor consists of a homogeneous cylindrical core surrounded by a reflector of pure water. The neutron flux distribution associated with a bare homogeneous reactor is changed markedly by the addition of a reflector because of the slowing down of neutrons (that occurs in the reflector adjacent to the core). A typical thermal flux shape resulting from the addition of a reflector is illustrated in Exhibit 3-25.

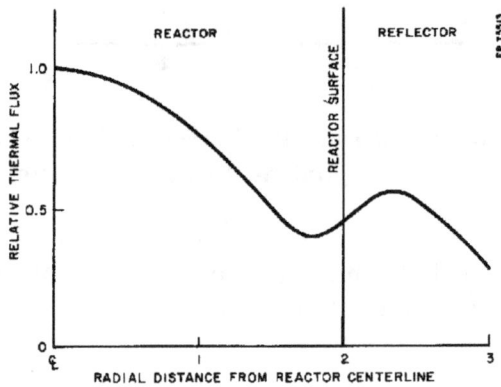

Exhibit 3-25 — Relative Neutron Flux Versus Radial Distance from a Reflected Homogeneous Cylindrical Reactor's Centerline

The leakage of thermal neutrons back into the core corresponds to a thermal neutron flux shape which slopes downward from the reflector, across the core surface, and into the core. The slope of the thermal neutron flux at the core's surface in Exhibit 3-25 is opposite the slope at the core's surface in Exhibit 3-23. This is represented by assigning a negative value to the core's thermal buckling.

The reflector's presence has only a small effect upon the fast flux inside the reactor. The buckling for the fast energy range may therefore be considered to be unaffected by the presence of the reflector. By contrast, the fast flux outside the reactor is affected strongly by the reflector. Rather than slowly decrease, as seen in Exhibit 3-23 and Exhibit 3-24, the fast flux from a reflected reactor decreases quicker in a linear fashion to zero within a short distance from the reactor's surface.

Section 3.12 — Multiplication Factor

3.12.1 — Criticality

A core is said to be **critical** when it can sustain a neutron chain reaction with no net change of free neutrons in the core. The average number of neutrons produced in each generation is exactly balanced by the average number of neutrons lost in each generation through absorption and leakage. Under these conditions, core power remains constant. If more neutrons are generated than needed (production exceeds loss), the core is

supercritical and power increases over time. If too few neutrons are generated to sustain the chain reaction (loss exceeds production), the core is **subcritical** and power decreases over time.

In Paragraph 3.10.2, k_{eff} was defined as the ratio of the number of neutrons in one generation to the number of neutrons in the preceding generation.

Equation 3-145

$$k_{eff} = \frac{neutrons\,in\,generation\,(i+1)}{neutrons\,in\,generation\,i}$$

For a critical system where neutron production equals neutron loss (leakage plus absorption):

Equation 3-146

$$k_{eff} = 1 = \frac{neutron\,production}{leakage + absorption}$$

When the core is supercritical,

Equation 3-147

$$k_{eff} > 1$$

When the core is subcritical,

Equation 3-148

$$k_{eff} < 1$$

3.12.2 — Reactivity

The term reactivity, sometimes symbolized with δk (read "delta k"), is a convenient means of specifying the extent that a reactor departs from criticality. It is defined by Equation 3-149.

Equation 3-149

$$\delta k = \frac{k_{eff} - 1}{k_{eff}}$$

Reactivity is positive for a supercritical reactor, negative for a subcritical reactor, and zero for a critical reactor.

A more physical interpretation of reactivity occurs when Equation 3-145 and Equation 3-149 are combined. Reactivity may be defined as the fractional change in neutron population per generation relative to its final value:

Equation 3-150

$$\delta k = \frac{\left(\begin{array}{c}\text{Population of}\\\text{Generation}(i+1)\end{array}\right) - \left(\begin{array}{c}\text{Population of}\\\text{Generation}\,i\end{array}\right)}{\text{Population of Generation}(i+1)}$$

δk is usually stated in units of 10^{-4} as:

   1. $\delta k = +10 \times 10^{-4}\ \delta k$
   2. $\delta k = -20 \times 10^{-4}\ \delta k$

δk is commonly reported in percent obtained by multiplying by 100:

Equation 3-151

$$\delta k(\%) = 100\ \delta k$$

Some examples are:

   1.   plus one-tenth percent $\delta k (\delta k = +10 \times 10^{-4}\ \delta k = 0.1\%)$
   2.   minus two percent $\delta k (\delta k = -200 \times 10^{-4}\ \delta k = -2.0\%)$

A change in reactivity can be referred to as $\Delta\delta k$ (read, "delta delta k"). For example,

Equation 3-152

$$\Delta\delta k = -300 \times 10^{-4}\ \delta k = -3.0\%\ \delta k$$

Table 3-3 — Relationship Between Effective Multiplication Factor, Reactivity, And Criticality State Of A System

Multiplication	Reactivity	State of Criticality
$k_{eff} > 1.0$	$\delta k > 0$	supercritical
$k_{eff} = 1.0$	$\delta k = 0$	critical
$k_{eff} < 1.0$	$\delta k < 0$	subcritical

The reactivity of a given core is determined by many factors. Some of these factors, such as core size and shape, number of control rods, amount of fuel and poison present in the core at the beginning of life, operating temperature and pressure, etc. are fixed by the designer. The remaining factors fall into two categories. The first category consists of those factors which reflect the changes that take place in the core as a result of energy generation. The amount of fuel and poison depleted and the accumulation, depletion, and radioactive decay of fission products such as ^{135}Xe are examples of these factors. Changes in reactivity due to these factors are more gradual and noticeable on time scales of hours to days. The second category contains those factors over which the operator has control, such as average primary coolant temperature and control rod position. Changes in reactivity due to these factors are much quicker and noticeable on time scales of seconds.

3.12.3 — Definitions Utilizing Reactivity

As a means of characterizing reactivity in certain situations, various parameters and concepts are used. The following subsections describe some of the more common ones.

3.12.3.1 — Excess Reactivity

At any time in life, the core has an excess of positive reactivity that varies with conditions such as xenon concentration, moderator temperature, etc. The amount of positive reactivity in a core decreases with lifetime due primarily to fuel and poison depletion and fission product buildup. When this positive reactivity no longer exists, the core cannot operate. This positive reactivity must be compensated for with the negative reactivity of the control rods.

Excess reactivity is defined as:

Equation 3-153

$$\delta k_{ex} = \text{core (positive) reactivity with } \underline{\text{all}} \underline{\text{rods}} \underline{\text{out}}$$

3.12.3.2 — Reactivity Margin

Reactivity margin is defined as the excess reactivity (for the core) under hot, maximum xenon plus samarium conditions.

3.12.3.3 — Excess Multiplication Factor, k_{ex}

Similar to δk_{ex} is the **excess multiplication factor**, defined as:

Equation 3-154

$$k_{ex} = k_{eff} - 1.0$$

k_{ex} is often stated in percent; for example, k_{ex} is 0.02 or 2.0 percent.

3.12.3.4 — Shutdown Margin, K_{SD}

It is important that there be enough negative reactivity in the control rods to assure complete shutdown of the chain reaction in the reactor at all times during core life. In other words, insertion of the control rods must ensure subcriticality ($k_{eff} < 1.0$). In practice, a **shutdown margin** of several percent in reactivity is desirable. **Shutdown margin** is the amount of negative reactivity when compared to the critical condition of $k_{eff} = 1.0$ attained by the core when control rods are fully inserted. Usually, a margin on the order of 1 to 2 percent δk is desired.

Section 3.13 — Temperature Coefficient Of Reactivity

3.13.1 — Core Reactivity And α_T

As a core is constructed of components, so is the core's reactivity δk.

Equation 3-155

$$\delta k = \delta k_{fuel} + \delta k_{bp} + \delta k_{rods} + \delta k_{fpp} + \delta k_{Xe} + \delta k_{Sm}$$

where:

δk_{fuel}	= reactivity of the fuel
δk_{bp}	= reactivity of the burnable poison
δk_{rods}	= reactivity of the control rods
δk_{Xe}	= reactivity of xenon
δk_{Sm}	= reactivity of the samarium
δk_{fpp}	= reactivity of the fission product poisons (excluding xenon and samarium).

Each of these reactivity terms is dependent on the six multiplication factors that determine k_{eff}. Because these multiplication factors (except for η) are affected by moderator temperature, the value of each of the terms in Equation 3-155 depends upon the primary coolant temperature. It is customary to reference each term of Equation 3-155 to the **normal operating temperature** (NOT) of the primary coolant and then account for any difference between <u>actual</u> average coolant temperature and NOT by another reactivity term called the temperature reactivity, $\delta k_{T_{ave}}$. It is necessary to base temperature reactivity upon the <u>average</u> coolant temperature, T_{ave} because the temperature of the moderator in an operating reactor varies from point to point depending upon local power density and heat transfer. With the addition of the temperature reactivity term, Equation 3-155 becomes:

Equation 3-156

$$\delta k = \delta k_{fuel} + \delta k_{bp} + \delta k_{rods} + \delta k_{fpp} + \delta k_{Xe} + \delta k_{Sm} + \delta k_{T_{ave}}$$

and is the general equation for determining core reactivity.

A change in any of the terms on the right-hand side of Equation 3-156 can cause the core reactivity δk to change, resulting in the addition of positive or negative reactivity. The reactivity balance equation,

Equation 3-157

$$\Delta \delta k = \Delta \delta k_{fuel} + \Delta \delta k_{bp} + \Delta \delta k_{rods} + \Delta \delta k_{fpp} + \Delta \delta k_{Xe} + \Delta \delta k_{Sm} + \Delta \delta k_{T_{ave}}$$

expresses the change in core reactivity resulting from a change in one or more of the individual reactivity terms. For example, if δk is the core reactivity at a given average primary coolant temperature, T_{ave}, then the reactivity addition due solely to the average primary coolant temperature Δ_{Tave} is $\Delta \delta k = \Delta \delta k_{Tave}$. The amount of temperature reactivity added by $\Delta \delta k_{Tave}$ is related to the primary coolant temperature change ΔT_{ave} by the relation

Equation 3-158

$$\Delta \delta k_{T_{ave}} = \alpha_T \Delta T_{ave}$$

where α_T is called the **temperature coefficient of reactivity**. Since a change in temperature reactivity results in an equal change in core reactivity $\Delta \delta k$, it is customary to write Equation 3-158 in the form

Equation 3-159

$$\Delta \delta k = \alpha_T \Delta T_{ave}$$

or

Equation 3-160

$$\alpha_T = \frac{\Delta \delta k}{\Delta T_{ave}}$$

The temperature coefficient of reactivity α_T is defined as the amount of reactivity added per degree F change in T_{ave}. α_T is usually expressed in units of 10^{-4} $\delta k/F$ and is not constant but varies with moderator temperature.

3.13.2 — Importance Of α_T

The response of a reactor to a change in temperature depends on the algebraic sign of α_T. If α_T is positive, $\Delta \delta k / \Delta T$ is positive and an increase in T leads to an increase in δk. An increase in δk leads to an increase in the power level of the reactor, giving rise to a further increase in the temperature, another increase in δk, and so on. Thus, with α_T positive, an increase in temperature leads to a continuous cycle until either the reactor is shutdown by outside intervention or an accident results.

If α_T is positive, a decrease in T leads to a decrease in δk. This reduces the reactor power, which reduces the temperature, giving a further decrease in δk, and so on, until the reactor eventually shuts down.

The situation is quite different when α_T is negative. With $\Delta \delta k / \Delta T$ negative, an increase in T gives a decrease in δk. A decrease in δk leads to a decrease in power, which tends to decrease the temperature and return the reactor to its original state. Furthermore, a decrease in T results in an increase in δk, so that if T goes down, the power goes up and the reactor tends to return to its original state.

A reactor with a positive α_T is inherently unstable to changes in core temperature while a reactor possessing a negative α_T is inherently stable. However, if α_T is too negative or not negative enough, undesirable consequences can result. If α_T is not sufficiently negative, the change in reactor power in response to a change in steam demand may be too sluggish. Yet a too negative α_T can be a liability under abnormal conditions. For example, a steam line rupture in which a large steam demand causes T_{ave} to decrease and add positive reactivity due to temperature, can cause reactor power to increase to a dangerous level.

3.13.3 — Equation For α_T

Rewriting Equation 3-160 as:

Equation 3-161

$$\alpha_T = \frac{\delta k_2 - \delta k_1}{T_2 - T_1}$$

and applying Equation 3-149, the following expression for α_T results.

Equation 3-162

$$\alpha_T = \frac{1}{k_{eff}^2} \bullet \frac{dk}{dT}$$

Using Equation 3-123 for the six-factor formula in Equation 3-162 yields for α_T:

Equation 3-163

$$\alpha_T = \frac{1}{k_{eff}} \left[\frac{1}{\varepsilon}\frac{\partial \varepsilon}{\partial T} + \frac{1}{p}\frac{\partial p}{\partial T} + \frac{1}{f}\frac{\partial f}{dT} + \frac{1}{\eta}\frac{\partial \eta}{\partial T} + \frac{1}{_f}\frac{\partial_{\ f}}{\partial T} + \frac{1}{_{th}}\frac{\partial_{\ th}}{\partial T} \right]$$

With some simplifying assumptions, Equation 3-163 can be written as:

Equation 3-164

$$\alpha_T = \frac{\beta^{mod}}{k_{eff}} \left[\left(\frac{\varepsilon-1}{\varepsilon}\right) - (1-p) + \left(\frac{\Sigma_a^{mod}}{\Sigma_a^T} - \frac{\Sigma_a^{lp}}{2\Sigma_a^T}\right) - (1 - {}_f) - (1 - {}_{th}) \right]$$

β_{mod} is the volumetric coefficient of expansion, the fractional change in coolant density resulting from a one °F change in T_{ave}. Σ_a^{lp} is the macroscopic thermal absorption cross section in the lumped poison. Σ_a^{mod} is the total macroscopic absorption cross section of the moderator. Σ_a^T is the total macroscopic absorption cross section of the core.

While Equation 3-164 involves many simplifying assumptions and cannot be used to calculate α_T in response to changes in the individual multiplication factors. In particular, variation in the temperature coefficient over core life may be estimated using Equation 3-164.

3.13.4 — Behavior With Increasing Temperature

The temperature coefficient of reactivity α_T is a consequence of the effect of a T_{ave} change upon the individual factors comprising the six-factor formula. α_T is equal to the sum of the reactivities added by the change in each of these six factors due to one °F increase in T_{ave}.

As discussed in Paragraph 3.9.3, L_s^2 and L^2 increase with increasing temperature due to a decrease in density with temperature. As a consequence, $_f$ and $_{th}$ decrease (with temperature), contributing negative terms to α_T.

In a simple reactor with no lumped poisons, f increases slightly as temperature increases since there is less moderator absorption as the density decreases. Thus,

$$\frac{\partial f}{\partial T}$$

is slightly positive.

In a reactor with lumped poisons, the effectiveness of the lumped poison depends not just upon the lumped poison's macroscopic absorption cross section but also upon the thermal diffusion length of the region just outside the poison lump. First, all (fuel and poison alike) absorption cross sections are decreasing as the average neutron thermal energy increases with increasing temperature. Second, the absorption rate is determined by:
1.   how much of the lump is exposed to incoming thermal neutrons (determined by the surface area) and
2.   the rate at which thermal neutrons can reach its surface, which is controlled by the thermal diffusion length

As L increases with T_{ave}, more neutrons are able to reach the lump, increasing the absorption rate. Overall, this means that lumped poisons are not as self-shielded at higher temperatures and absorb more neutrons. Therefore, the fraction of neutrons absorbed by the fuel decreases and

$$\frac{\partial f}{\partial T}$$

is negative.

Since L_s^2 increases with an increase in temperature, the fast fission factor ε increases because the neutrons travel further, spending more time in the resonance region, and increasing the probability of fast fission.

Since the neutrons travel farther while slowing down at higher temperature, they have a higher probability of leaking from the core. Therefore, the resonance escape probability p decreases.

The reproduction factor η is independent of T_{ave} since the average number of neutrons liberated in fission is independent of temperature and changes in macroscopic cross sections cancel one another. Therefore, its $\Delta\delta k/F$ is zero and does not contribute to α_T

Table 3-4 — Parameter Trends As Temperature Increases

T ↑		
$_f \downarrow$	$\varepsilon \uparrow$	$_{th} \downarrow$
$f \uparrow$	$p \downarrow$	$\eta \leftrightarrow$
$\alpha \downarrow$		

Since those factors from the six-factor formula that decrease with increasing temperature out-weigh those factors that increase, α_T decreases or becomes more negative as temperature increases.

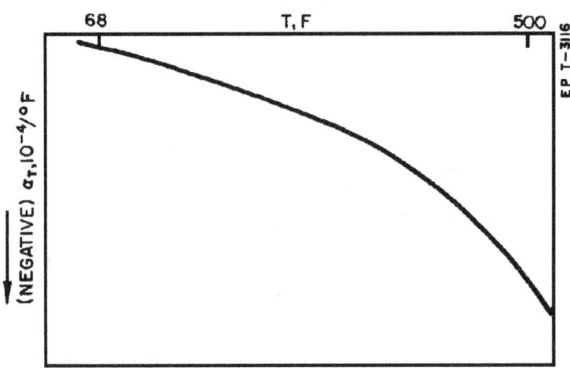

Exhibit 3-26 — General Shape Of α_T With Temperature

3.13.5 — Behavior With Lifetime

The following effects cause α_T to increase (become slightly less negative) over the life of the core:

1. The reproduction factor η does not change appreciably.
2. The fast fission factor ε decreases because the macroscopic resonance fission cross section decreases.
3. Thermal utilization increases because lumped poison burns out.
4. The fast non-leakage factor, $_f$ increases as a result of two competing effects. First, the slowing down length L_S increases with core life, which tends to decrease $_f$. Second, the rods are near the top of the core which decreases buckling, tending to increase $_f$. The effect of the buckling is dominant and $_f$ increases.
5. The thermal non-leakage factor $_{th}$ decreases. This is primarily due to the ratio of thermal flux in the core to thermal flux in the reflector increasing with core life. This makes diffusion of neutrons from the core to the reflector more likely, increasing the thermal leakage and making $_{th}$ decrease.

6. The resonance escape probability p increases. This is due to less absorption of fast neutrons in the resonance region of the fuel and lumped poisons. Therefore, there is an increased probability that a fast neutron slows down through the resonance region without being absorbed.

3.13.6 — Behavior With N^{Mod}/N^{U235}

In a simple but large reactor with low buckling B^2 and little fuel, it is possible for the core to have a positive temperature coefficient at low temperatures. Under these conditions a high N^{Mod}/N^{U235} results and α_T gains a positive contribution from

$$\frac{\partial f}{\partial T}.$$

Section 3.14 — Fission Products

During the course of reactor operation, fission fragments and their many decay products accumulate. Some that are stable or possess reasonably long half-lives also have thermal absorption cross sections. These nuclei act as reactor poisons and affect the multiplication factor, chiefly by decreasing thermal utilization. The concentration of fission product poisons in a reactor is related to the thermal neutron flux. Consequently, when the reactivity is changed, so that there is an accompanying change in the flux, the concentration of fission products are affected, although at a much slower rate, which in turn influences the reactivity. Two especially important fission products poisons are ^{135}Xe and ^{149}Sm.

3.14.1 — Xenon – 135

^{135}Xe is produced directly from fission with a yield of $\gamma^{Xe} \approx 0.3\%$ and by the radioactive decay of ^{135}I, which for practical purposes has a direct yield form a fission of $\gamma^I \approx 6.1\%$. ^{135}Xe is unstable and decays with a half-life of 9.2 hours to ^{135}Cs. It can also be removed by thermal absorption. Its thermal absorption cross section (at 0.0253 eV) of 2.7×10^6 barns is the largest among thermal absorbers and is the chief reason that ^{135}Xe is a good fission product poison.

The differential equations governing the concentrations of xenon and iodine at any time in the reactor are:

Equation 3-165

$$\frac{dN^{Xe}}{dt} = \gamma^{Xe} \Sigma_f^{U235} \phi + N^I \lambda^I - \Sigma_a^{Xe} \phi - N^{Xe} \lambda^{Xe}$$

and

Equation 3-166

$$\frac{dN^I}{dt} = \gamma^I \Sigma_f^{U235} \phi + N^I \lambda^I$$

These equations can be interpreted as:

Equation 3-167

$$\begin{pmatrix} \text{xenon time} \\ \text{rate of change} \end{pmatrix} = \begin{pmatrix} \text{production from} \\ \text{U-235 fission} \end{pmatrix} + \begin{pmatrix} \text{production from} \\ \text{iodine decay} \end{pmatrix} - \begin{pmatrix} \text{destruction by} \\ \text{absorption} \end{pmatrix} - \begin{pmatrix} \text{destruction by} \\ \text{decay to Cs-135} \end{pmatrix}$$

and

Equation 3-168

$$\begin{pmatrix} \text{iodine time} \\ \text{rate of change} \end{pmatrix} = \begin{pmatrix} \text{production from} \\ \text{U} - 235 \end{pmatrix} - \begin{pmatrix} \text{destruction by} \\ \text{decay to Xe} - 135 \end{pmatrix}$$

3.14.1.1 — Equilibrium Concentrations

Equilibrium conditions exist when the time rate of change for the concentration is zero (or when sum of the production terms equal the sum of the destruction terms). By setting the left hand sides of Equation 3-165 and Equation 3-166 equal to zero, expressions for the equilibrium concentrations of iodine and xenon can be determined. Therefore,

Equation 3-169

$$N_{Eq}^{Xe} = \frac{\left(\gamma^I + \gamma^{Xe} \right) \Sigma_f^{U235} \phi}{\lambda^{Xe} + \sigma_a^{Xe} \phi}$$

and

Equation 3-170

$$N_{Eq}^I = \frac{\gamma^I \Sigma_f^{U235} \phi}{\lambda^I}$$

where:

N_{Eq}^{Xe} = the ^{135}Xe equilibrium concentration

N_{Eq}^I = the ^{135}I concentration.

The value of the equilibrium xenon concentration is power dependent. In the denominator of Equation 3-169, the λ^{Xe} term dominates the low fluxes from here N_{Eq}^{Xe} increases with ϕ. As ϕ increases to moderate values, the

$\sigma_a^{Xe}\phi$ term in the denominator dominates as the flux destroys the xenon faster than it decays. As ϕ approaches higher values, N_{Eq}^{Xe} saturates to a value of:

$$((\gamma^I + \gamma^{Xe})\Sigma_f^{U235})/\sigma_a^{Xe}$$

Equation 3-170 shows that the equilibrium concentration of iodine increases linearly as the flux increases. The equilibrium concentration of xenon and iodine as a function of flux (power) is shown in Exhibit 3-28.

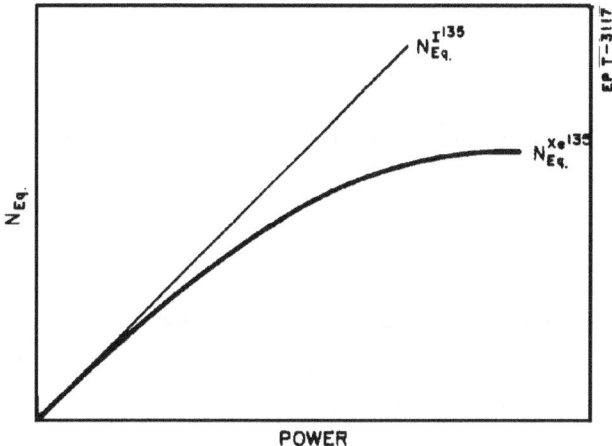

Exhibit 3-27 — Equilibrium Concentrations Of ^{135}I And ^{135}Xe

The buildup to an equilibrium xenon concentration requires 40 to 50 hours. This buildup behavior and the xenon equilibrium concentration value as a function of power are illustrated in Exhibit 3-28.

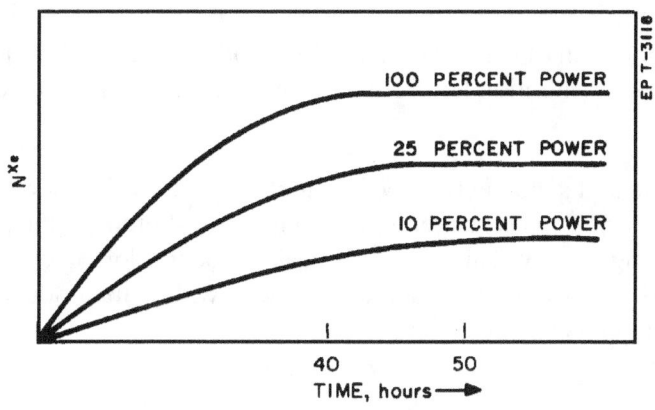

Exhibit 3-28 — Power And Time Dependence Of ^{135}Xe Concentration

3.14.1.2 — Xenon Concentration After Shutdown

When the reactor is shutdown so there is no power (flux) production, Equation 3-165 reduces to

Equation 3-171

$$\frac{dN^{Xe}}{dt} = N^I \lambda^I - N^{Xe} \lambda^{Xe}$$

The general solution of this equation is:

Equation 3-172

$$N^{Xe}(t) = \frac{\lambda^I}{\lambda^I - \lambda^{Xe}} N_0^I \left(e^{-\lambda^{Xe}t} - e^{-\lambda^I t} \right) + N_0^{Xe} e^{-\lambda^{Xe}t}$$

where:

N_0^I = the initial concentration of iodine

N_0^{Xe} = the initial concentration of xenon.

When Equation 3-172 is used to analyze shutdown concentrations after the reactor established equilibrium concentrations of iodine and xenon, it becomes

Equation 3-173

$$N^{Xe}(t) = \frac{\lambda^I}{\lambda^I - \lambda^{Xe}} N_{Eq}^I \left(e^{-\lambda^{Xe}t} - e^{-\lambda^I t} \right) + N_{Eq}^{Xe} e^{-\lambda^{Xe}t}$$

The first term represents production of (indirect) xenon from iodine and its subsequent decay. The last term represents the (direct) xenon concentration left from the decay of the equilibrium xenon originally present at shutdown.

If equilibrium levels of iodine and xenon existed before shutdown, the value of N^I at shutdown in Equation 3-173 is larger than the value of N^{Xe}. Therefore, immediately after shutdown, the production rate of xenon is greater than the loss rate through decay. This leads to a peak concentration of xenon called peak xenon (max xenon if the reactor was at 100% power before shutdown), after which the production of xenon diminishes since the concentration of iodine is reduced.

The first term of Equation 3-173 dominates at high powers when N_{Eq}^I is much larger than N_{Eq}^{Xe}. Therefore, the magnitude of the xenon peak is a function of N_{Eq}^I / N_{Eq}^{Xe}. The time to reach peak xenon is a function of the flux before shutdown. As the flux increases, the time to peak asymptotically approaches 11.1 hours.

Exhibit 3-29 shows, as a function of reactor power, the xenon concentration building to equilibrium, peaking after shutdown, and decaying exponentially.

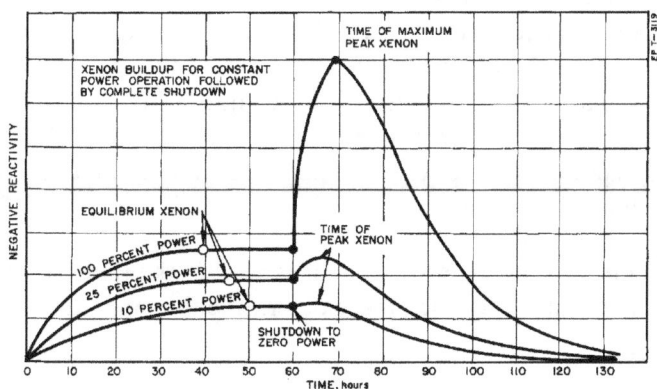

Exhibit 3-29 — Typical ^{135}Xe Buildups And Decays

3.14.1.3 — Variation With Lifetime

From Equation 3-170, it can be seen that N_{Eq}^{I} increases linearly with power.

Equation 3-174

$$N_{Eq}^{I} = \text{constant} \times \text{power}$$

If power is held constant over core lifetime, N_{Eq}^{Xe} remains constant.

For higher flux operations, the N_{Eq}^{I} concentration is given by:

Equation 3-175

$$N_{Eq}^{Xe} = \frac{\left(\gamma^{I} + \gamma^{Xe}\right)\Sigma_{f}^{U235}}{\sigma_{a}^{Xe}}$$

Over an entire core lifetime, N^{U235} (hence Σ^{U235}) may decrease by as much as 50%; the saturation value of N_{Eq}^{I} tends toward 50% of its initial value. At lower fluxes, the decrease in N_{Eq}^{I} is less dramatic since Equation 3-175 is no longer appropriate.

3.14.1.4 — Reactivity Considerations

The variation in equilibrium xenon reactivity over core life is small compared to the variation in peak xenon reactivity. For practical purposes, equilibrium xenon reactivity can be considered constant. However, equilibrium xenon reactivity tends to increase slightly (become more negative) with core life since the concentration of ^{235}U is decreasing faster than the concentration of ^{135}Xe.

The xenon reactivity following a shutdown is the sum of the direct and the indirect components. At the time of shutdown, the direct component is equal to the equilibrium xenon reactivity just prior to shutdown. This component decays with a half-life of 9.09 hours as the shutdown continues. The indirect component is initially zero at the time of shutdown and arises from the decay of the equilibrium concentration of iodine present in the core at the time of shutdown. This indirect component rises to a maximum at about 11.11 hours after shutdown and then decays to essentially zero over the next 50 hours. Direct and indirect components together trend to reduce the time to peak xenon after shutdown.

The peak xenon reactivity attained following a shutdown, as well as the time to reach the peak, depends on the ^{135}Xe and ^{135}I concentrations at shutdown, since these determine the magnitudes of the direct and indirect components.

The indirect reactivity that develops from the shutdown ^{135}I is inversely proportional to the amount of fuel remaining in the core. For this reason, the indirect xenon reactivity following shutdown from a given power increases over core life. In turn, this increases the magnitude of the peak xenon reactivity and lengthens time after shutdown.

3.14.2 — Samarium – 149

^{149}Sm is not produced directly is fission, but from the radioactive decay of ^{149}Pm (half-life of 53.1 hour) which possesses a direct fission yield of 1.07%. ^{149}Sm is stable and can only be removed by neutron absorption. Its thermal absorption cross section (at 0.0253 eV) 4.08×10^4 barns.

The differential equations governing the concentrations of samarium and promethium at any time in the reactor are:

Equation 3-176

$$\frac{dN^{Sm}}{dt} = \lambda^{Pm}N^{Pm} - \Sigma_a^{Sm}\phi$$

and

Equation 3-177

$$\frac{dN^{Pm}}{dt} = \gamma^{Pm}\Sigma_f^{U235}\phi - \lambda^{Pm}N^{Pm}$$

These equations can be respectively interpreted term by term as:

Equation 3-178

$$\begin{pmatrix} \text{samarium time} \\ \text{rate of change} \end{pmatrix} = \begin{pmatrix} \text{production from} \\ \text{promethium} \end{pmatrix} - \begin{pmatrix} \text{destruction by} \\ \text{absorption} \end{pmatrix}$$

and

Equation 3-179

$$N^{Sm} = N_{Eq}^{Sm} + N_{Eq}^{Pm}$$

3.14.2.1 — Equilibrium Concentrations

By setting the left hand sides of Equation 3-176 and Equation 3-177 to zero, expressions for the equilibrium concentrations of samarium and promethium can determined. Therefore,

Equation 3-180

$$N_{Eq}^{Sm} = \frac{\gamma^{Pm} \Sigma_f^{U235}}{\sigma_a^{Sm}}$$

and

Equation 3-181

$$N_{Eq}^{Pm} = \frac{\gamma^{Pm} \Sigma_f^{U235} \phi}{\lambda^{Pm}}$$

where:

N_{Eq}^{Pm} = the ^{149}Pm equilibrium concentration.

N_{Eq}^{Sm} = the ^{149}Sm equilibrium concentration

Equation 3-180 shows that the equilibrium concentration of promethium is power dependent. However, the equilibrium concentration of samarium is power independent. Examination of Equation 3-181 shows that there is no flux dependence. The ratio $\sigma_f^{U235}/\sigma_a^{Sm}$ is fairly constant over core life. Therefore, N_{Eq}^{Sm} is proportional to the fuel concentration and decreases throughout core life.

3.14.2.2 — Samarium Concentration After Shutdown

When the reactor is shutdown so that there is no power (flux) production, Equation 3-176 reduces to

Equation 3-182

$$\frac{dN^{Sm}}{dt} = N^{Pm}\lambda^{Pm}$$

The general solution of this equation is:

Equation 3-183

$$N^{Sm}(t) = N_0^{Pm}(1 - e^{-\lambda^{Pm}t}) + N_0^{Sm}$$

where:

N_0^{Pm} = the initial concentration of promethium

N_0^{Sm} = the initial concentration of samarium.

When Equation 3-183 is used to analyze shutdown concentrations after the reactor established equilibrium concentrations of promethium and samarium, it becomes,

Equation 3-184

$$N^{Sm}(t) = N_{Eq}^{Pm}(1 - e^{-\lambda^{Pm}t}) + N_{Eq}^{Sm}$$

The first term represents build up of promethium. At times long after shutdown (greater than 20 days), Equation 3-184 simplifies to:

Equation 3-185

$$N^{Sm} = N_{Eq}^{Sm} + N_{Eq}^{Pm}$$

Since samarium does not decay, this concentration exists until the reactor resumes power operations and samarium is depleted with the neutron flux.

Exhibit 3-30 shows typical behavior for ^{149}Sm concentration as it builds up from BOL, reaches saturation, and saturates a second time following an increase after reactor shutdown.

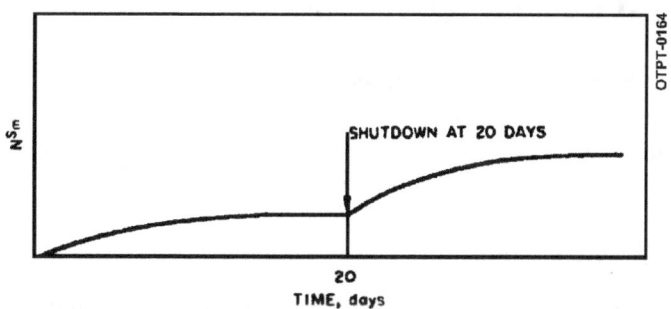

Exhibit 3-30 — Time Dependence Of ^{149}Sm

3.14.2.3 — Reactivity Considerations

Equilibrium samarium reactivity remains relatively constant over core life.

^{149}Sm is a stable isotope and is removed from the core only by depletion. Following a shutdown from 100% power, the magnitude of samarium reactivity increases to its maximum value and remains there. Subsequent power operation depletes the resulting ^{149}Sm until the equilibrium samarium reactivity is restored. The time to maximum samarium is about 20 days and is independent of core age.

3.14.3 — Other Fission Products

Fission yield, radioactive decay, and neutron reaction data exist for almost one thousand fission products. Of these, about 180 are explicitly considered in the process of core design in order to accurately represent the fission product poison inventory and its effect on core reactivity. These fission product poisons do not have a rapid transient behavior.

The effect of all fission product poisons, other than ^{135}Xe and ^{149}Sm, can be represented approximately by a single gross or aggregate fission product microscopic neutron absorption cross section, σ_a^{fpp}. This cross section has units of barns/fission rather than the usual units of barns/atom. The corresponding macroscopic cross section (Σ_a^{fpp}) is therefore formed by multiplying this microscopic cross section by the cumulative number of fissions per cm^3 (N_f^{U235}) that have occurred at a given time in life and core location.

Equation 3-186

$$\Sigma_a^{fpp} = \sigma_a^{fpp} N_f^{U235}$$

Equation 3-186

shows that Σ_a^{fpp} increases with core lifetime.

The fission product poisons contribute negative reactivity to the core. This reactivity increases (becomes more negative) as core life increases due to the increase in Σ_a^{fpp} and the decrease in fuel concentration (Σ_a^{U235}).

Section 3.15 — General Reactor Kinetics Equations

Reactor kinetics by its nature deals with changes in reactor power. Power changes cause changes in delayed precursor concentrations which in turn act to modify further power changes. Any accurate treatment of reactor kinetics must therefore account for the rate of change of both reactor power and delayed neutron precursor concentrations.

3.15.1 — The First Kinetics Equation

The First Kinetics Equation expresses how the change in total fission rate εR_f per second is related to core reactivity and to the parameters which characterize delayed neutrons and non-fission neutron sources.

The First Kinetics Equation,

Equation 3-187

$$\ell^* \frac{d}{dt}(\varepsilon R_f) = (\delta k - \bar{\beta})\varepsilon R_f + \bar{C}\lambda_{eff} + \bar{S}$$

can be interpreted term-for-term as:

Equation 3-188

$$
\begin{pmatrix} \text{Change in Total} \\ \text{Fission Rate per} \\ \text{Generalization} \end{pmatrix} = \begin{pmatrix} \text{Effect of} \\ \text{Prompt Neutrons} \\ \text{upon Change} \end{pmatrix} + \begin{pmatrix} \text{Effect of} \\ \text{Delayed} \\ \text{Neutrons} \\ \text{upon Change} \end{pmatrix} + \begin{pmatrix} \text{Effect of} \\ \text{Non-Fission} \\ \text{Sources} \\ \text{upon Change} \end{pmatrix}
$$

where:

ℓ^* $= \ell_p/k_{eff} \bullet \ell_p$ is the prompt neutron lifetime. The prompt neutron lifetime is used because it properly represents the average time between generations (~ 26 μsec).

δk = reactivity

εR_f = total fission rate per second

$\bar{\beta}$ = effective delayed neutron fraction. The fraction of thermalized fission neutrons that were born delayed. It has a constant value of ~0.0065.

$\bar{C}$ = effective (fission equivalent) delayed neutron precursor concentration. C is the total concentration of the precursors or fission products that emit delayed neutrons. $\overline{C}$ is the effective concentration, which accounts for the greater effectiveness of delayed neutrons in causing thermal fission and includes a $1/v$ multiplier, where v is the average number of neutrons born per fission, to compensate for the fact that each precursor decay yields only one neutron.

λ_{eff} = the effective decay constant for the total concentration of precursors, $dC/dt = -\lambda C(t)$. The value of λ_{eff} depends on the mix of concentrations of the six precursor groups. For a constant reactivity insertion, a curve similar to that shown in Exhibit 3-31 may be used to determine λ_{eff}. To understand the behavior of λ_{eff} in Exhibit 3-31, the student should recall that λ_{eff} is an effective decay constant for all six groups of delayed neutron emitters as in Equation 3-189

Some precursor groups have shorter half-lives (larger λ) than other groups. When the reactor is at constant power, the precursor concentrations are at equilibrium and λ_{eff} has a value of approximately 0.077 sec $^{-1}$. However, during changes in power level, the short-lived precursors exist due to the most recent seconds of power while the long-lived precursors exist due to the last several hundred seconds of power. A step addition of reactivity causes an up-power transient during which there are proportionately more short-lived precursors whose decay constants are larger and λ_{eff} decreases.

$\overline{S}$ = effective source of neutrons from non-fission sources per cm^3 per second that includes a $1/v$ multiplier. Source neutrons can be ignored above 10^{-6} % power.

Equation 3-189

$$\lambda_{eff}(C_1 + C_2 + C_3 + C_4 + C_5 + C_6)$$

$$= \lambda_1 C_1 + \lambda_2 C_2 + \lambda_3 C_3 + \lambda_4 C_4 + \lambda_5 C_5 + \lambda_6 C_6$$

The prompt neutron terms $(\delta k - \overline{\beta})\epsilon R_f$ follows the total fission rate very closely and can vary rapidly with time. It is also the largest term and is always negative since the condition $\delta k \geq \overline{\beta}$ never occurs in a properly operated and protected reactor. Because $\delta k - \overline{\beta}$ is always negative, the core fission rate would eventually decrease to zero were it not for the effect of the delayed neutrons and non-fission sources of neutrons.

The delayed neutron term $\overline{C}\lambda_{eff}$ expresses the contributions to the fission rate by the neutrons born of delayed neutron precursor decay. This term varies slowly compared to the prompt term because the average precursor lifetime $1/\lambda_{eff}$ is on the order of seconds rather than microseconds as in the prompt neutron lifetime. The magnitude of this term depends upon the fission rate that existed up to about a minute before. This term is always positive.

The source term $\overline{S}$ expresses the contribution of the non-fission sources to the fission rate. This term varies slowly, compared with the other terms in the equation, can be considered a constant under all normal conditions. This term is always positive and is much smaller than the other two terms except when reactor power is in the source range (~10^{-6} % power) where non-fission sources become important.

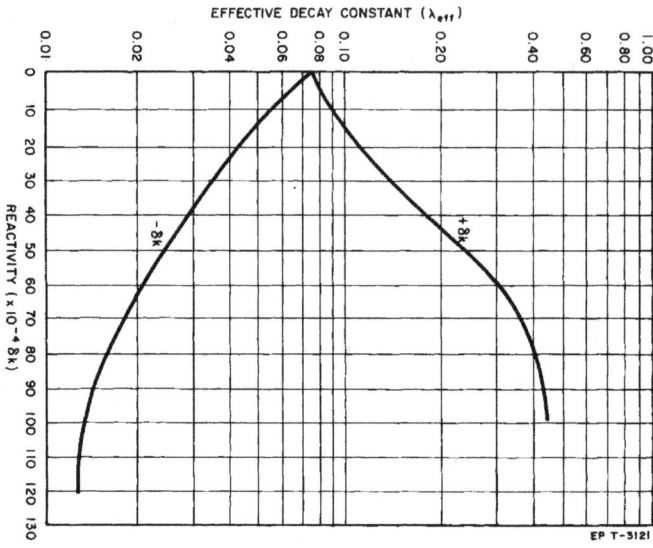

Exhibit 3-31 — λ_{eff} Versus Reactivity

3.15.2 — The Second Kinetics Equation

The Second Kinetics Equation expresses how the rate of change of the fission-equivalent delayed neutron precursor concentration, $\overline{C}$, is related to the fission rate and the rate of precursor decay.

The Second Kinetics Equation,

Equation 3-190

$$\frac{d\overline{C}}{dt} = \overline{\beta}\varepsilon R_f - \lambda_{eff}\overline{C}$$

can be interpreted term-for-term as:

Equation 3-191

$$
\begin{pmatrix}
\text{Rate of Change of} \\
\text{Fission Equivalent} \\
\text{Precursor Concentration}
\end{pmatrix}
=
\begin{pmatrix}
\text{Effective Production} \\
\text{Rate of Delayed} \\
\text{Neutron Precursors} \\
\text{due to Fission}
\end{pmatrix}
-
\begin{pmatrix}
\text{Effectiv Loss} \\
\text{Rate of Delayed} \\
\text{Neutron Precursors} \\
\text{due to Beta-Decay}
\end{pmatrix}
$$

For the same reasons given at the end of Paragraph 3.15.1, the production term $\overline{\beta}\varepsilon R_f$ can vary rapidly with time, while the loss term $\lambda_{eff}\overline{C}$ generally varies more slowly.

3.15.3 — Reactor Period And Startup Time

Reactor period T is a convenient method of expressing how fast reactor power (neutron population) changes. The reactor period is defined as:

Equation 3-192

$$
T \equiv \frac{P(t)}{dP(t)/dt} \equiv \frac{n(t)}{dn(t)/dt}
$$

where:

P(t)   = reactor power as a function of time

n(t)   = neutron population as a function of time.

The period T has dimensions of time with its units usually expressed in seconds.

If the reactor period T is constant, P(t) and n(t) can be expressed as follows:

Equation 3-193

$$
P(t) = P_0 e^{t/T} \qquad \text{or} \qquad n(t) = n_0 e^{t/T}
$$

where:

P_0   = power at time $t = 0$

n_0   = neutron population at time $t = 0$

Physically, the reactor period T is the time the reactor takes for power (neutron population) to change by a factor of "e" (~2.72).

When $\delta k < 60 \times 10^{-4}$ and $P > 10^{-6}\%$, the simplified equation for T is:

Equation 3-194

$$T = \frac{\bar{\beta} - \delta k}{\lambda_{eff} \delta k + d\delta k/dt}$$

An alternate method of expressing how fast power changes is startup rate, SUR. The reactor startup rate is defined as the decades (powers of ten) by which the reactor power changes in 1 minute. SUR has dimensions of inverse time, but are always expressed as decades per minute, DPM.

The relationship between startup rate and reactor period is given by

Equation 3-195

$$\frac{P(t)}{P_0} = e^{t/T} = 10^{SUR \cdot t'}$$

where:

t = time in seconds

t' = time in minutes.

Solving Equation 3-195 for SUR results in

Equation 3-196

$$SUR = 26.06/T$$

Using a reactor period T in units of seconds in Equation 3-196 produces a value for SUR with units of DPM.

It is convenient to express the startup rate as DPM since certain reactor operations such as startup and shutdown may require the operator to monitor reactor power over a range from as low as 10^{-10}% power to vicinity of 100% power. For this reason, reactor instrumentation is calibrated in powers of ten (decades).

3.15.4 — Prompt Criticality

Prompt criticality is defined as the reactivity condition wherein the reactor would be just critical in the absence of both source neutrons and delayed neutrons. In the First Kinetics Equation, the absence of source and delayed neutrons is equivalent to the condition $\bar{C}\lambda_{eff} + \bar{S} = 0$ and criticality implies a constant fission rate $[d(\varepsilon R_f)/dt = 0]$. Therefore, the condition of prompt criticality is

Equation 3-197

$$\delta k = \bar{\beta}$$

A reactor is said to be prompt supercritical or prompt subcritical when $\delta k > \overline{\beta}$ or $\delta k < \overline{\beta}$, respectively.

In a prompt critical reactor, the fission rate is such that the resulting prompt neutrons alone are sufficient to offset neutron losses due to absorption and leakage, and maintain power. Thus, the delayed neutrons and source neutrons constitute an excess of neutrons whose multiplication in the neutron life cycle results in a very rapid power increase. In fact, the startup rate for a prompt critical reactor is typically in the hundreds of DPM. Such startup rates are beyond human control and must be guarded against by proper plant design and operation.

3.15.5 — The Transient Startup Rate Equation

From the First and Second Kinetics Equations, the Transient Startup Rate Equation can be developed. This equation accurately describes the rate at which reactor power changes in response to a given behavior of core reactivity.

Equation 3-198

$$SUR = 26.06 \left(\frac{\lambda_{eff} \delta k + \dot{\delta k} + \dfrac{\lambda_{eff} \overline{S}}{\varepsilon R_f}}{\overline{\beta} - \delta k} \right)$$

The term $\dot{\delta k}$, appearing in the numerator of Equation 3-198, represents the rate at which reactivity is added to the core and is usually expressed in 10^{-4} δk/sec. Thus, $\dot{\delta k}$ has the same meaning as $\delta k/dt$.

The startup rate is given by Equation 3-198 is the same everywhere in the core because δk and its variation with time, $\dot{\delta k}$, is the same throughout the core. This follows from the fact that the neutron life cycle at any given location in the core is intimately connected to every other location in the core by fast and thermal leakage, so that δk and $\dot{\delta k}$ represent the entire core as a whole.

3.15.6 — Power Turning

From the definition of criticality, the rate at which the total fission rate changes with time, $\left[d(\varepsilon R_f)/dt \right]$, must be equal to zero for a critical reactor ($\delta k = 0$). But $d(\varepsilon R_f)/dt$ can also be zero when δk is not equal to zero.

The significance of $d(\varepsilon R_f)/dt = 0$ is that reactor power is neither increasing or decreasing. For $\delta k \neq 0$, this can occur at the point when reactor power has just risen to a maximum and is beginning to turn down (decrease), or when decreasing reactor power has just reached its lowest point and begins to turn upward (increase). Such maximum and minimum are called power turning points and are marked "PT" in Exhibit 3-32.

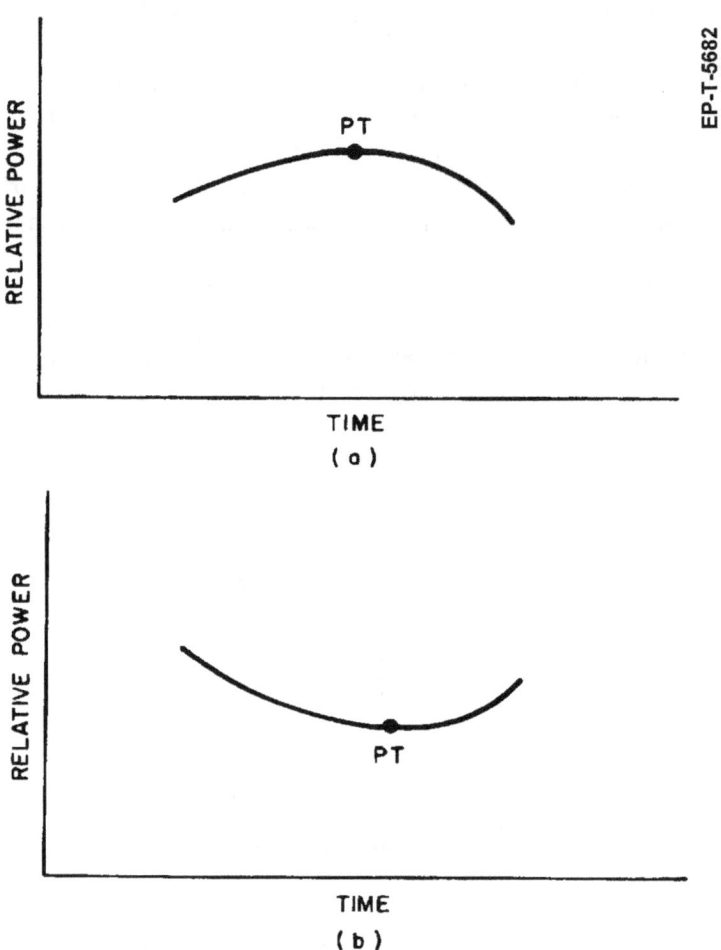

EP-T-5682

Exhibit 3-32 — Power Turning Points For Increasing (a) And Decreasing (b) Reactor Power

Given a power curve as in Exhibit 3-32, the startup rate can also be determined by taking the slope of the logarithm (base 10) of the power curve at any time. In each of the curves illustrated in Exhibit 3-32, the power turning point is the point at which the slope of the reactor power is just changing sign. The slope therefore is zero at the instant of power turning. A zero reactor power slope versus time is equivalent to a zero startup rate; therefore, SUR = 0 at any point of power turning. The fact that SUR = 0 at all power turning points provides the information necessary to predict the conditions which result in power turning.

Reactor startup rate can be zero only if the numerator of the startup rate equation, Equation 3-198, is zero.

Equation 3-199

$$\lambda_{eff}\delta k + \dot{\delta k} + \lambda_{eff}\overline{S}/\varepsilon R_f = 0$$

There are only three physically realistic conditions under which Equation 3-199 is satisfied. The first condition,

Equation 3-200

$$\delta k = \dot{\delta k} = \lambda_{eff}\overline{S}/\varepsilon R_f = 0$$

corresponds to a critical reactor with power $> 10^{-6}\%$ (so that the source term is negligible).

The second condition,

Equation 3-201

$$\dot{\delta k} = \lambda_{eff}\delta k + \lambda_{eff}\overline{S}/\varepsilon R_f = 0$$

corresponds to a subcritical reactor at constant power.

The third condition

Equation 3-202

$$\lambda_{eff}\delta k + \dot{\delta k} = 0$$

is possible when the fission rate is constant, $\left[d(\varepsilon R_f)/dt = 0\right]$, and the reactor is above $10^{-5}\%$ power, not critical, and core reactivity is changing $\dot{\delta k} \neq 0$. Equation 3-202 is referred to as the power turning condition.

3.15.7 — Time To Power Turning

A time delay, known as the time to power turning Δt_{PT}, occurs between the time corrective reactivity insertion begins and the time power actually turns.

For a given $\dot{\delta k}$, power turning occurs when the power turning condition is met, Equation 3-202. So the power turning reactivity δk_{PT} is given by:

Equation 3-203

$$\delta k_{PT} = -\frac{\dot{\delta k}}{\lambda_{eff}}$$

The time required for the core reactivity to change from its initial value δk_0 to the value

$\delta k_{PT} = \delta k_0 + \Delta\delta k$ required for power turning is equal to the time required to add the reactivity

Equation 3-204

$$\Delta\delta k = \delta k_{PT} - \delta k_0$$

Since $\dot{\delta k}$ is the amount of reactivity added per second, the amount of reactivity added during the time interval Δt_{PT} between the beginning of the reactivity insertion and the moment power turns is

Equation 3-205

$$\Delta \delta k = \dot{\delta k} \Delta t_{PT}$$

Substituting Equation 3-204 into Equation 3-205 and rearranging yields:

Equation 3-206

$$\Delta t_{PT} = \frac{\delta k_{PT} - \delta k_0}{\dot{\delta k}}$$

Substituting the expression δk_{PT} from Equation 3-203 into Equation 3-206 yields:

Equation 3-207

$$\Delta t_{PT} = -\left(\frac{1}{\lambda_{eff}} + \frac{\delta k_0}{\dot{\delta k}} \right)$$

where:

$$\Delta t_{PT} = t_{PT} - t_0$$

t_{PT} = the time at which power turns

t_0 = the time at which δk_0 exists.

The time to power turning and the actual power level at the power turn are important factors in determining the reactors response in a casualty situation. To reduce Δt_{PT} and the probability of a larger power spike, the reactivity addition rate must be sufficiently negative. This $\dot{\delta k}$ will determine the mode of automatic protection for the reactor (i.e. scram, cut back or fast insertion).

Section 3.16 — Subcritical Multiplication

In a subcritical reactor, the number of neutrons lost as a result of leakage and non-fission capture in each generation depends on the neutron flux (reactor power) in the reactor. When an operating reactor is shut down, the reactor's neutron population (power) decreases until the number of neutrons lost per generation is equal to the number of source neutrons produced per generation. At this point, the source neutron production rate just offsets the rate of neutron loss, establishing an equilibrium so that reactor power stabilizes. **Subcritical multiplication** is the process by which a subcritical reactor achieves this equilibrium power level in the source range. The resulting reactor power is called the **subcritical equilibrium power**.

When the reactor is subcritical and the neutron source level is constant, the left-hand sides of the First and Second Kinetics Equations equal zero. The resulting equations, when solved, give

Equation 3-208

$$\varepsilon R_f = -\overline{S}/\delta k$$

where:

εR_f = fission rate in fissions per cm^3–second

$\overline{S}$ = S/v, and S is given in neutrons per cm^3–second.

This is the number of fission events produced per second by a fixed source after many generations.

As Equation 3-208 shows, the fission rate is only dependent on the fission equivalent source strength and the shutdown reactivity (this is a negative quantity, hence the negative sign in the equation). $\overline{S}$ changes with core EFPH and time after shutdown because it is equal to the sum of the intrinsic, photoneutron, and transuranic fission equivalent source strengths, all of which vary.

The subcritical multiplication process is illustrated with the following example. A reactor is shutdown with k_{eff} = 0.8, correlating to $\delta k = 2500 \times 10^{-4}\delta k$. This subcritical reactor has a fission equivalent source strength of 100 neutrons/cm^3–second born from fission in generation N. In subsequent generations, k_{eff} **(fission neutrons + source neutrons)** are born from fission. As this process is applied from generation N on, the neutron population approaches a constant number born in each generation from fission.

Equation 3-209

$$0.8(500 + 100) = 480$$
$$0.8(480 + 100) = 464$$
$$0.8(464 + 100) = 451$$
$$0.8(451 + 100) = 441$$
$$0.8(441 + 100) = 433$$
$$\bullet$$
$$\bullet$$
$$\bullet$$
$$0.8(400 + 100) = 400$$

Exhibit 3-33 shows how the neutron population converges to the subcritical equilibrium level.

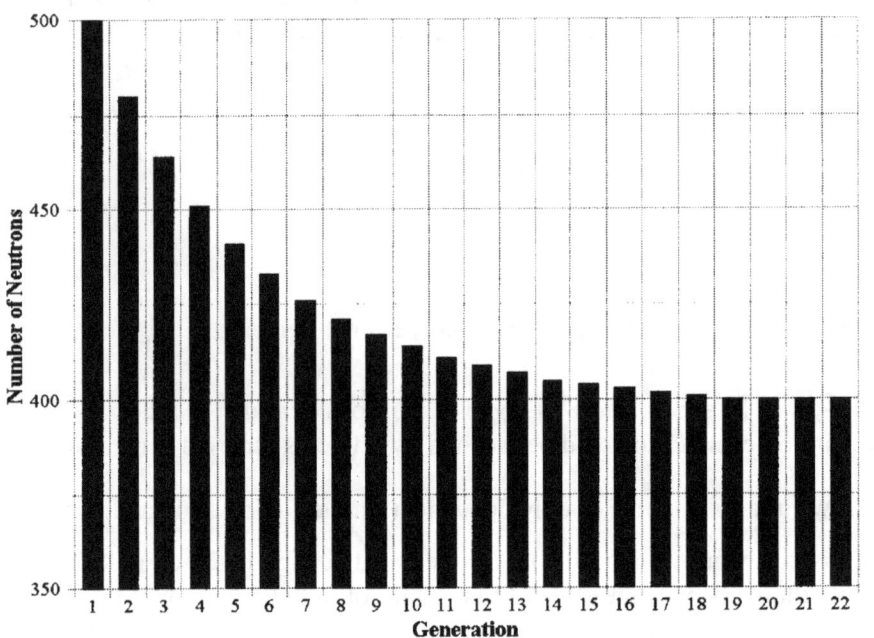

Exhibit 3-33 — The Subcritical Multiplication Process As The Neutron Level Approaches Equilibrium For The Given Example

This result can be confirmed with Equation 3-208

Equation 3-210

$$\varepsilon R_f = -\left(\frac{100}{-0.25}\right) = 400$$

Section 3.17 — Gamma Attenuation

3.17.1 — Photon Fundamentals

Photons are quanta of the electromagnetic field. They are regarded as discrete particles possessing zero mass and no electric charge, but carry energy and momentum. Gamma rays and x-rays are generically referred to as photons since they behave as packets or bundles of energy moving with the speed of light.

The difference between gamma rays, x-rays, light or any other type of electromagnetic radiation centers around the values of the frequency and wavelength. All electromagnetic waves are governed by the following formula:

Equation 3-211

$$C = \nu\lambda$$

where:

    C   = the speed of light

    v = the frequency

    λ = the wavelength.

Since C is a constant, an increase in the frequency requires a decrease in the wavelength. Visible light occupies a narrow band in the electromagnetic spectrum: 7.6×10^{-5} cm (3.9×10^{14} hertz) to 3.8×10^{-5} cm (7.9×10^{14} hertz). X-rays typically occupy that region of the electromagnetic spectrum from 10^{-6} (3×10^{16} hertz) to 10^{-8} cm (3×10^{18} hertz). Gamma rays typically occupy the region from 10^{-8} cm (3×10^{18} hertz) and higher.

The terms x-ray and gamma ray are used almost synonymously. Usually electromagnetic waves coming from nuclei are called gamma rays, while those of high energy coming from atoms are called x-rays. At the same frequency they are physically indistinguishable, no matter what the source.

The photon energy is given by the following formula:

Equation 3-212

$$E = hv = h\frac{C}{\lambda}$$

where:

    C   = the speed of light

    v = the frequency

    λ = the wavelength.

    h   = Planck's constant

    E   = Energy

Since Planck's constant is fixed, the photon's energy increases if its frequency increases or its wavelength decreases.

An electromagnetic wave can be characterized by the frequency, wavelength, or energy of one of its photons. All three of these quantities are in common use with each suited to a particular domain of the electromagnetic spectrum. X-rays and gamma rays, detected individually as single photons, are most often specified by their energy. The electron volt (eV) and its multiples, the keV (10^3 ev), the MeV (10^6 eV), and the GeV (10^9 eV), are common units of photon energy. The upper limit on atomic x-rays is typically 40 keV, while the upper limit on nuclear gamma rays is typically 40 MeV.

3.17.2 — Photoelectric Effect

A gamma ray or gamma can be absorbed, losing all of its energy, by a bound electron with the removal of the electron from the atomic shell or orbit if the energy of the gamma meets or exceeds the electron's binding energy. The kinetic energy of the ejected electron is given by:

Equation 3-213

$$T_e = h\nu - W$$

where:

h	= Planck's constant
ν	= the frequency
T$_e$	= the electron's kinetic energy
W	= the work function.

The work function, W, is equivalent to the electron's binding energy to the atom. W is sometimes referred to as the ionization energy of the electron. This value depends upon Z for the nuclide and the orbital shell populated by the electron. K-shell binding energies are larger than L-shell binding energies. For heavy nuclei, 80% of the photoelectrons come from the K-shell. The appearance of the work function here makes this a threshold process where a minimum energy (or frequency) is required for it to happen.

Removal of an electron by the photoelectric effect momentarily creates a vacancy or hole in the atomic electron shells. This is an unstable arrangement for the atom which reaches stability through the emission of fluorescent x-rays or Auger electrons.

This process cannot take place with a free electron. With a bound electron and its atom, conservation of linear momentum is assured when the atom recoils in a direction opposite to that of the escaping photoelectron. The tighter an electron is bound, the larger is the probability for the process to occur. (Electron binding energies increase with Z). As the gamma energy decreases, the difference between hν and W decreases so that the electron appears (to the gamma) to be more and more tightly bound to the atom. When this occurs the interaction probability increases until it reaches maximum when hν equals W.

The interaction probability for this process is given by

Equation 3-214

$$\mu_{pe} \propto \frac{Z^m}{E_p^{\ n}}$$

where:

μ$_{pe}$	= the photoelectric effect attenuation coefficient

E_p = gamma energy

Z = atomic number

m varies from 4.0 (at 100 keV) to 4.6 (at 3 MeV)

n varies from 3 (for $E\gamma < 150$ keV) to 1 (for $E\gamma > 150$ keV)

The interaction probability is generically known as the macroscopic cross section. By tradition, gamma ray cross sections are called attenuation coefficients.

3.17.3 — Compton Scattering

Compton scattering is the elastic scattering of a photon by an electron in which both energy and momentum are conserved. As shown in Exhibit 3-34, the incident photon with energy E_p and wavelength λ interacts with an electron and scatters through an angle θ while the electron recoils through an angle ϕ. To conserve momentum in the interaction, energy is given up by the photon so that $E_p' < E_p$ and $\lambda' > \lambda$. As a result, the electron gains energy that is equal to the amount of energy given up by the photon.

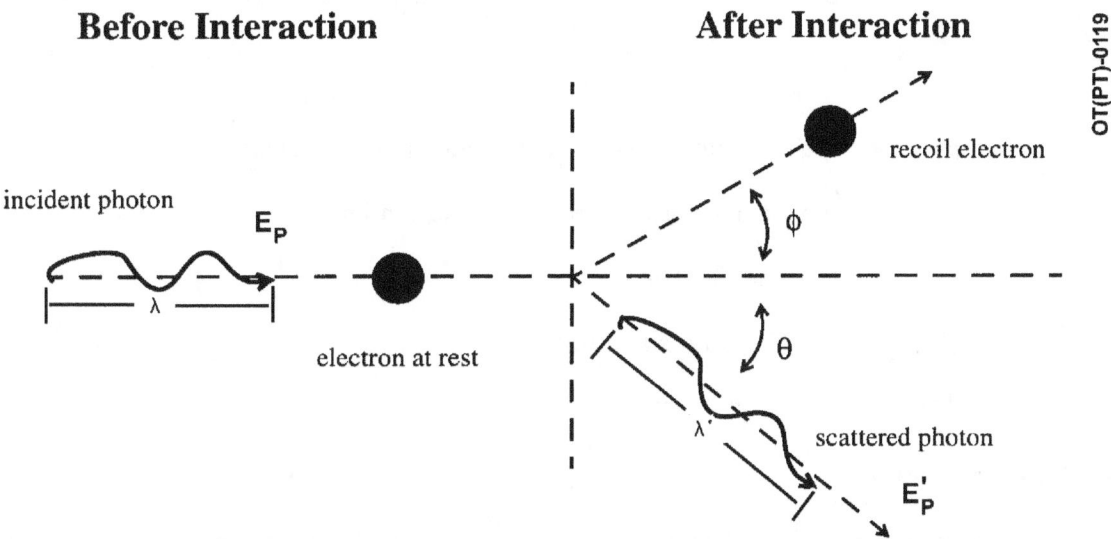

Exhibit 3-34 — Photon Compton Scattering From An Electron

Given two incident photons of the same energy, the photon scattering through the larger angle loses the most energy. The maximum energy loss occurs for backward scattering where the energy of the scattered photon is given by:

Equation 3-215

$$\left| \Delta E_{p,max} \right| = \frac{E_p}{\dfrac{m_e c^2}{2E_p} + 1}$$

Low energy photons are scattered with only a moderate energy change, but high energy photons suffer a very large change in energy.

For Compton scattering, the photon interacts with individual electrons, whether they are free or bound. It is possible to define a Compton cross section per electron, $_e\sigma_C$. The magnitude of this cross section decreases with increasing photon energy with the fall off going as $1/E_p$ when the photon's energy is greater than 0.5 <eV. For atoms containing Z electrons, the Compton scattering cross section is given by:

Equation 3-216

$$\sigma_{CS} = Z \bullet {}_e\sigma_{CS}$$

The attenuation coefficient for Compton scattering is:

Equation 3-217

$$\mu_{CS} = N \bullet \sigma_{CS} = N \bullet Z \bullet {}_e\sigma_{CS} = \frac{\rho N_a}{GAW} \bullet Z \bullet {}_e\sigma_{CS}$$

where:

μ_{CS}	= the Compton scattering attenuation coefficient
N	= the number of atoms per cubic centimeter
ρ	= material density
N_a	= Avogadro's number
GAW	= Gram Atomic Weight

3.17.4 — Pair Production

Above incident photon energies of 1.022 MeV, pair production occurs. In this process, the photon is absorbed by the electric field of a nucleus and an electron-positron pair is created. The nucleus is necessary so that there is a system where momentum is conserved before the creation of the electron-positron pair. During and after pair creation, the incident photon' energy is shared among the nucleus, electron, and positron with the nucleus receiving only a small amount. After creation, the nucleus recoils slightly. This nuclear motion in conjunction with the motion of the electron and positron is such that momentum is conserved after the pair is created. Therefore, with three particles involved with conserving momentum, it is possible for both the electron and positron to travel in the forward direction (relative to the direction of photon travel).

This process has a threshold of 1.022 MeV since this is the minimum amount of energy needed to create two particles with masses equal to the rest mass of an electron, 0.511 MeV. The photon energy in excess of the threshold value of 1.022 MMeV primarily appears as kinetic energy of the electron-positron pair.

After creation, the electron and positron separately lose energy by ionization. After the positron has lost a considerable amount of energy, it may form a kind of atom with one of the electrons from the medium. This so-called positronium atom is like hydrogen except that an electron and positron are spinning around a center-of-mass instead of an electron and a proton. The positronium atom does not last long. The electron and positron come together and annihilate one another and create two gammas each with an energy of 0.511 MeV. The annihilation gammas travel away from their creation site in opposite directions to conserve momentum.

The interaction probability for this process is proportional to the charge producing the electric field.

Equation 3-218

$$\mu_{pp} \propto Z^2$$

where:

μ_{pp} = the pair production attenuation coefficient

The interaction probability is a function of energy increasing as the energy of the incident photon increases beyond the threshold.

3.17.5 — Interaction Probability Curve

The total interaction probability for photon interactions is the sum of the individual interaction probabilities.

Equation 3-219

$$\mu_{tot} = \mu_{PE} + \mu_{CS} + \mu_{pp}$$

Each of the photon interaction mechanisms has a dependence on the atomic number, Z, and incident photon energy, E_p. As a result of this, given Z, each interaction mechanism has an energy rage over which it dominates. These results are summarized in Table 3-5

Table 3-5 — Attenuation Coefficients And Their Energy Ranges

Attenuation Coefficient	Z Dependence	Most Probable Energy Range
μ_{PE}	$\sim Z^4$	< 0.5 MeV
μ_{CS}	Z	0.5 to 5 MeV
μ_{PP}	Z^2	> 5 MeV

Exhibit 3-35 illustrates the behavior of the linear attenuation coefficients for lead. A consequence of the individual attenuation coefficients' variation with energy is a minimum in the total attenuation coefficient between 1 and 10 MeV, typically, for all elements. This energy range is often referred to as the "gamma window" since gamma rays with these energies traverse materials quite easily compared to other energies and are the most difficult to attenuate.

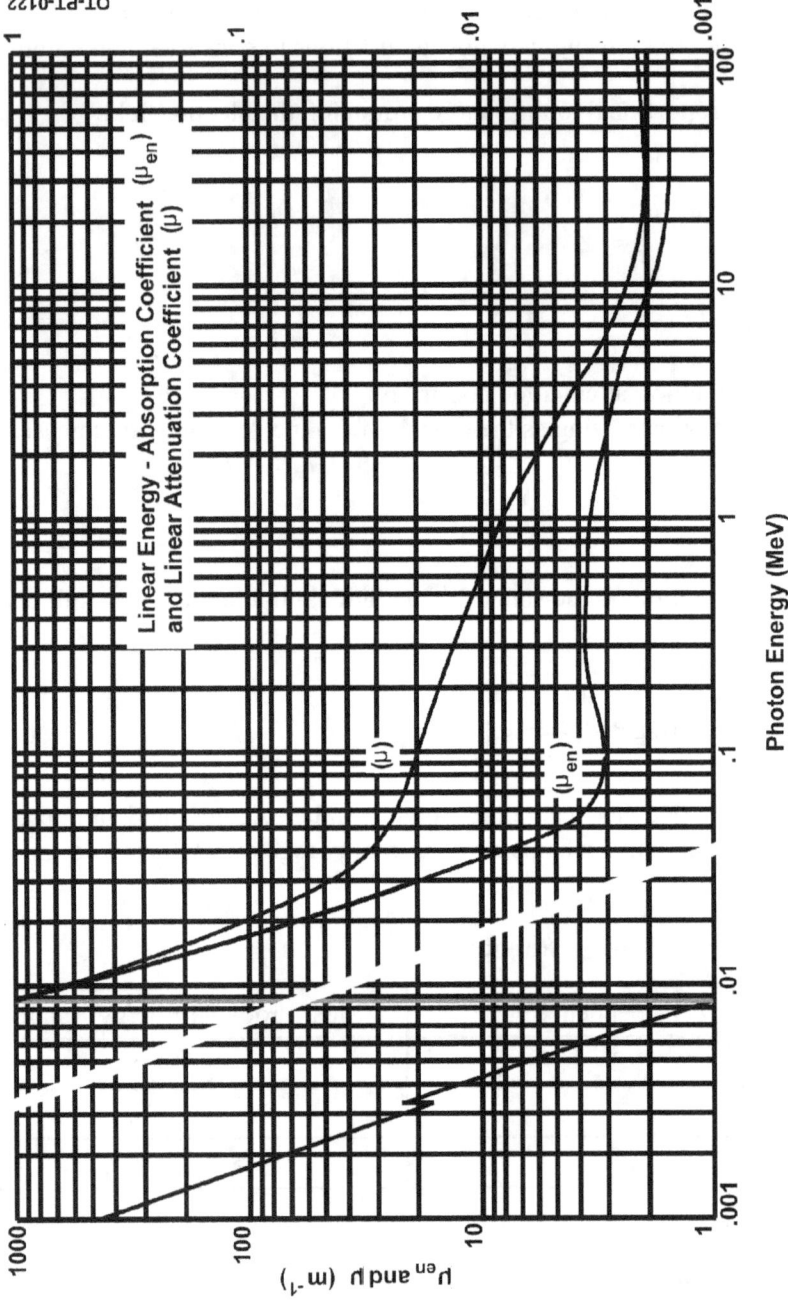

Exhibit 3-35 — Linear Attenuation Coefficients For Photons In Lead

Section 3.18 — Neutron Sources

The neutron-induced fission process makes possible a self-sustaining nuclear chain reaction which produces virtually all of the neutrons in an operating reactor. However, to initiate this chain reaction, a neutron source which does not depend on neutron-induced fission is required. The major reason for concern with neutrons produced by reactions other than neutron-induced fission is the role they play in the reactor startup.

Neutron-producing reactions are possible with the materials contained in the reactor core. These reactions are initiated by the particles available from radioactive decay. The principal means of producing neutrons are nuclear reactions induced by charged particles (α, n) or by photons (γ, n). Normally of less importance because of their smaller neutron yields are spontaneous fission, photofission, induced by fission product gamma-rays, and cosmic ray interactions.

3.18.1 — Intrinsic Neutron Sources

Intrinsic source neutrons are those produced by the natural alpha decay of the nuclear fuel present in a reactor through the (α, n) reaction. All of the naturally occurring isotopes of uranium plus ^{236}U are alpha emitters; however, ^{234}U and ^{238}U have comparable alpha emission rates in natural uranium. ^{234}

U alphas produced greater neutron yields because of their higher kinetic energies. In enriched uranium, most of the alpha activity is due to ^{234}U.

The alpha particles emitted by the fuel have a very short range and interact only with the oxygen present in the fuel mixture to produce neutrons. Of the naturally occurring oxygen isotopes, ^{18}O with an abundance of 0.204 a/o has a low enough threshold for the (α, n) to proceed.

Equation 3-220

$$ _2^4\alpha + {}_8^{18}O \rightarrow {}_{10}^{21}Ne + {}_0^1n $$

The resulting neutrons possess a kinetic energy of ~3 MeV. ^{17}O does not have a threshold for (α, n) reactions, it should be noted that a small part (a few percent) of the intrinsic source is due to spontaneous fission of the uranium field.

3.18.2 — Photoneutrons

Photoneutrons are the exiting particles from (γ, n) reactions that result when gamma rays of sufficient energies are captured by a nucleus and neutrons are emitted. Gamma rays of various energies are given off by fission products, but only those with sufficient yields and energies need be considered.

The gamma-ray energy required is set by the binding energy of the last neutron in the target nucleus; hence (γ, N) reactions are threshold reactions. Two isotopes with the lowest binding energies for the last neutron are ^2H (2.23 MeV) and ^9Be (1.67 MeV). Beryllium is rarely found in reactor cores unless specifically

placed there as a photofission source. Since it comprises 0.015 a/o of all hydrogen, $_1^2H$ is found in water moderated cores as D_2O. ^2_1H is the major source of photoneutrons in water moderated reactors as shown with the following reaction:

Equation 3-221

$$\gamma + {}^2_1H \rightarrow {}^1_1H + {}^1_0n$$

Because the neutron level in reactors is dominated by delayed neutrons during the first few minutes after shutdown, only those fission products whose controlling half-lives are greater than or equal to 15 minutes are considered. When energy and yield requirements are added to that for the half-life, there are ~12 fission products which emit gammas with sufficient energy to produce photoneutrons.

The most important gamma emitting fission product for short time scales is ^{140}La, the beta-decay daughter of ^{140}Ba. ^{140}Ba is produced from fission with a yield of over 6% and decays to ^{140}La with a half-life of 12.8 days. ^{140}La has a shorter half-life of 40 hours, so that after ~2 weeks, transient equilibrium exists between parent and daughter. When ^{140}La finally beta-decays to ^{140}Ce, gammas with sufficient energy to induce photoneutron production are emitted.

The most important gamma emitting fission product for long time scales is ^{106}Rh, the beta-decay daughter of ^{106}Ru. ^{106}Ru is produced from fission with a yield of 0.4%, and slowly decays with a half-life of ~1 year to ^{106}Rh. ^{106}Rh has a shorter half-life of ~30 seconds, so that after a few minutes, secular equilibrium (λ daughter $\gg \lambda$ parent) exists between parent and daughter. When ^{106}Rh beta-decays to ^{106}Pd, gammas with sufficient energy to induce photoneutron production are emitted.

After the first 15 minutes of shutdown, photoneutron sources are the major source of neutrons. This source is dependent upon the reactor power level before shutdown, (primarily the last 100 hours before shutdown and decreases as the shutdown time increases.

3.18.3 — Transuranic Sources

During reactor operation, about 19% of all ^{235}U absorptions result in the production of ^{236}U by the (n, γ) reaction. The ^{238}U present may also capture a neutron to become ^{239}U. Further neutron absorption and subsequent beta decay lead to unstable transuranics such as ^{238}Pu, ^{242}Cm, and ^{244}Cm, to name a few. To relieve Coulomb repulsion, these transuranics decay by spontaneous fission producing neutrons directly or by emitting alpha particles which undergo (α, n) reactions, as discussed in Paragraph 3.18.1.

After thousands of hours of reactor operation, significant amounts of the transuranics exist and continue to increase with core life. After shutdown, the transuranic source decreases with time, but not as quickly as the photoneutron source.

Chapter 4 — MECHANICAL REVIEW

Section 4.1 — Steam Thermodynamics

4.1.1 — Energy Changes

Most stored or **latent energy**, a form of internal energy, is released in the form of thermal energy or heat. Coal, fuel oil, and nuclear power are examples of stored energy that is released primarily in the form of heat. The problem is to convert the thermal energy into useful work. Gas engines, steam generating systems, and thermo-electric convertors are examples of methods whereby heat energy is converted into useful work.

We are interested in the **steam cycle** and how it is used to convert thermal energy into useful work. First, heat is added to water, which increases the internal energy and converts the water to steam. The steam then passes through a nozzle where the internal energy is converted into kinetic energy. The high velocity steam leaving the nozzle is then deflected by the turbine blade and its kinetic energy is absorbed by the blade, which exerts a torque on the turbine shaft. The steam is then condensed and returned to be reconverted into steam.

4.1.2 — General Energy And Mass Equations For Fluids

In a steady-flow open system the rate of heat transfer, rate of doing work, rate at which the fluid is flowing, and conditions at each point in the fluid and equipment all remain at steady-state values. In other words, energy enters and leaves the system in equal amounts with no change in the amount of stored energy within the system. The **energy balance equation** for the system may be stated as

Equation 4-1

$$\frac{v_1^2}{2g_cJ} + \frac{gz_1}{g_cJ} + h_1 + q_{NET} = \frac{v_2^2}{2g_cJ} + \frac{gz_2}{g_cJ} + h_2 + w_{NET}$$

where:

v	= Average Linear Velocity of Fluid, ft/sec
g_c	= Unit Conversion, 32.17 lbm • ft/lbf • sec^2
J	= Unit Conversion, 778.2 ft • lbf/Btu
g	= Standard Acceleration of Gravity, 32.17 ft/sec^2
z	= Height Above Reference, ft
h	= Specific Enthalpy, Btu/lbm
q_{NET}	= Net Heat Transferred into (+) or out of (-) each lbm of Fluid through the System, Btu/lbm

w_{NET} = Net Work Done on (-) or by (+) each lbm of Fluid Flowing through the System, Btu/lbm.

The general energy equation is derived from the law of conservation of energy.

In a closed system the mass flow rate entering the system must equal the mass flow rate exiting the system. Based on this balance, the following equivalence can be determined:

Equation 4-2

$$\rho_1 A_1 v_1 = \rho_2 A_2 v_2$$

where:

ρ = Fluid Density, lbm/ft^3

A = Cross-sectional Area of Flow Path, ft^2

v = Linear Velocity of Flow, ft/sec or ft/hr.

This equivalence is simply a statement of the law of conservation of mass.

4.1.3 — Nozzles

Nozzles are used in a steam turbine to transform the enthalpy of steam into kinetic energy. This is accomplished by shaping the nozzle (Exhibit 4-1) to cause an increase in the speed of the steam as it expands from a high-pressure region to a low-pressure region. Basically, a **nozzle** is nothing more than a smooth-shaped hole in a wall that separates a high-pressure region from a low-pressure region. The high-pressure steam expands through the nozzle and emerges at the low-pressure side as a high-speed jet. Nozzles may have many forms, but all are similar in principle of operation. They all consist of an **entering section**, a **throat**, and a **mouth**. In addition to their primary function of energy conversion (thermal to kinetic), nozzles also serve to direct the exiting high-speed jet of steam tangentially onto the moving blades where the final energy conversion takes place (kinetic to mechanical).

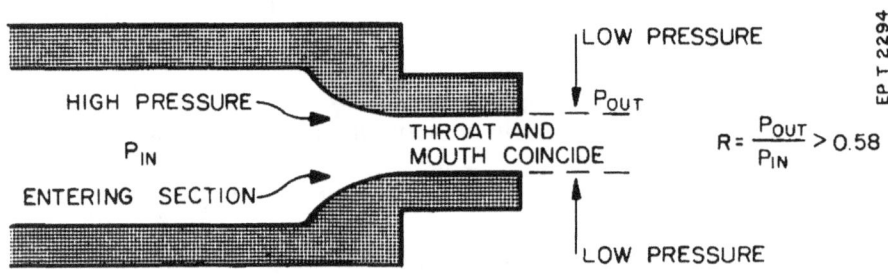

Exhibit 4-1 — Convergent Nozzle

Exhibit 4-1 shows the simple **convergent** nozzle, which is employed when the exit pressure is no less than approximately one-half of the initial pressure. When the pressure differential between entering and exiting steam is greater than this, there is a tendency for the steam to expand rapidly in all directions upon leaving the throat, creating turbulence. To remedy this turbulent expansion, a section of gradually increasing area is added after the throat. This results in the steam emerging as a uniform steady flow. This type of nozzle, illustrated in Exhibit 4-2, is called a **convergent-divergent** nozzle. A graphical representation of changes in volume, speed, and pressure across a convergent-divergent nozzle is also shown in Exhibit 4-2. Every nozzle must be specifically designed, as to throat opening and divergent section length, for the pressure ratio at which it will be operated. Operation at any pressure ratio other than the design ratio will cause a decrease in nozzle efficiency.

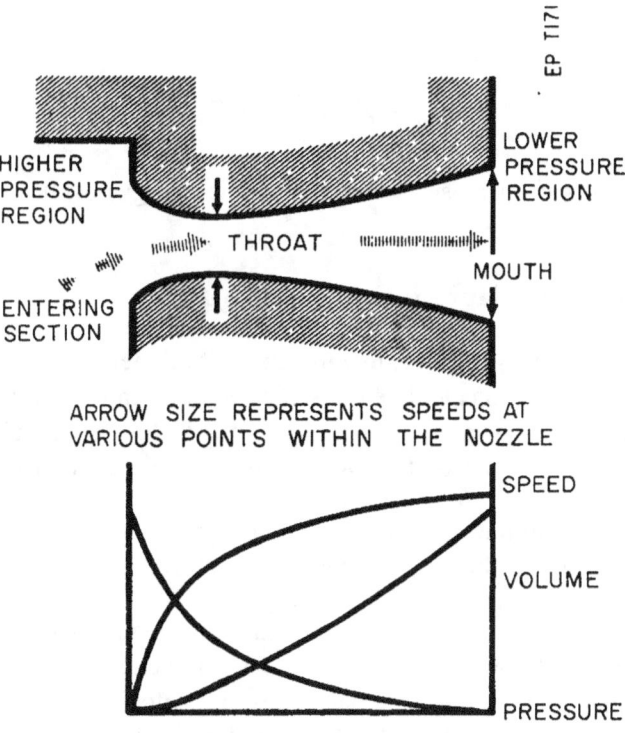

Exhibit 4-2 — Convergent-Divergent Nozzle

The general energy equation from Paragraph 4.1.2 may be applied to derive an equation for the exit speed of steam from the nozzle. By eliminating nonapplicable terms and disregarding nozzle friction losses and initial steam speed (both are usually negligible), the equation reduces to:

Equation 4-3

$$v_{exit}^2/2g_c + (h_1 - h_2)J$$

where:

v_{exit} = Exit Speed of Steam, ft/sec

g_c = Constant, 32.17 lbm-ft/lbf-sec^2

h_1 = Inlet Enthalpy, Btu/lbm

h_2 = Outlet Enthalpy, Btu/lbm

J = Unit Conversion, 778.2 ft-lbf/Btu.

Solving the equation for exit speed and substituting in the constants yields

Equation 4-4

$$v_{exit} = 223.7(h_1 - h_2)^{0.5}$$

It can be seen that the nozzle acts to increase the speed (kinetic energy) of the steam jet at the expense of enthalpy.

4.1.4 — Turbine Blade Velocity Diagram For Impulse Turbine

The purpose of the **velocity diagram** is to determine the amount of energy that can be removed from the steam by the turbine blades. The first step is to calculate the steam velocity in the steam nozzle and then draw the velocity diagram. Assume that 200 psia saturated steam is in the turbine steam chest, and that the condenser vacuum is 25 inches Hg or 5 inches Hg absolute. For the moment neglect the fluid friction of steam.

The energy taken by the turbine from the steam is obtained by first converting the internal energy of the steam into kinetic energy and then absorbing the kinetic energy in the turbine blades.

The internal energy is converted into kinetic energy by expanding steam through a nozzle. The velocity of the steam is determined by the enthalpy change of the steam. The total energy leaving the nozzle is equal to the total energy entering the nozzle. Neglecting frictional losses, the velocity of the steam leaving the nozzle can be determined by applying Equation 4-4. The value for enthalpy can be found either from steam tables or a Mollier diagram. Substituting the enthalpy values into Equation 4-4 yields:

$$V_{exit} = 223.7(1198 - 896)^{0.5}$$

$$V_{exit} = 223.7(17.4) = 3892 \, \text{ft/sec}.$$

The easiest way to calculate the work absorbed by the blades is to select a suitable scale and lay out the nozzle angle, steam velocities, and turbine blade speed on graph paper as in Exhibit 4-3. Exhibit 4-3 assumes a blade speed of 2000 ft/sec and a 20° nozzle angle.

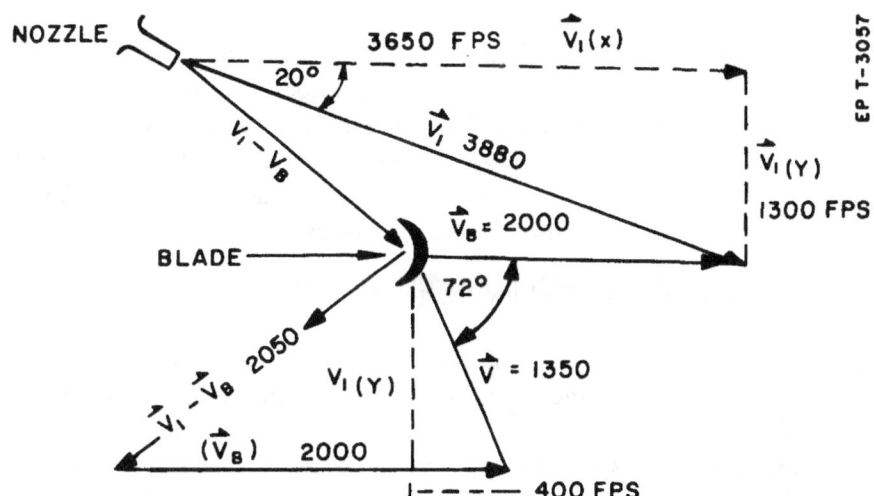

Exhibit 4-3 — Velocity Diagram

The force on the turbine blade is the difference in velocity between v_1 and v_2 in the direction of the blade multiplied by the mass flow rate of steam:

Equation 4-5

$$f = \frac{\dot{m}}{g}(v_1 - v_2)$$

For the flow of 1 lbm/sec of steam, the force is

$$f = \frac{(1\,\text{lbm/sec})(3650\,\text{ft/sec} - 400\,\text{ft/sec})}{32.17\,\text{lbm-ft/lbf-sec}^2} = 101\,\text{lbf}.$$

The work is force times distance. Power is the rate of doing work or work per unit of time, so

Power 　　= (force) × (blade velocity)

　　　　　　= (101 lbf)(2000 ft/sec)

　　　　　　= 202,000 ft-lbf/sec (1 HP/(550 ft-lbf/sec))

　　　　　　= 367 HP

　　　　　　= (367 HP) (0.746 KW/HP) = 274 KW.

It should be noted that the quantities selected for the above problem were for illustration only and do not necessarily reflect values that would exist in a marine steam turbine.

4.1.5 — Internal Turbine Losses

In each of the previous examples an ideal case of "no losses" was assumed. Actually, several types of internal losses reduce the efficiency of energy conversion. Their general effect is to increase the entropy and enthalpy of the exiting steam and, in all cases, to somewhat reduce the efficiency of the turbine. The turbine is, however, a highly effective means of converting thermal energy into useful work. A turbine operates within the range of 80 percent efficiency. Paragraph 4.1.5.1 and Paragraph 4.1.5.2 briefly describe some of the internal turbine losses that reduce turbine efficiency, and throttling losses are discussed in more detail in Paragraph 4.1.8.4.

4.1.5.1 — Internal Losses In An Impulse Turbine Stage

1. Windage Losses - The disc or wheel on which the moving blades are mounted is rotating in a steam atmosphere. Friction of the steam on the nonworking part of that wheel or disc causes turbulence, which consumes power.
2. Suction Losses - Suction losses only occur in partial admission stages, such as in an impulse stage, where the nozzles only go partially around the periphery of the stage. Suction losses take place in the inactive portion of the arc. This is where the blades push the steam back to the active section of the arc to allow the steam to exhaust to the next stage.
3. Nozzle End Losses - In a partial admission stage, when the blade reaches the admission arc, or the first nozzle in the arc, the relatively slow moving steam in the blade must be pushed out of the blade by the incoming steam.
4. Moisture Losses - At some point in the turbine, the steam reaches its saturation pressure and continues to expand, and moisture droplets begin to form. These slow velocity water droplets impinge on the blades and cause a braking action.
5. Diaphragm Packing Losses - The small amount of steam leaking from one stage to the next through the diaphragm packing is not available for work because it does not pass through the nozzle.
6. Leaving Losses - The kinetic energy left in the steam as it exits the turbine (Paragraph 4.1.8).
7. Throttling Losses - A drop in pressure without the performance of work, such as the pressure drop when steam passes through a steam admission valve (Paragraph 4.1.2).
8. Blade Losses - The frictional losses as the steam passes over the blades.
9. Nozzle Losses - The frictional losses due to the steam passing over the walls of the nozzle.

4.1.5.2 — Internal Losses In A Reaction Stage

1. Windage Losses - Friction of the steam on the nonworking portions of the moving blades and the rotor.
2. Tip Leakage Losses - Unused energy in that part of the steam which leaks past the blade tips without performing work.
3. Moisture Losses - At some point in the turbine, where the steam reaches its saturation pressure and continues to expand, moisture droplets begin to form. These slow velocity water droplets impinge on the blades and cause a braking action.
4. Leaving Losses - The kinetic energy left in the steam as it exits the turbine (Paragraph 4.1.8).
5. Throttling Losses - A drop in pressure without the performance of work, such as the pressure drop when steam passes through a steam admission valve (Paragraph 4.1.8).
6. Blade Losses - The frictional losses from the steam passing over the blades.

7.   Nozzle Losses - The frictional losses from the steam passing over the walls of the nozzle.

4.1.6 — External Turbine Losses

There are also external turbine losses referred to as mechanical losses. Mechanical losses are those associated with friction in the turbine bearings and the shaft packing. Heat thus produced is carried away by the circulating lube oil and is not available for conversion to work. However, this heat loss is small.

4.1.7 — Temperature-Entropy And Mollier Diagrams

Temperature-entropy (T-s) and Mollier (h-s) diagrams are useful for visualizing the thermodynamic variations in a steam cycle. Exhibit 4-4 illustrates the basic steam cycle plotted on a temperature-entropy diagram and the relationship of the plotted points.

This steam cycle uses 1000 psig saturated vapor and the turbine exhausts to a 25 inch Hg vacuum. We will start at Point 1 and explain the cycle.

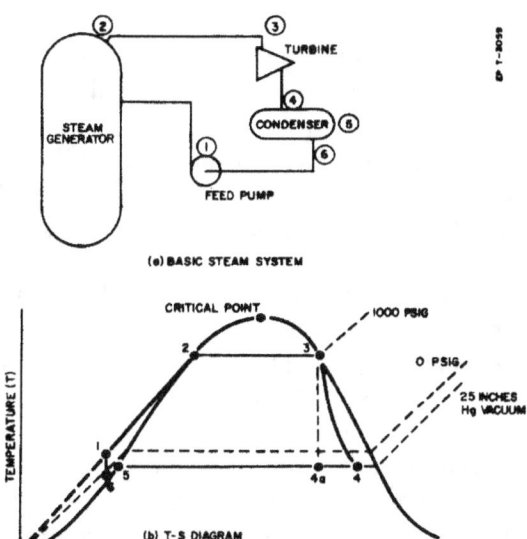

Exhibit 4-4 — Basic Steam Cycle Plot On T-s Diagram And Sketch Of Associated Steam Cycle

1.   Points 1-3.
   a.   Point 1 is where the liquid enters the steam generator.
   b.   Point 2 is where the liquid in the steam generator has been increased in temperature to its saturation point.
   c.   Point 3 is where the saturated liquid in the steam generator at Point 2 has been changed to saturated vapor (addition of latent heat of vaporization).
2.   Points 3-4.
   a.   Point 3 also indicates where the saturated vapor enters the turbine.

      b.   Point 4 indicates wet vapor exhaust from the turbine (4a is for an ideal turbine with a constant entropy process). The energy loss from Point 3 to Point 4 is that energy converted to mechanical energy to propel the ship.

3.   Points 4-6.
      a.   Point 4 also indicates the entrance of the wet vapor into the condenser
      b.   Point 5 indicates the point where there is a complete change from a wet vapor to a saturated liquid.
      c.   Point 6 shows a decrease in temperature (constant pressure) of the liquid (subcooling), which is called **condensate depression**.

4.   Points 6-1.
      a.   Point 6 also shows the suction of the feed pump.
      b.   Point 1 shows the point of discharge from the feed pump, where there is a slight increase in temperature with a large increase in pressure.

Another bit of information obtainable from this diagram is that the heat of vaporization decreases as temperature and pressure are increased until the **critical point** is reached, where the saturated liquid changes to a saturated vapor without further heat addition.

Now we will briefly discuss the Mollier diagram (Exhibit 4-5 and Exhibit 4-6) as it applies to a steam cycle. In this case we are using a steam generator operating at 320 psig and exhausting into a condenser at 28 inches of Hg vacuum.
    1.   The line between Points 2 and 3 represents the addition of heat in the steam generator.
    2.   The line between Points 3 and 4 represents the removal of heat by the turbine.
    3.   The line between Points 4 and 1 represents the removal of heat by the condenser.

As you can see from Exhibit 4-5 most of the heat energy that we add to our system is removed by the condenser. The more perfectly we design a turbine the closer we come to having a constant entropy process from the turbine, which means more heat is removed by the turbine. Therefore, if we can remove a greater percent of the heat energy with the turbine, we can increase the efficiency of the system.

The primary use of the Mollier Diagram is in the analysis of turbine operation. Particularly for the **ideal turbine approximation** (constant entropy expansion in the turbine), the turbine exhaust enthalpy is readily determined when the turbine exhaust pressure is known. Exhibit 4-6 compares ideal turbine behavior (constant entropy) with real turbine behavior on T-s and h-s diagrams. For real turbines $s_{real} > s_{ideal}$.

4.1.8 — Vapor Cycles (Steam Cycle)

The steam (Rankine) cycle can best be described using the T-s diagram in Exhibit 4-7. Heat is added to the water in the steam generator between Points 1 and 2. Additional heat added between Points 2 and 3 converts the water into steam. The steam passes through the turbine (Point 3 to Point 4), where energy is removed by the turbine. Additional energy is removed by the condenser (Point 4 to Point 1) converting the steam to water and completing the steam cycle. The amount of heat supplied by the steam generator is represented by the area under the curve 1-2-3. The heat rejected or removed by the condenser is represented by the area under the line 1-4. The net remaining area represents the net available for useful work.

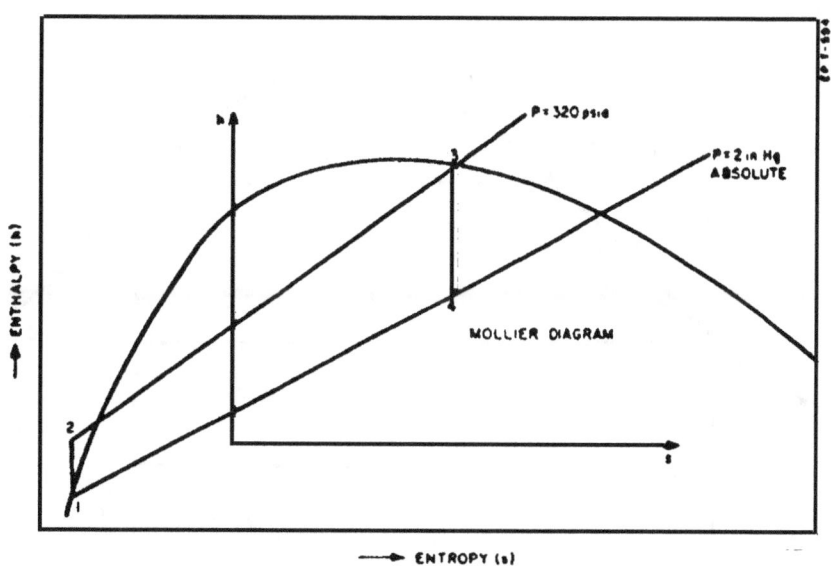

Exhibit 4-5 — Mollier (h-s) Diagram

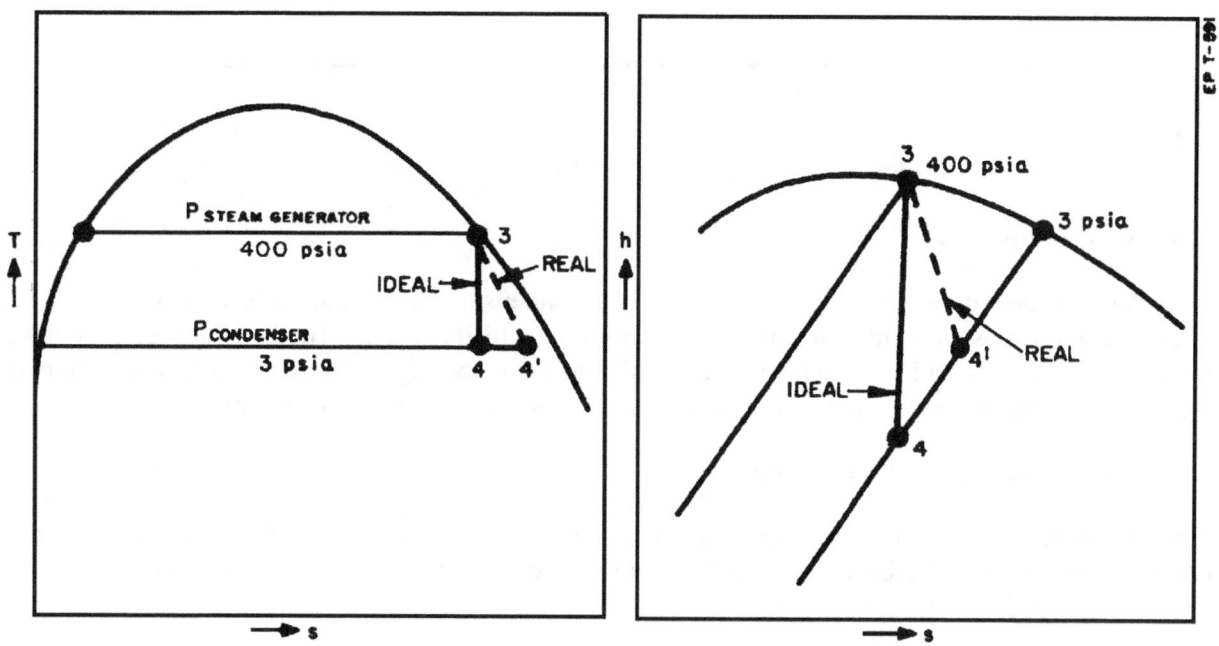

Exhibit 4-6 — Comparison Of Ideal And Real Turbines Behavior On T-s And h-s Diagrams

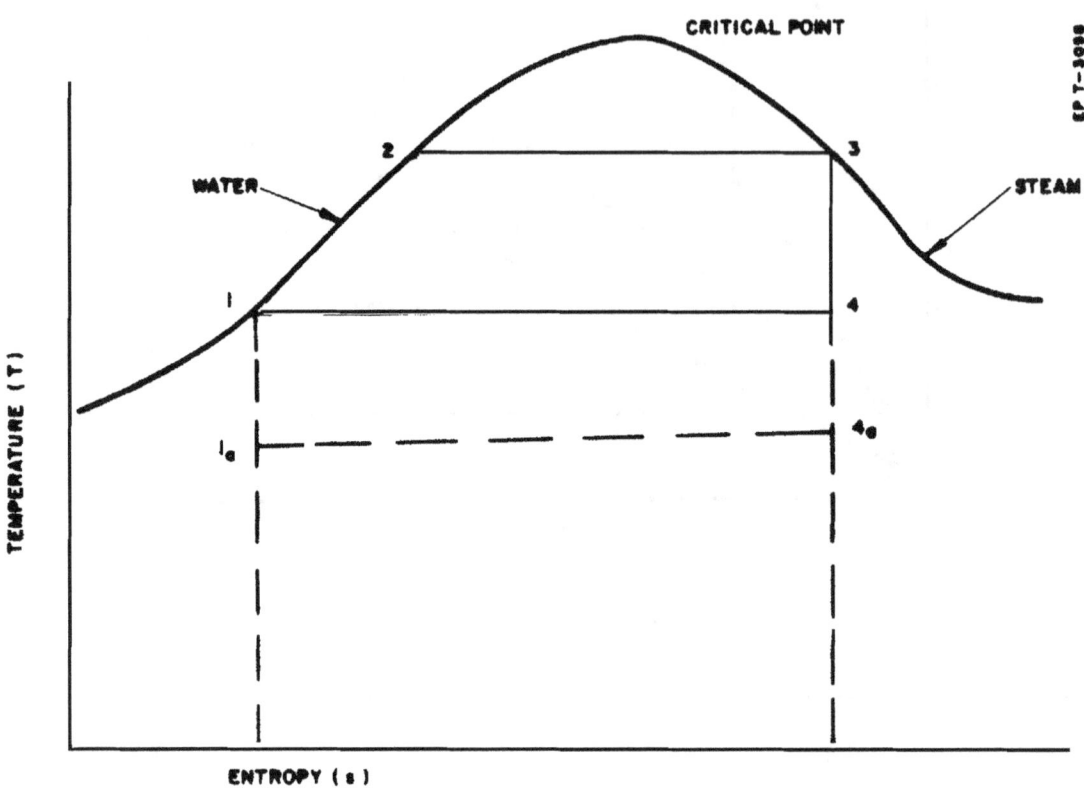

Exhibit 4-7 — Temperature-Entropy (T-s) Diagram

4.1.8.1 — Sink Temperature

As can be demonstrated by reference to Exhibit 4-7, if the condensing temperature is lowered (line to 4a), more energy is available for useful work. The lowest temperature at which the steam can be condensed is known as the sink temperature. It is set by the temperature of the condenser cooling water. At the prototype plants it is the spray pond or cooling tower temperature, and on ships it is the seawater temperature.

4.1.8.2 — Steam Cycle Thermal Efficiency

The thermal efficiency of a steam cycle is the percentage of the total heat added in the steam generator that is converted into useful work. Referring again to Exhibit 4-7, the efficiency, expressed in terms of the specific enthalpy at Points 1, 2, 3, and 4, is

Equation 4-6

$$\text{EFFICIENCY}\,(\eta) = (h_3 - h_4)/(h_3 - h_1)$$

4.1.8.3 — The Effect Of Condenser Pressure In The Steam Cycle

Reducing the pressure in the condenser by reducing the condensing temperature (Exhibit 4-7) increases the thermal efficiency. To illustrate this point consider a steam cycle which produces saturated steam (Point 3) at 310 psia (T_3 = 420.36°F). If the exhaust pressure of the turbine is 12.5 psia and the condenser produces no subcooling then $T_1 = T_4 = T_{sat}$ for 12.5 psia or 203.95°F. To obtain the efficiency for this steam cycle we proceed as follows:

1. <u>Specific Entropy and Enthalpy of Saturated Steam at 310 psia s_3 and h_3</u> (point 3) - From the steam tables we find that the specific entropy of saturated steam at 310 psia = s_3 = 1.5076 Btu/lbm °F. The specific enthalpy = h_3 = 1203.2 Btu/lbm.

2. <u>Specific Enthalpy at Turbine Exhaust (P_4 = 12.5 psia)</u> - Assuming that entropy is unchanged from point 3 to point 4, from the Mollier Diagram we determine that the specific enthalpy at point 4 = h_4 = 974 Btu/lbm, which is the specific enthalpy for s_3 = 1.5076 and P_3 = 12.5 psia.

3. <u>Specific Enthalpy at Condenser Exhaust (h_1)</u> - Assuming no subcooling, the specific enthalpy at the condenser exhaust (h_1) is that for saturated fluid at 12.5 psia, h_1 = 172.06 Btu/lbm.

4. <u>Efficiency (η)</u> - We can now estimate the thermal efficiency:

$$\eta = (h_3 - h_4)/(h_3 - h_1)$$

$$= (1203 - 974)/(1203 - 172) = 0.222.$$

We now compare the efficiency for P_4 = 12.5 psia (22.2%) with that obtained when the turbine exhaust pressure is 2 inches of mercury absolute or 0.982 psia.

Repeating Steps 2, 3, and 4 above we determine that h_4 = 841 Btu/lbm and h_1 = 69 Btu/lbm. The efficiency is:

$$\eta = (h_3 - h_4)/(h_3 - h_1)$$

$$= (1203 - 841)/(1203 - 69) = 0.319.$$

The thermal efficiency for a typical steam plant with a condenser injection temperature of 60 °F is in the range of 25 percent.

4.1.8.4 — Loss Of Work Due To Throttling

The available work per pound of steam will decrease if the governor or throttle valve is partially closed, reducing the pressure at the nozzle entrance. For example, assume the steam entering a turbine is throttled so that steam chest pressure is 200 psig, all other conditions remaining the same. To determine the effect of throttling we shall compare the efficiency obtained assuming that $P_4 = P_1$ = 2 inches of Hg absolute and, therefore, h = 69 Btu/lbm. Assuming that throttling reduces pressure without changing the enthalpy, the specific entropy at the turbine entrance is that for h_3 = 1203 Btu/lbm and a pressure of 215 psia. From the Mollier diagram we obtain s_3 = 1.544 Btu/lbm °F. A constant entropy expansion in the turbine to pressure of 2 inches of Hg absolute yields h_4 = 862 Btu/lbm. The efficiency is then given by

$$\eta = (h_3 - h_4)/(h_3 - h_1)$$

$$= (1203 - 862)/(1203 - 69) = 0.301 (30.1\%),$$

versus 0.319 (31.9%) percent in the absence of throttling. Efficiency losses from throttling effects (drop in steam pressure without the performance of work) are small; throttling losses are minimized by the throttling configuration of a series of consecutively opening **poppet valves**, which minimize the substantial pressure drop the steam would experience across a single, partially open throttle valve.

Section 4.2 — Propulsion Plant Equipment

4.2.1 — Turbines

A steam turbine is a heat engine in which thermal energy in steam is converted to kinetic energy and then to work. In the turbine, the steam's velocity is increased as it passes through a nozzle, which converts the steam's thermal energy to kinetic energy by expanding the steam from a higher to a lower pressure. The **impulse** (push) of the steam is directed via the nozzle to **blades** attached to a rotating **wheel** and forces the wheel (**rotor**) to move.

There are two basic types of turbines used, **impulse** and **reaction**. In the impulse turbine category are the **simple impulse**, **velocity-compounded**, **pressure-compounded**, and **pressure-velocity compounded** subtypes.

4.2.1.1 — Simple Impulse (Rateau Stage) Turbine

This turbine has one set of nozzles and one **row**, or **wheel**, of moving blades (Exhibit 4-8). To get maximum work from a turbine we want to operate it at its **ideal blade speed (IBS)**, which for a simple impulse wheel is:

Equation 4-7

$$IBS = \frac{v \cos \alpha}{2}$$

where:

v velocity of entrance steam, ft/sec

α angle of steam entry (i.e., nozzle angle), degrees.

At the ideal blade speed, the steam leaving the blade has a zero component of absolute velocity in the direction of blade motion, i.e., the speed that yields the optimal efficiency of converting kinetic to mechanical energy.

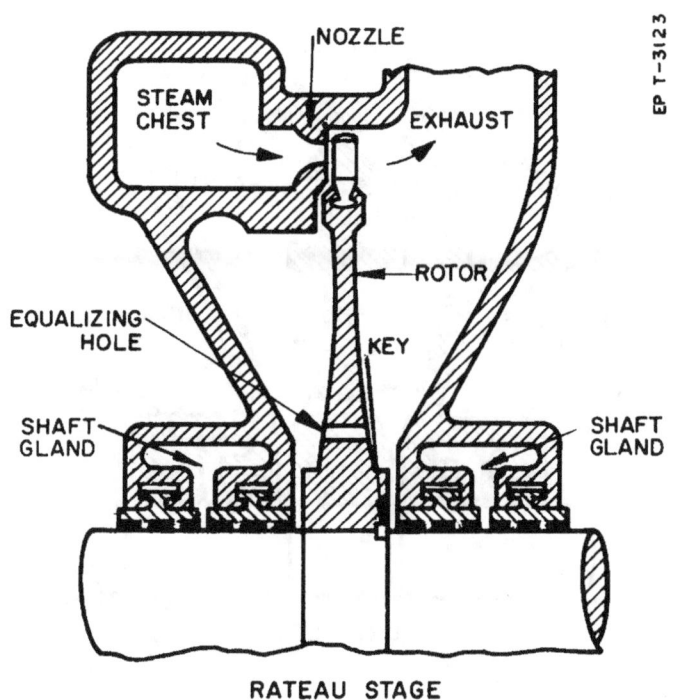

EP T-3123

RATEAU STAGE

Exhibit 4-8 — Simple Impulse (Rateau Stage) Turbine

Because of the limitations of the various turbine materials, the simple impulse turbine cannot be built to operate at its ideal blade speed; consequently, it must operate at less than maximum efficiency. Although inherently inefficient, simplicity of design and construction make it well-suited as a small auxiliary turbine. For a more complete discussion of ideal blade speed refer to Reference 6.

4.2.1.2 — Velocity-Compounded (Curtis Stage) Turbine

In the **velocity-compounded**, or **Curtis stage** impulse turbine, we have at least two sets of moving blades for each set of nozzles (Exhibit 4-9) as opposed to only one set of blades in the simple impulse. The advantage of this turbine is that ideal blade speed is lower for the same velocity and entrance angle. Ideal blade speed for a Curtis wheel that has two rows of moving blades is half that of a simple impulse wheel; therefore, for a constant blade speed, the pressure drop across a nozzle in a Curtis stage is greater than the pressure drop across a nozzle in a Rateau stage. With a Curtis stage first, for a constant blade speed, the pressure and temperature at the inlet to the Rateau stage blading is less, which allows the casing to be designed with less strength and at a lower cost.

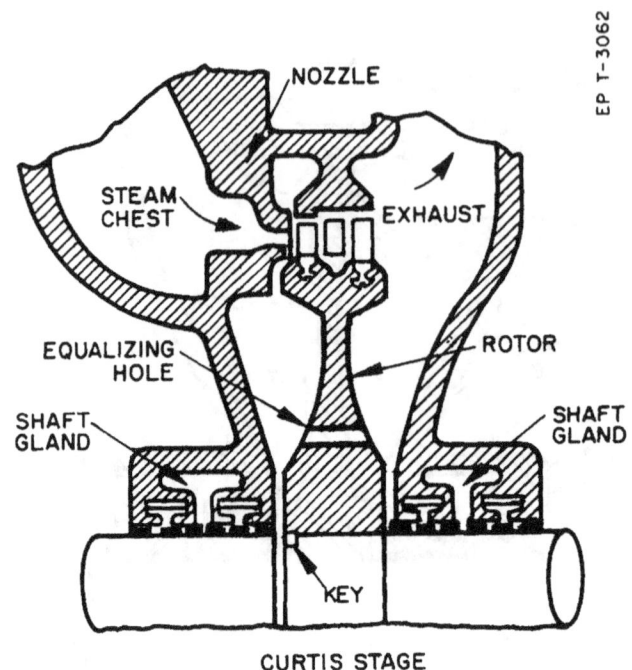

EP T-3062

Exhibit 4-9 — Velocity-Compounded (Curtis Stage) Turbine

With a lower temperature at the inlet of a Rateau stage, blade **creep** is minimized. This allows the blade design strength and cost to be reduced. Creep conditions may still exist at full power operation but, since full power is utilized only a small percentage of the time, and since creep is a function of time, the amount of creep is small.

4.2.1.3 — Pressure-Compounded Impulse Turbine

Another means of reducing ideal blade speed is to allow the pressure reduction to take place in steps rather than to cause the total pressure drop to occur only in one set of nozzles. The result is a number of simple impulse stages connected in series. The steam leaving the first stage, instead of going to the exhaust, flows through a second set of nozzles, then through a second row of moving blades, and so on, depending upon the number of simple impulse stages in series. In the pressure-compounded turbine the fixed blades between alternate rows of moving blades are **vanes** attached to the casing and are shaped as convergent nozzles. Since a pressure drop occurs in each set of nozzles, each combination of a set of nozzles and its succeeding row of moving blades constitutes a stage. Therefore, the pressure-compounded turbine is a multi-stage turbine (Exhibit 4-10).

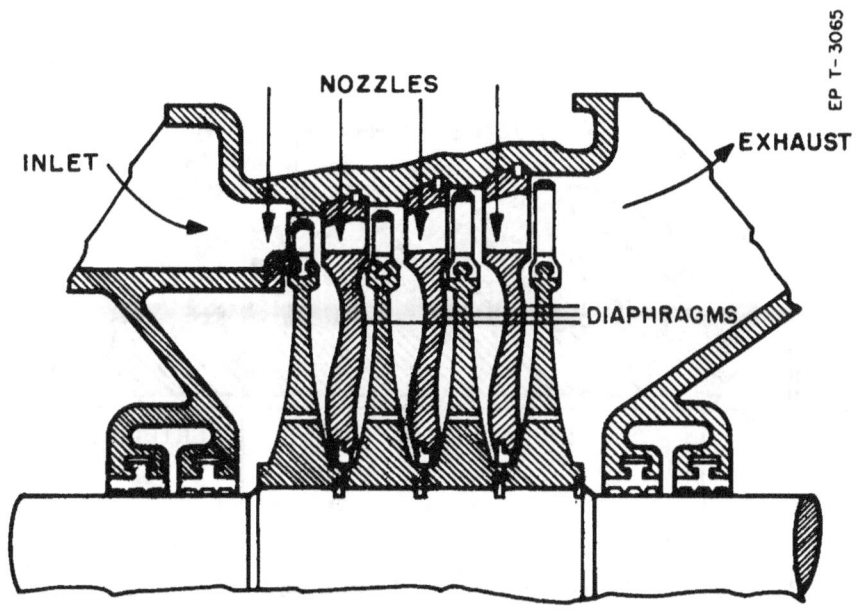

Exhibit 4-10 — Pressure-Compounded Impulse Turbine

In the pressure-compounded impulse turbine there is no pressure drop across the moving blades; in all impulse stages the only pressure drop occurs across the nozzles.

The steam is expanded by the first stage nozzles to produce a high-speed jet. It then passes through the first row of moving blades and gives up part of its kinetic energy; thus, its speed is reduced. From these blades the steam enters the second set of nozzles, and the process is repeated. The total pressure drop, from steam chest to exhaust, is divided into as many steps as there are stages. This results in a relatively low pressure drop in each set of nozzles and a relatively low steam entrance speed for each stage. If a sufficient number of pressure stages is used, an ideal blade speed within the practical turbine material limitations can be obtained.

The simple impulse stage is sometimes called a Rateau stage, and a pressure-compounded turbine, which is a series of Rateau stages, is known as a **Rateau turbine**.

4.2.1.4 — Pressure-Velocity-Compounded Impulse Turbine

The ideal blade speed for a Curtis turbine is one-half that of a Rateau turbine for the same steam velocity, so to use these turbines in combination we place the Curtis stage ahead of the Rateau turbine. The steam velocity to the Rateau would be less than to the Curtis, so turbines could operate in series and more closely approach the ideal blade speeds of both turbines. In naval turbine installations it is common to have the high-pressure turbine constructed of a Curtis stage followed by a series of Rateau stages all on a common shaft. A cross-section of a pressure-velocity compounded turbine is shown in Exhibit 4-11.

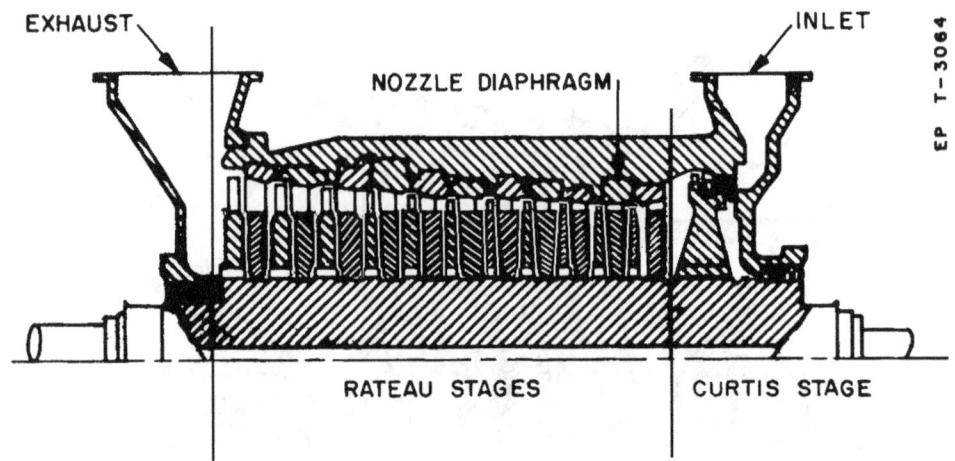

EP T-3064

Exhibit 4-11 — Pressure-Velocity-Compounded Impulse Turbine

4.2.1.5 — Reaction Turbine

If you hold a garden hose while water is flowing through it, you feel the push, or **reaction**, of the hose in the opposite direction of water flow. Similarly, in a reaction turbine the moving blade is shaped like a nozzle and there is a push against the blade from the steam expansion as the steam leaves the blade. Also in a reaction turbine there is some force on the blade due to the impulse of the steam on the blades, but most of the driving force is due to reaction or expansion of the steam.

As previously observed, both the moving blades and the fixed blades are shaped like nozzles; therefore, for simplicity in discussing reaction turbines, nozzles will be referred to as fixed blades and the word nozzles will not be used. You can now see that a reaction turbine could have just one set of fixed blades and one set of moving blades. But this would not be practical.

The ideal blade speed for a reaction turbine is $\mathbf{v}\ \mathbf{cos}\ \alpha$, so for high steam velocities a simple reaction turbine would have too high a blade speed to operate at its most efficient point; therefore, reaction turbines are not normally used as high-pressure turbines in the Navy. Normally, reaction turbines are used where there is low velocity steam, such as in low-pressure turbines, because they can operate closer to ideal blade speed. Since there is a lower pressure and temperature the turbine can be constructed of lighter and less expensive materials.

In the impulse turbine there is no pressure drop across the moving blades so there is no axial thrust. However, since the reaction turbine has a pressure decrease, there is a thrust on the blades and rotors in the direction of steam flow. The axial thrust is created by the drop in pressure across the moving blades and the unequal steam pressure acting on the two ends of the drum. As shown in Exhibit 4-12, there are two ways to cancel out this thrust.

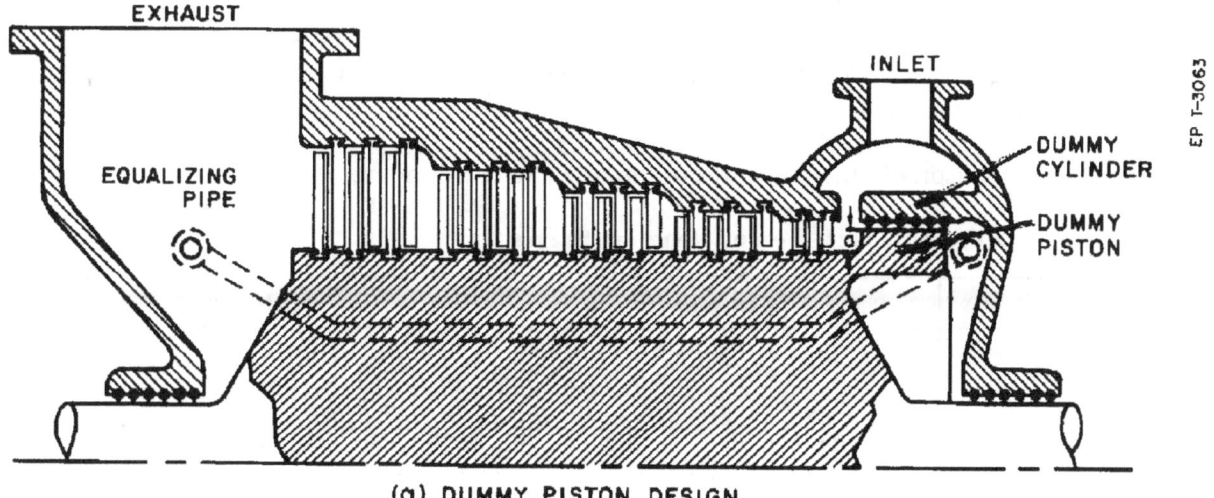

(a) DUMMY PISTON DESIGN

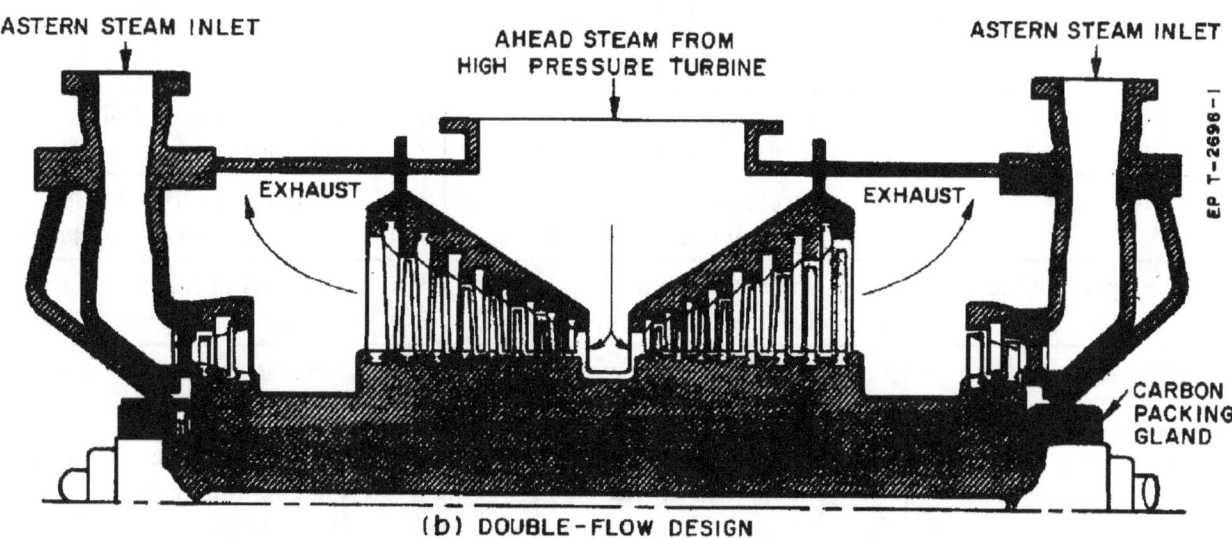

(b) DOUBLE-FLOW DESIGN

Exhibit 4-12 — Reaction Turbine Designs Created In The Other Section

Low-pressure turbines are frequently of the pure reaction type. The usual practice in low-pressure turbines is to design the turbine so that the steam enters the middle of the turbine and there divides to flow in opposite directions toward exhaust chambers at each end. Such a turbine is called a **double-flow turbine** and is shown in Exhibit 4-12. This construction allows the axial thrust created by the pressure drop across the moving blades in one section of the turbine to counter the axial thrust created in the opposite section of the turbine.

Another technique is to construct a turbine using a dummy piston. The **dummy piston** is an extension of the rotor drum having a step-like cross section. Surrounding this piston is a part of the casing known as the **dummy cylinder**. Labyrinth packing, usually referred to as **dummy packing**, is installed between the dummy piston and cylinder. An **equalizing pipe**, which connects the exhaust end of the turbine with the area between the dummy packing and shaft gland at the high pressure end, equalizes the pressure in these two areas. As can be seen in design (a) of Exhibit 4-12, the shoulder area on the rotor is under full inlet steam pressure while the corresponding area on the other side of the dummy piston is under exhaust pressure. This difference in pressure causes a thrust that partially counterbalances the thrust in the opposite direction caused by the pressure drop across the turbine blades.

Table 4-1 presents a comparison of salient features of impulse and reaction turbines.

Table 4-1 — Salient Features: Impulse Versus Reaction Stages

	Impulse	Reaction
Blading and Velocity Diagram		
Steam Enthalpy	1. Decreases in nozzles only 2. Constant in moving blades	1. Decreases in fixed blades 2. Decreases in moving blades
Steam Pressure	1. Decreases in nozzles only 2. Constant in moving blades	1. Decreases in fixed blades 2. Decreases in moving blades
Absolute Speed of Steam	1. Increases in nozzles 2. Decreases in moving blades.	1. Increases in fixed blades 2. Decreases in moving blades
Relative Speed of Steam	1. Increases in nozzles 2. Constant in moving blades	1. Increases in fixed blades 2. Increases in moving blades
Nozzles	1. Convert entire enthalpy drop in stage to kinetic energy 2. Direct steam into moving blades	1. Convert part of enthalpy drop in stage to kinetic energy. 2. Direct steam into moving blades
Fixed Blades or Guide Vanes	In velocity-compounded stages only; serves only to redirect (guide) the steam; no nozzle action.	Not used in reaction stages
Ideal Blade Speed	$v_{ibs}= (v_1 \cos \alpha)/2$ in an impulse stage	$v_{ibs}= (v_1 \cos \alpha)$ in a pure reaction stage
Stage Types	1. Simple impulse 2. Velocity compounded 3. Pressure compounded 4. Pressure velocity compounded.	1. Pure reaction Pressure compounded.

4.2.2 — Reduction Gears

To operate at maximum efficiency turbines need to operate at a high speed; propellers, or screws, need to operate at low speeds. **Reduction gears** reduce the speed of the turbines to a slower speed for main shaft and propeller operation. Although reduction gears are of the single or double reduction types, reduction gears in naval ships are almost always of the double reduction type.

In naval main propulsion installations, the reduction gear **wheels** and **pinions** are of the **double-helical** or **herringbone** type. A **single helical** or **twisted spur gear** has teeth cut at an angle. They are helical in form and become long-pitch screw threads that run very smoothly with little noise and with even pressure distribution along the entire length of the tooth. This twist introduces an end thrust in the shaft, which is undesirable. When a **double-helical gear** is used, in which each tooth cuts half a right-hand helix and half a left-hand helix, the two halves introduce end thrusts in opposite directions which neutralize each other. A constant stream of lubricating oil, supplied to these gears from the main engine lubrication system, is directed via supply nozzles between the gears at the point where they mesh.

Various types of reduction gears are illustrated in Exhibit 4-13 through Exhibit 4-16; Exhibit 4-13 shows an early type of single reduction gear, illustrating the high-pressure and low-pressure turbine connections through separate shafts; Exhibit 4-14 shows a simple double-reduction gear used in modern cargo and passenger ships; and Exhibit 4-15 shows the **nested** arrangement of a simple double-reduction gear in which the first reduction gears are nested between the two helices of the main gear to save space.

Most modern combatant ships are equipped with the **locked-train** type of double-reduction gear shown in Exhibit 4-16. Because the power is transmitted to the main gear through four pinions instead of two, the tooth pressures are lower. This design also provides for a more compact unit. To assist in equalizing the loads on the gear teeth a **quill shaft** with flexible couplings connects the first reduction gear to the second reduction pinion. A quill shaft assembly consists of a shaft within a shaft; it allows for some axial and radial misalignment and provides more flexibility than if the gears were connected solidly to each other.

The term **articulated** means that the gear set has quill shafts. The term **locked train** indicates that the power is distributed equally between the two quill shafts on each side of the assembly. The floating member (quill shaft) is relatively long and passes through the hollow centers of both gears. It is supported only at the ends where its teeth mesh with the teeth of the shaft rings. This arrangement allows the pinion to be placed close to the driven gear, resulting in a shorter reduction gear casing. A quill shaft is shown in Exhibit 4-17. The quill shaft assembly also provides radial flexibility. In summary, the quill shaft assembly provides the axial and radial flexibility required between the pinion and driven gear of high-speed reduction gears.

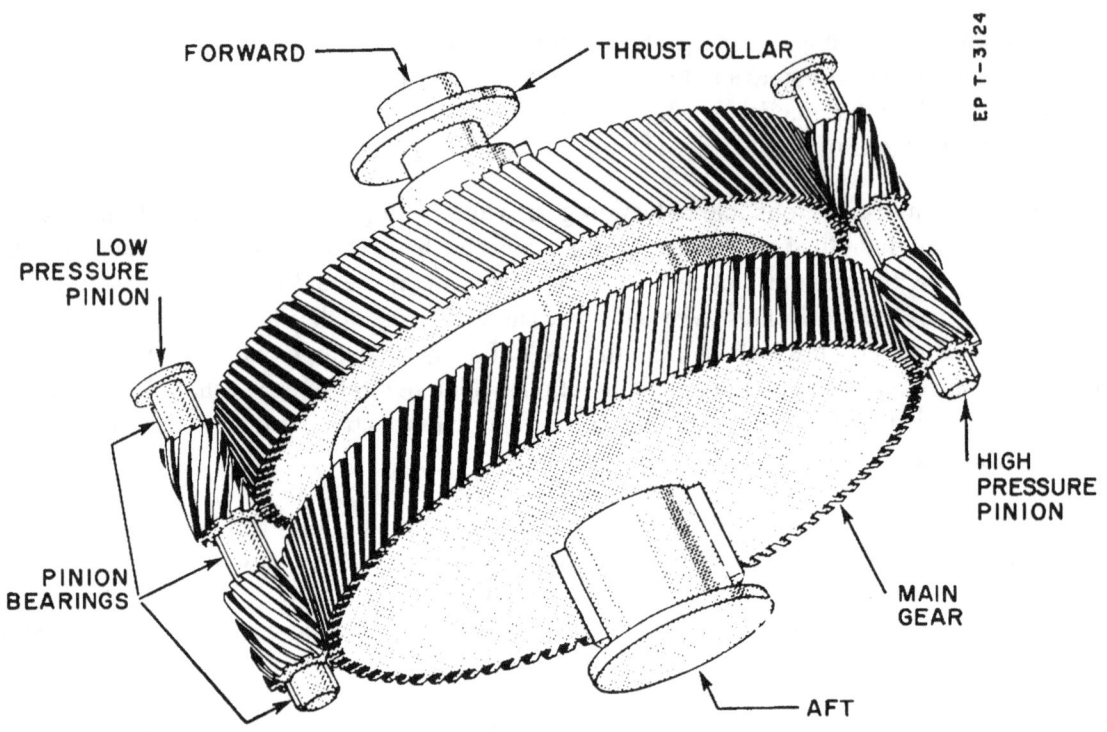

Exhibit 4-13 — Reduction Gear Arrangement, Single-Reduction Type

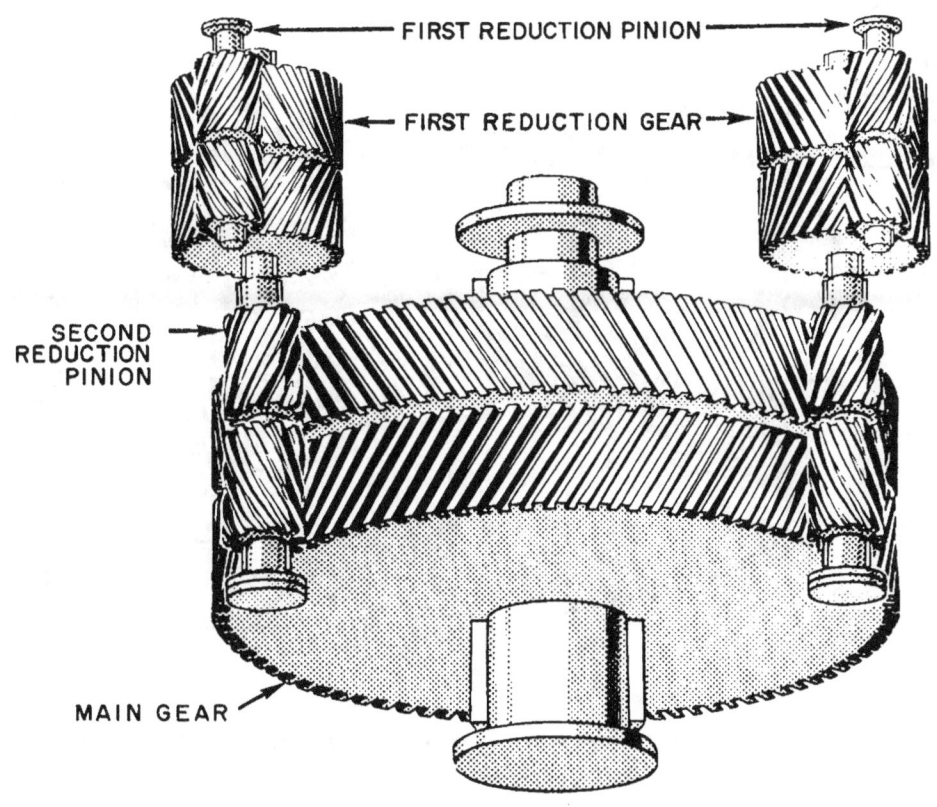

Exhibit 4-14 — Reduction Gear Arrangement, Simple Double-Reduction Type

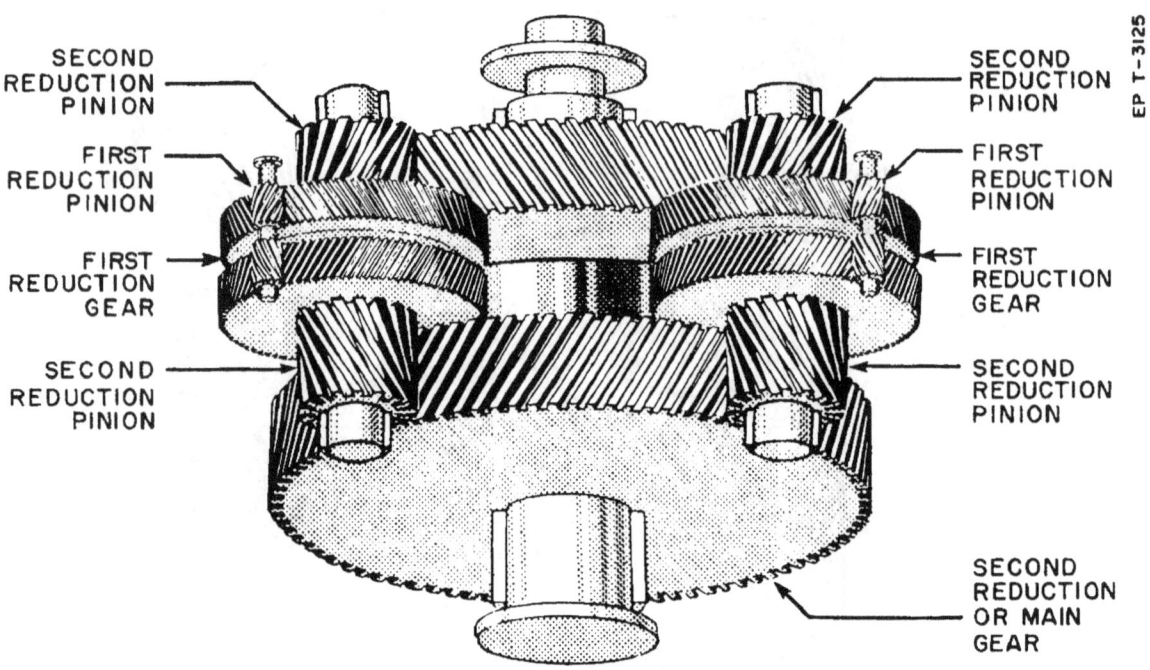

EP T-3125

Exhibit 4-15 — Reduction Gear Arrangement, Nested Double-Reduction Type

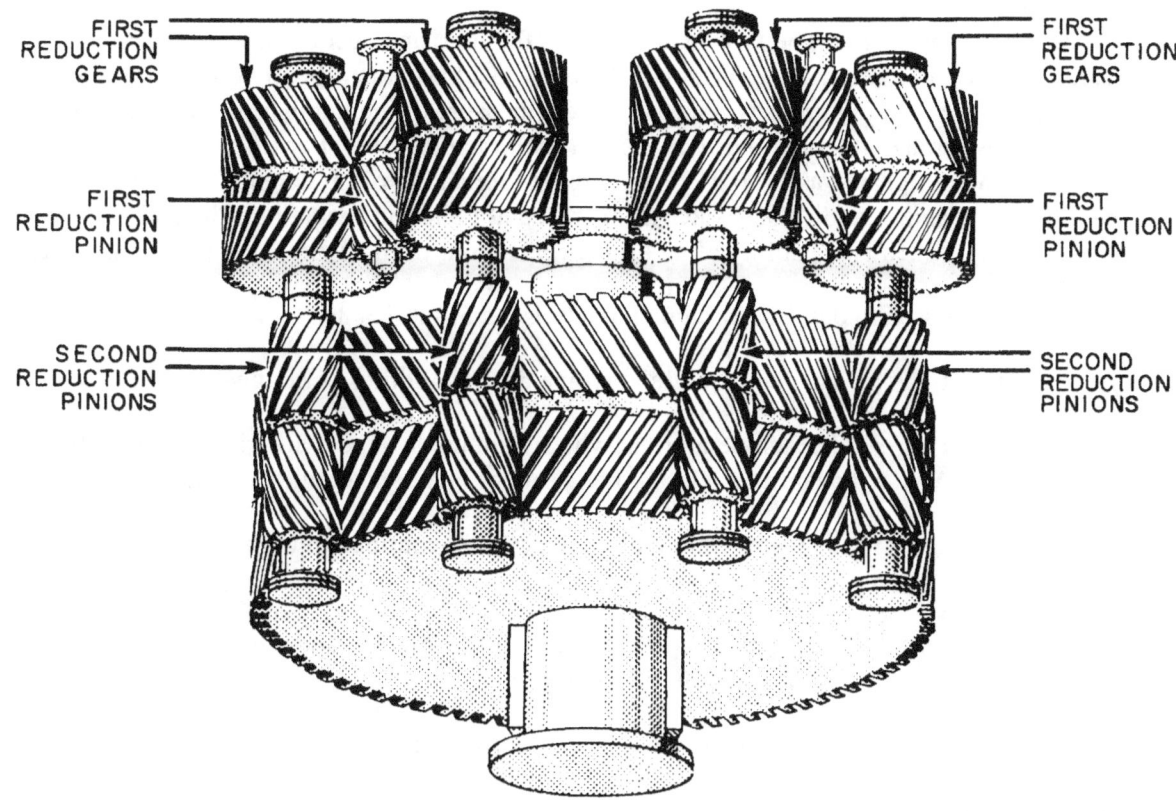

Exhibit 4-16 — Reduction Gear Arrangement, Locked-Train Double Reduction Type

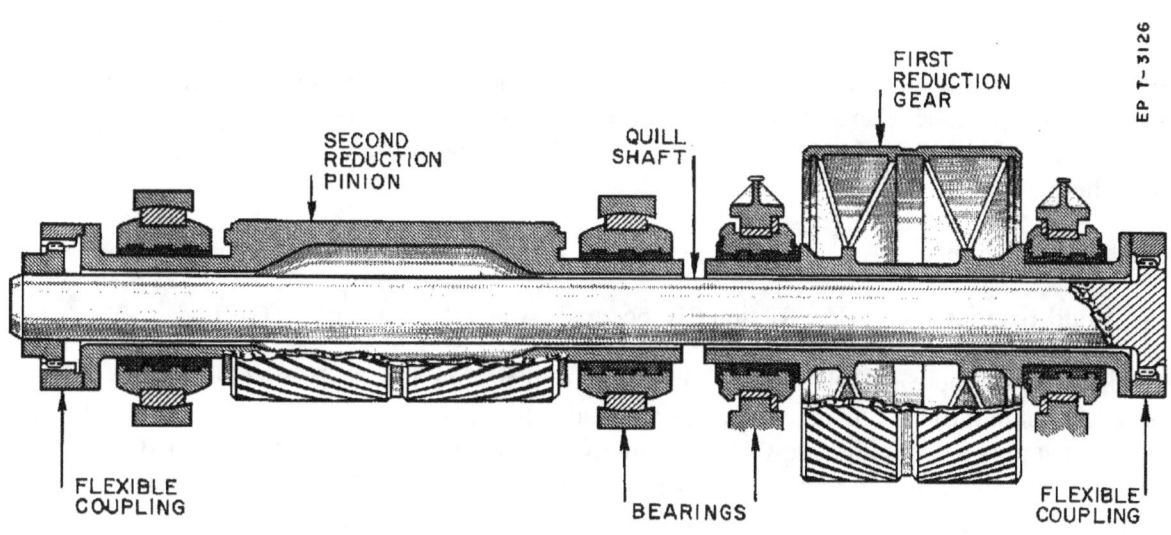

Exhibit 4-17 — Quill Shaft

4.2.3 — Lubricants (Oil)

Lube oil symbols have four digits. The first digit classifies the lubricant according to use and the last three digits indicate viscosity. The letters H, T, TH, or TEP added to a basic identification number indicate that the oil contains additives for special purposes. Viscosity is the measure of the number of seconds it takes a given quantity of oil to flow through a standard orifice at a particular temperature. For turbine oil, this temperature is 130 °F. Lubrication oil designated as 2190 TEP is a turbine lubricating oil. It contains additive materials that increase the lubricant's ability to displace water from steel and to inhibit oxidation. The additives have little effect on demulsibility, which is the ability of an oil to separate out from an emulsion of oil and water.

The following terms are used to describe certain characteristics of lubricants:

Pour Point	The temperature at which an oil solidifies or congeals.
Flash Point	The temperature at which flammable vapors are given off in sufficient quantity to flash when brought into contact with a flame.
Fire Point	The temperature at which oil will continue to burn when ignited.
Autogenous Ignition Temperature	The temperature at which the flammable vapors given off will burn without the application of a spark or flame.
Viscosity (body)	A measure of the internal friction of the fluid.
Neutralization Number	A measure of the acid content of an oil: defined as the number of milligrams of potassium hydroxide (KOH) required to neutralize 1 gram of the oil.

4.2.4 — Bearings

Friction is defined as the resistance to motion. As the resistance to motion increases, the amount of heat generated increases and the chance of damage to moving parts increases; consequently, prime consideration is given to the reduction of friction in the moving parts of equipment. There are two types of friction to be considered here:

Sliding friction	The friction that occurs between two solid surfaces sliding over each other.
Fluid friction	The friction that occurs between the molecules of a fluid.

Fluid friction is much more desirable than sliding friction; therefore, it is desirable to substitute fluid friction for sliding friction in turbines and other equipment. Also, a lot of heat is generated at the turbine bearings, so it is necessary to supply a continuous flow of lubricant to remove this heat. The heat generated comes from two places in a turbine; some from fluid friction within the lubricant, but mostly from heat conducted along the shaft from the rotor, which is heated to a high temperature by the steam flowing through the turbine.

There are two main types of **bearings** in turbine applications under consideration, **journal** and **thrust** bearings. In these bearings there are two distinct **lubricating films** established, a **boundary film** and a **fluid film**. The boundary film is the same in both journal and thrust bearings. It is a result of a physical and chemical union between the metal and lubricant, is only a few molecules thick, and has characteristics of both the metal and the lubricant. Boundary films are important during starting, stopping, and temporary interruptions of the oil supply. The film is very strong and is the last boundary between metal-to-metal contact when the metals are being pressed together.

4.2.4.1 — Journal Bearings

Exhibit 4-18 illustrates how the fluid film for a journal bearing is established. **Journal bearings** maintain alignment and provide support for the shaft or rotor by absorbing radial thrust, which is the thrust at right angles to the axis of the shaft. The rotating shaft is called the **journal**, and the metal surrounding the shaft is the **bearing**. In Exhibit 4-18 the clearances in a journal bearing, the region where oil is supplied, are exaggerated to make the lubricating principle more visible. When at rest, the journal is in contact with the bearing. There is a slight separation provided by the boundary film layer. When journal rotation is started, the journal tends to roll toward a new contact point. As the journal rotational speed increases, oil adheres to the journal and is carried into the film, which increases the film thickness and lifts the journal. As journal speed continues to increase, steady-state **hydrodynamic** lubrication is established as more oil is drawn into the wedge between journal and bearing, and a full fluid film is built up, as shown in Exhibit 4-18.

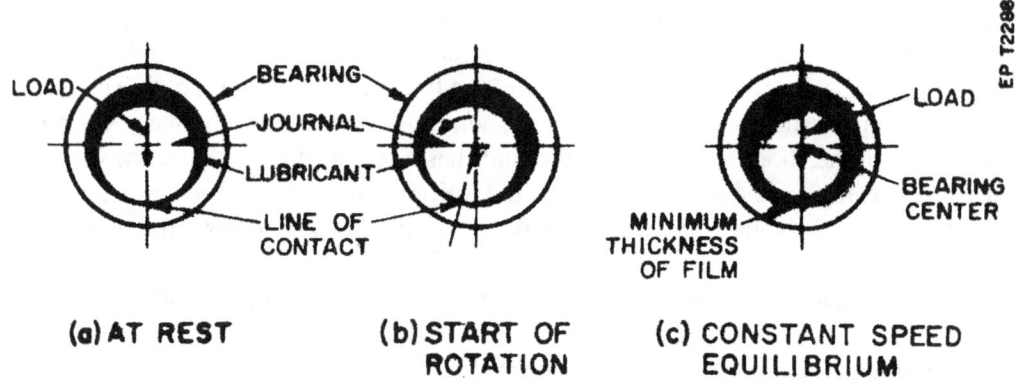

Exhibit 4-18 — Shift Of Journal Position During Changes In Speed

The **journal bearing shell** is made of steel and is split horizontally; the **bearing lining** is made of tin-base **babbitt**. The shell backs are doweled together, and the bearing is doweled to the case to prevent rotation of the bearing. The top of the bearing babbitt is relieved to receive the oil supply and to spread an oil film over the length of the bearing. An **annular groove** is machined into the bearing babbitt on the inner face to distribute oil evenly across the lubricated areas. Radial drain passages in the lower half of the groove drain the oil back to the pump. **Thermocouples** are used to measure bearing temperature. They are inserted to a point 1/64 inch below surface of the bearing babbitt at the point of greatest load. All turbine bearing temperatures are monitored remotely at a temperature monitoring (TM) panel.

Oil seals consist of rings of labyrinth-type seals made in halves. The lower half of the oil seal has vertical drains through it. Oil creeping along the shaft is collected by the oil seal and drained through the lower seal to be collected in a pocket in the lower half. This pocket forms a vapor block that prevents oil vapor from passing through the seal and into the ship.

A typical shaft penetration, starting inside the turbine case and working out consists of:

Gland seal	Three labyrinth seals with a steam supply and a leak-off.
Slop drain	Space between the gland seal and the oil seal, which collects leakage from gland seal and oil seal and drains it to the bilge.
Bearing	A bearing with an oil seal on each side.

4.2.4.2 — Thrust Bearings

A **thrust bearing** also employs the principle of oil adhering to a turning component and thus creating a fluid film between the moving and stationary components. The thrust bearing differs from a journal bearing in that it is used to support weight or take up thrust that is transmitted axially or parallel to the shaft, or rotor, of the component. The main applications of thrust bearings of the **Kingsbury (pivoted shoe)** type are for turbine and main shaft installations with, of course, the larger of the two being main shaft thrust bearings. The principle of the Kingsbury thrust bearing is that an **oil wedge** is formed when one surface is moved past another thus separating the surfaces with a film of oil. This is accomplished by dividing the stationary bearing element into segments called **shoes** (Exhibit 4-19). These shoes are supported so that they are free to tilt slightly. The rotating element (or **collar**) is attached to, and turns with, the turbine or propeller shaft. The entire assembly is submerged in oil. When the shaft and the collar rotate, some oil is dragged in between the collar and the shoe at the leading edge of the shoe. The thrust on the shaft has a tendency to squeeze the oil out again. Since the oil can enter from one edge and exit at three edges the trailing edge oil passage can be much narrower than the leading edge. Thus, the shoe is tilted and the wedge established. The essential elements of the Kingsbury thrust bearing are:

1. Thrust collar transmits thrust from the shaft through the oil film to the shoes. The thrust collar is keyed to the turbine or propeller shaft.
2. Shoes maintain the proper oil wedge and transmit the thrust to the leveling plates.
3. Leveling plates equalize the thrust load among the shoes. The shoes rest on the upper leveling plates and the upper leveling plates rest on the lower leveling plates.
4. Base ring holds the leveling plates in place and transmits the thrust to the ship's structure. The lower leveling plates rest upon the base ring.

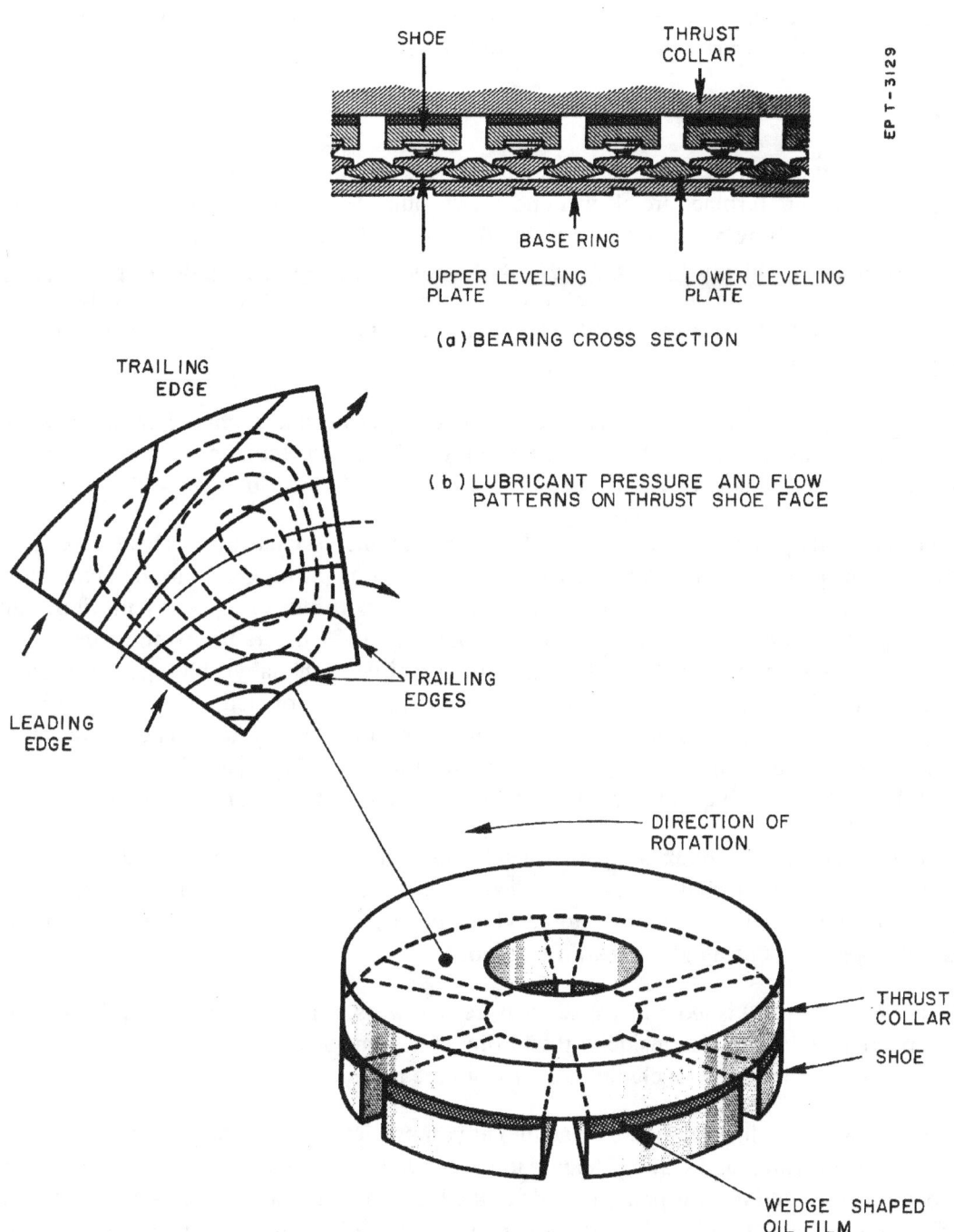

Exhibit 4-19 — Kingsbury Thrust Bearing

Exhibit 4-19 shows the face of a thrust shoe with arrows indicating oil entering the leading edge and oil leaving the trailing edge. The solid lines indicate oil flow and the dotted lines indicated equal pressure areas with the highest pressure area in the center. Both double- and single-acting thrust bearings are used. **Double-acting**

bearings have shoes and leveling plates on both sides of the collar; **single-acting bearings** have shoes on only one side and are used where thrust will be exerted in only one direction.

4.2.5 — Seals

In the high-pressure end of high-pressure turbines a pressure exists that is considerably above atmospheric pressure. Most low-pressure turbines are always under a vacuum. When the astern turbine (usually located in the same casing with the low pressure ahead turbine) is in use, the ahead turbines run idle in a vacuum. When engines are stopped with condensers in operation, as when standing by, both turbines are under a vacuum. Wherever a turbine rotor shaft passes through its casing there exists a pressure differential between the inside and outside of the turbine casing. **Shaft glands** are installed to restrict the steam along the shaft from flowing out of a turbine.

There are two general types of shaft glands, **labyrinth-packing glands** and **carbon-packing glands**. Labyrinth packing, in its simplest form, consists of machined packing strips or fins mounted on the casing surrounding the shaft so as to make a very small clearance between the shaft and the strip.

The principle of labyrinth packing is that as steam leaks through the very narrow spaces between the packing strips and the shaft, its pressure drops. As the steam passes from one packing strip to the next, its pressure is gradually reduced and any velocity that it might gain through **nozzling** is lost by **eddy motion**. Labyrinth packing is used on all high-pressure glands where high-pressure and temperature would damage carbon packing. Carbon-packing glands, which employ the same principle as the labyrinth type, are used in glands where low pressures and temperatures exist. Carbon packing rings are mounted around the shaft in segments that are held together by means of springs and are prevented from turning by keys. These block segments are formed so that when the ends of the segments are forced together by springs, there is a close clearance between the blocks and the shaft. A small quantity of steam leaks between the blocks and the shaft.

Neither the labyrinth nor the carbon packing completely stops the flow of steam out of the turbine, nor will they stop the flow of air into a turbine. Because air leaking into a turbine will find its way into the condenser and destroy the vacuum it is vitally important to provide a positive stop to prevent air in-leakage through the shaft glands. This is done by using **gland sealing steam**.

Steam at about 2 psig (17 psia) is led into a space between two sets of gland packing. As its pressure is greater than the atmosphere and the pressure inside of the turbine, the gland steam flows both into the casing and out into the engineroom thus positively excluding all air from the turbine.

Since even high pressure turbines are under a vacuum at various times (when standing by or backing down) all turbine glands are fitted with gland sealing steam. However, when operating at high speeds, the turbine internal pressures at both the high and the low pressure ends of the high pressure turbine are usually considerably above atmospheric pressure, while at both ends of the low pressure turbine the pressure is below atmospheric. Suppose now that the pressure inside the turbine, instead of being 1 psia, is 40 psia and, after having passed through the labyrinth packing into the chamber where the gland steam is admitted, its pressure has been reduced to 18 psia. The steam will then flow in two directions. Part of it will continue to flow out of the turbine gland through the carbon packing, but most of it, being at a higher pressure than the gland sealing steam, will flow into the gland sealing connection reversing the direction of the gland steam. Usually the gland sealing line will be closed at such times, in which case a **gland leak-off connection** is provided to lead the excess steam to a lower stage of the turbine or to the condenser.

Exhibit 4-20 shows a typical gland installed on the high-pressure end of a modern high pressure turbine. In addition to the standard gland sealing connection this gland is also provided with two leak-off connections. These connections conduct part of the steam that leaks past the labyrinth packing back into the lower stages of the turbine. In this particular gland the steam is led back into the 8th and 12th stages of the high pressure turbine.

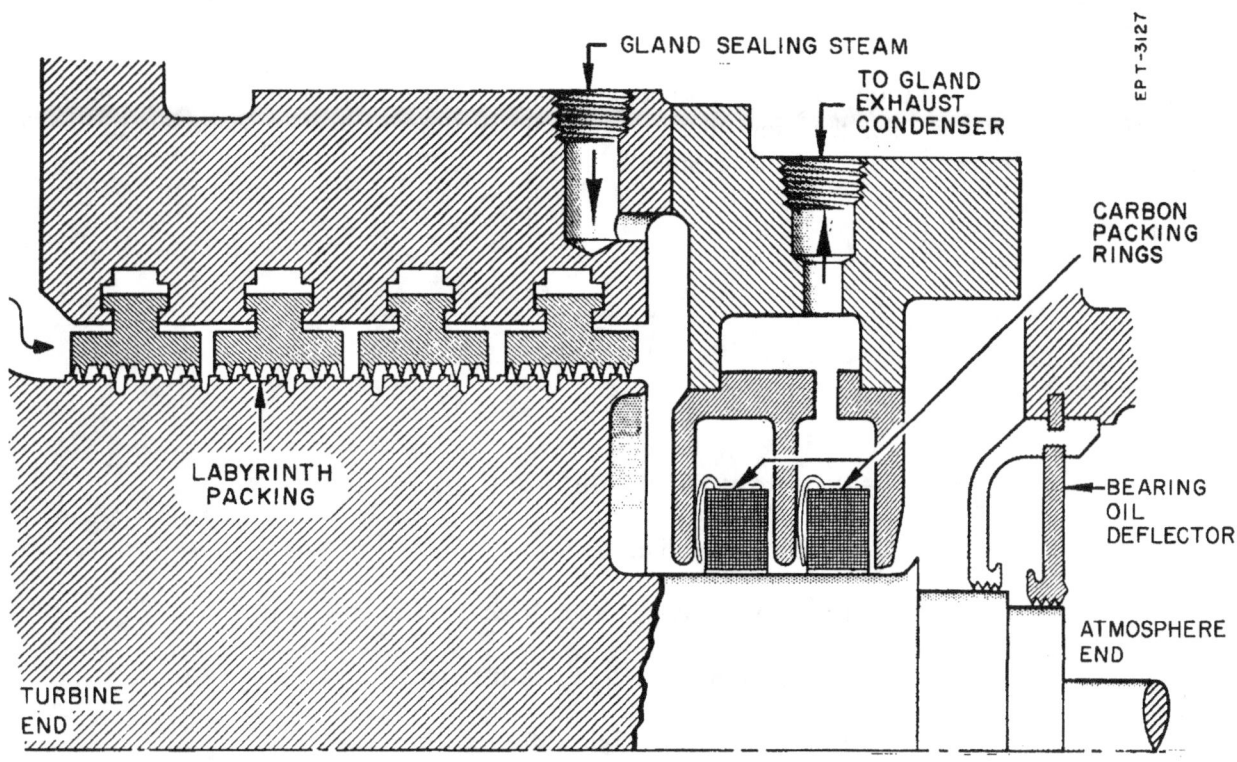

Exhibit 4-20 — Typical High-Pressure Turbine Gland

The labyrinth packing shown is made up of four **packing rings**. These six-segment rings are mounted in the **gland housing** and held in place by **carriage springs**. The dimensions of the segments are such that, when held inward toward the shaft by the springs, there is a small clearance between fins and shaft. The carbon packing rings are also made in segments; they are held in place by means of a metal strap, the ends of which are connected by a spring. The carbon blocks are cut to provide proper clearance between shaft and blocks. Both types of packing are prevented from turning by keys or pins. The use of carbon packing rings in high pressure turbines has been discontinued in favor of labyrinth packing in modern installations.

The shaft glands are fitted with a gland leak-off system by which excess steam and air are pulled into a small condenser by a gland exhaust compressor which maintains about a 1 inch Hg vacuum on the leak-off line. The gland leak-off connection in Exhibit 4-20 is between the first and second carbon packing rings. Steam is condensed and returned to the condensate system. Noncondensibles are discharged to the ship's atmosphere by the gland exhaust compressor.

Section 4.3 — Pumps

4.3.1 — Energy Relationships

The purpose of a pump is to raise fluid pressure so that a fluid may be transferred from one point in a system to another. While a pump may occasionally cause a change in fluid speed, its net effect is generally an increase in pressure at the pump discharge.

Pump head (H_P) is the **ideal pump work** performed on a unit mass of fluid as it flows through the pump. In other words, it is the energy which the pump adds to a unit mass of fluid, neglecting frictional effects within the pump. Pump head can be expressed mathematically as

Equation 4-8

$$H_p = \frac{P_2 - P_1}{\rho} + \frac{v_2^2 - v_1^2}{2g_c} + \frac{g(z_2 - z_1)}{g_c}$$

where:

P	= Pressure of Fluid, lbf/ft^2
ρ	= Fluid Density, lbm/ft^3
v	= Linear Velocity of Fluid, ft/sec
g_c	= Unit Conversion, 32.17 $lbm\text{-}ft/lbf\text{-}sec^2$
z	= Height Above Reference, ft
g	= Standard Acceleration of Gravity, 32.17 ft/sec^2.

The first term in Equation 4-8 accounts for the increase in specific flow energy across the pump; the second term accounts for the increase in specific kinetic energy across the pump; the last term accounts for the increase in specific potential energy across the pump. Refer to Reference 6 for a more detailed discussion.

For the common case in which the inlet and outlet diameters of the pump are equal ($v_2 = v_1$) and the height differential between the inlet and outlet is negligible ($z_2 = z_1$), the work done on the fluid by an ideal pump results in an increase in the flow energy of the fluid. This reduces Equation 4-8 to:

$$H_p = \frac{P_2 - P_1}{\rho}$$

The power required by an ideal pump is:

Equation 4-9

$$\dot{W}_p = H_p \cdot \dot{m}$$

where:

$\dot{W}_p$ = Ideal (Hydraulic) Pump Power, ft-lbf/sec

$\dot{m}$ = Mass Flow Rate, lbm/sec.

As fluid passes through a real pump, friction and turbulence result in energy losses, which appear as an increase in internal energy of the fluid or as in ambient heat loss from the system. The ratio of ideal power to real power is defined as the **mechanical efficiency** of the pump.

4.3.2 — Centrifugal Pumps

A centrifugal pump, as the name implies, depends on **centrifugal force** for operation. This force is generated by the high-speed rotation of the **impeller**, shown in Exhibit 4-21. Essentially, the impeller consists of a **shaft** with a radially mounted series of curved **vanes** and with suitable means provided for entry of the liquid at the **eye**, or center. When a positive supply of the liquid being pumped is initially introduced at the eye, the high speed rotation of the impeller causes the vanes to literally throw the liquid outward, imparting to it a high velocity head. The pressure head of the liquid at this stage is relatively low as compared to that at discharge. As a result, almost all of the energy that the moving liquid possesses at this point is kinetic.

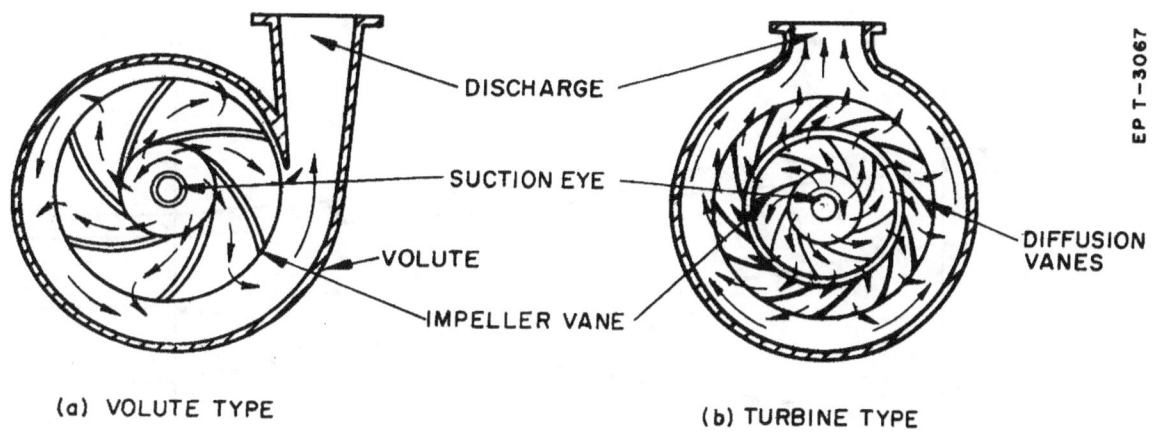

Exhibit 4-21 — Centrifugal Pumps

Upon leaving the curved vanes of the impeller, the high velocity liquid enters the stationary pump **casing** which encloses the impeller. This section of the casing is termed the **volute**. The clearance between the vanes and casing increases as the liquid approaches the discharge connection to the casing. The purpose of this volute is to collect the fluid being discharged from the impeller at high velocity and gradually bring it to a relatively low velocity, thus converting a large part of the velocity head to pressure head.

In this type of pump the casing is provided with suction, discharge, drain, and vent connections. The suction connection is designed to guide the liquid from the suction chamber of the casing to the eye of the impeller.

Principles of the straight diffuser illustrate the pressure-and-velocity changes that occur in a pump volute, as the volute is essentially a variation of the straight diffuser, which has been changed to a diverging nozzle whose axis is bent in a 360 degree circle.

Assumptions concerning Exhibit 4-22 will apply to a centrifugal pump volute. If a fluid is flowing through the diffuser from Point A to Point B and there is no change in density of the fluid and no work done on or by the fluid, then the energy possessed or exhibited by the fluid is the same at Point A as it is at Point B. This energy is the sum of the potential (pressure) energy and kinetic (velocity) energy. Thus, KE + PE at A equals KE + PE at B. If the mass flow rate equals the cross-sectional flow area times the velocity, and the area at B is larger than the area at A (density constant), then the velocity at B is less than the velocity at A. Therefore, the kinetic energy at B is less than the kinetic energy at A. Then the potential energy at B must be greater than the potential energy at A for the energy at point A to equal the energy at point B. Therefore, in a closed flowing fluid system, if the flow area is increased, the velocity decreases and the pressure increases. This is an example of the energy and mass balances that Equation 4-1 and Equation 4-2 utilize. The purpose of the pump volute is to decrease the velocity and increase the pressure of the fluid being pumped.

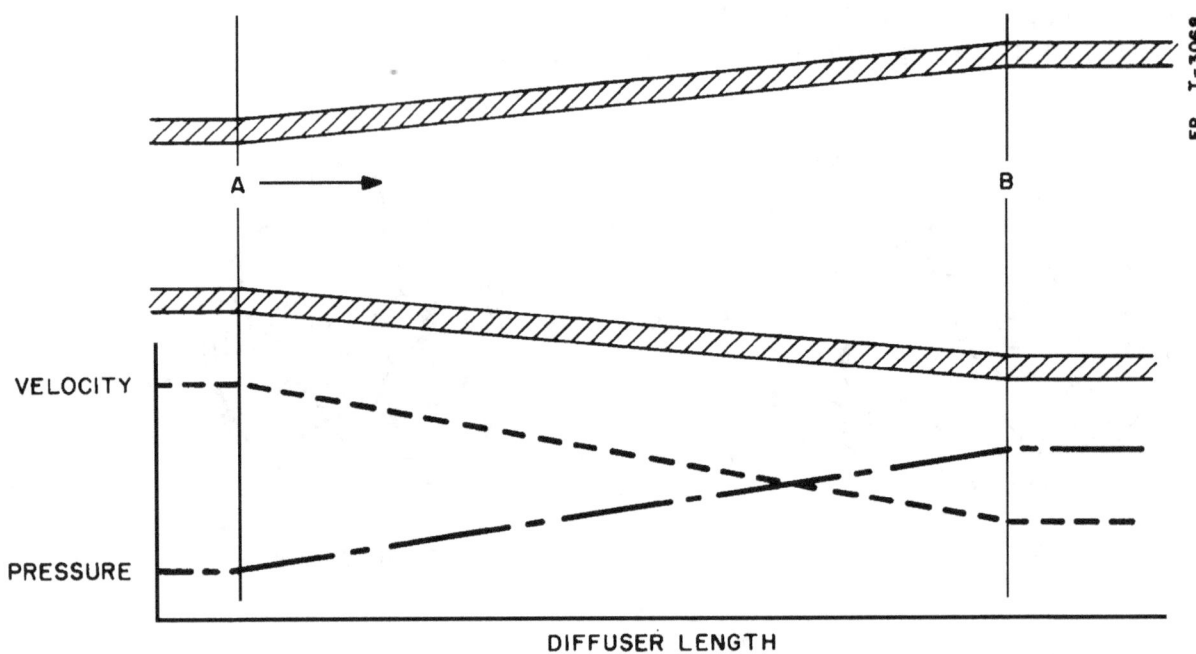

Exhibit 4-22 — Straight Diffuser And Associated Pressure And Velocity

A volute converts some of the kinetic energy of the high-velocity mass of fluid into potential energy or pressure in permitting the moving mass to slow down as it passes through the widening channel of the volute. As the passage widens toward the outlet, less linear velocity is required to pass the same mass of liquid in a given length of time. Now if, at the same time, the discharge is restricted so that the volute remains filled with fluid, it can be seen that the impact of the entering fluid upon the restrained mass of fluid tends to develop a static pressure. This discussion will help in understanding the following characteristics of the centrifugal pump:

1. Opening the discharge valve with no change in pump speed results in a decrease in discharge pressure. Similarly, closing the valve causes an increase in pressure (Exhibit 4-23).
2. If the discharge valve is fully shut, the pressure attains a certain value, called the **shutoff head**, after which the impeller churns and heats the fluid with no further increase in pressure.
3. Increasing or decreasing the impeller speed, with a constant valve opening, causes a corresponding increase or decrease in discharge pressure.
4. The pump is not self-priming; the pump casing and impeller must be full of fluid before the pump begins to function. To attain this **net positive suction head (NPSH)**, centrifugal pumps are located below the level of the source from which they take suction, or fluid is supplied to the pump suction nozzle under pressure supplied by another pump placed in the suction line. Mathematically, NPSH is determined to be

$$NPSH = \frac{P_{INLET}}{\rho} + \frac{V_{INLET}}{2g_c} - \frac{P_{SAT}}{\rho}$$

where:

P_{INLET} = Fluid Pressure at Pump Inlet, lbf/ft^2

ρ = Fluid Density, lbm/ft^3

V_{INLET} = Linear Velocity of Fluid at Pump Inlet, ft/sec

g_c = Unit Conversion, 32.17 $lbm\text{-}ft/lbf\text{-}sec^2$

P_{SAT} = Vapor Pressure of the Fluid, lbf/ft^2.

In some types of pumps, the volute is dispensed with and the liquid leaving the periphery of the impeller passes through radial diffusion vanes (Exhibit 4-21), which are shaped to convert the velocity to static pressure and deliver it to the outer periphery of the pump casing. This pump is a turbine-type centrifugal pump.

Centrifugal pumps are classified as follows:

Shaft Position — Horizontal or vertical, depending upon whether the pump shaft is horizontal or vertical.

Number of Stages — Single stage, with discharge pressures usually not exceeding 150 psig, or multistage pumps for pressures in excess of 150 psig. In the latter case, the first-stage impeller discharges to the suction eye of the second-stage impeller, the second to the third and so on.

Number of Inlets   Single suction impellers where water enters the eye in only one direction or double suction where water enters the eye from two directions.

Type of Impeller   Closed or open impellers. The open impeller uses the pump casing to confine flow through the vane channels; whereas, the closed impeller uses shrouds on the impeller. Impeller vanes can be either straight or francis vanes. Straight vanes are curved, while francis vanes are curved and twisted.

It is necessary to maintain a certain minimum suction head (inlet pressure) on an operating centrifugal pump to prevent the formation of vapor or gas bubbles from the gas dissolved in the pumped fluid. If the static pressure becomes lower than the saturation pressure of the fluid, these bubbles will form at points of high velocity within the pump and migrate to points of lower velocity. As these bubbles move along to points of lower velocity where the static pressure is higher than the saturation pressure of the fluid being pumped, they will collapse. This alternating bubble formation and collapse phenomenon is called **cavitation** and can result in flow decrease, vibration, noise, and erosion. In vertical centrifugal pumps it can lead to catastrophic bearing failure in a very brief time. The difference between total head at the pump inlet and saturation pressure head at the pump inlet is called **minimum net positive suction head (NPSH$_{MIN}$)**, which is determined to be

Equation 4-10

$$NPSH = \frac{P_{INLET} - P_{SAT}}{\rho}$$

where:

P_{INLET} = Fluid Pressure at Pump Inlet, lbf/ft^2

P_{SAT} = Vapor Pressure of the Fluid, lbf/ft^2.

ρ = Fluid Density, lbm/ft^3.

One means used to prevent cavitation and provide positive suction head to a centrifugal pump is to use a booster pump to keep a pressure at the suction of the pump. Another method of keeping the suction pressure sufficiently high is to locate the pump as far as possible below the level of the storage tank from which it is taking a suction. Other ways to prevent cavitation are to subcool the liquid being pumped prior to its entry into the pump suction, and to keep the amount of free gases in solution to a minimum.

Centrifugal pump flow, head, and power in a closed system can be interrelated by a series of equations known simply as the **pump laws**. These laws are:

1.   Volumetric flow rate ($\dot{V}$) is proportional to the speed of the pump (N), or ($\dot{V} \propto N$).
2.   Pump head (H$_P$) is proportional to the speed of the pump squared (N^2), or $\mathbf{H_P} \propto \mathbf{N^2}$.
3.   Pump power (P) is proportional to the speed of the pump cubed (N^3), or $\mathbf{P} \propto \mathbf{N^3}$.

The flow characteristics for centrifugal pumps can be readily depicted graphically on a H$_P$ versus $\dot{V}$ plot, commonly called a **pump curve**. Pump curves are commonly developed empirically by the pump manufacturer.

The flow characteristics for multiple pump combinations can be determined from the pump curves. The two main types of configurations are parallel and series.

Parallel   Since the outlets for parallel pumps have matching discharge pressures and the inlets have matching suction pressures, pumps in parallel cannot develop a higher H_P than a single pump. Two pumps can handle twice the flow rate. So for pumps in parallel the shutoff head is the same as for a single pump, but the flow rate is increased by the multiple of the number of pumps in parallel. Refer to Exhibit 4-23.

Series   When in series each pump cannot handle more volume than a single pump, but the pump's total head is added to the previous pumps. So for pumps in series, the maximum flow rate is the same as for a single pump, but the total head is increased by the multiple of the number of pumps in series.

The frictional forces within and on the working fluid of a system cause energy being used to maintain flow (mechanical and flow energy) to be converted into a less useful form of energy (internal energy). The converted energy is called the **headloss (h_L)** of the system. Headloss for a closed system can be defined as:

Equation 4-11

$$h_L = (\Sigma K) \bullet \dot{V}^2$$

where:

ΣK = Summation of Resistance Coefficients for Piping and Individual Components, $lbf\text{-}sec^2/lbm\text{-}ft^5$

$\dot{V}$ = Volumetric Flow Rate, ft^3/sec.

The curve for the headloss of the system intersects the pump curve at the **operating point (S)** of the system, as depicted in Exhibit 4-23.

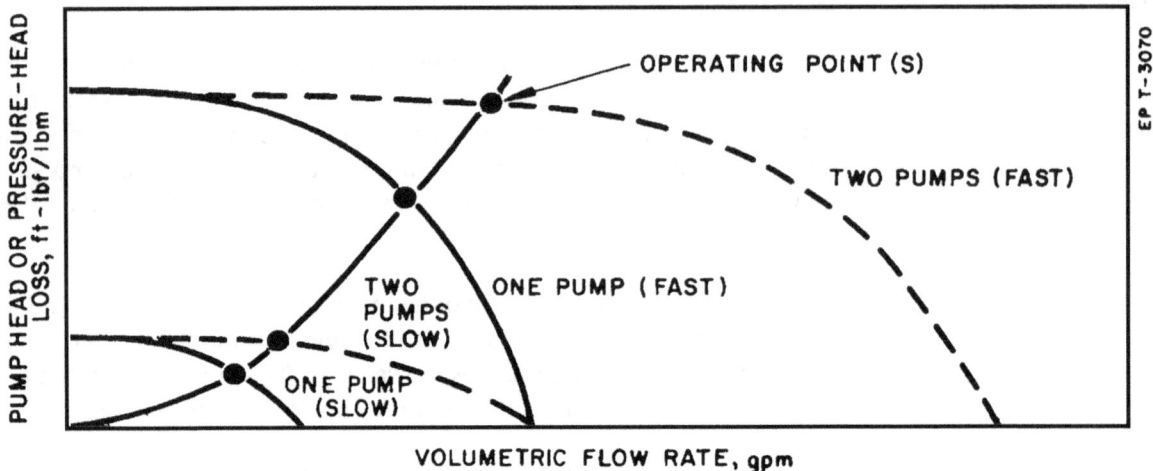

Exhibit 4-23 — Operating Points For Two 2-Speed Pumps

Centrifugal pumps are used in series to overcome a larger system head loss than one pump can compensate for individually. As illustrated in Exhibit 4-23, two identical pumps operating at the same volumetric flow rate contribute the same pump head. Since the inlet to the second pump is the outlet of the first pump, the head produced by both pumps is the sum of the individual heads. The volumetric flow rate from the inlet of the first pump to the outlet of the second remains the same.

Also, using two pumps in series does not actually double the resistance to flow in the system. The two pumps provide adequate pump head for the system and also maintain a slightly higher volumetric flow rate.

Consider operating two two-speed pumps in series in a closed system. We would have four operating points with two pumps in that system as shown in Exhibit 4-23. Note that the shut-off head (zero flow head) of the pumps does not change with the number of pumps operating, but the flow and head at the operating points does increase. Note also that the flow does not double, but only increases approximately 25 percent by doubling the number of operating pumps. Further increases in the number of pumps operating in a system will cause only small flow increases. Reference 6 explains in detail how pump and system curves are developed.

4.3.3 — Positive Displacement Pumps

One characteristic of all positive displacement pumps is that a constant volume output results when the pump is operated at a constant speed, regardless of the pressure or flow resistance. This characteristic makes these pumps valuable when a high-pressure and/or a metered (constant volume per unit time) output is necessary. A typical use for the positive displacement pump in the primary system is the coolant charging pump, which has a high-pressure and a metered output requirement. In the secondary system, the positive displacement lube oil pumps provide a metered output. Positive displacement pumps are generally classified as either **reciprocating** or **rotary**. Reciprocating positive displacement pumps have one or more pistons fitting snugly into cylinders (Exhibit 4-24). On one half-cycle, the chamber is filled through the **suction valve** with the fluid to be pumped. On the other half-cycle, the liquid is forced into the line through the **discharge valve**. This operation means that for each cycle the pump pushes exactly one cylinder volume of fluid into the line. The valve arrangement of this pump is such that on the suction half-cycle, the suction valve is drawn off its seat by the low pressure created when the piston moves out of the cylinder. Line pressure seats the discharge valve. On the pumping stroke, the suction valve is seated and the discharge valve lifts off its seat to allow the fluid to enter the line.

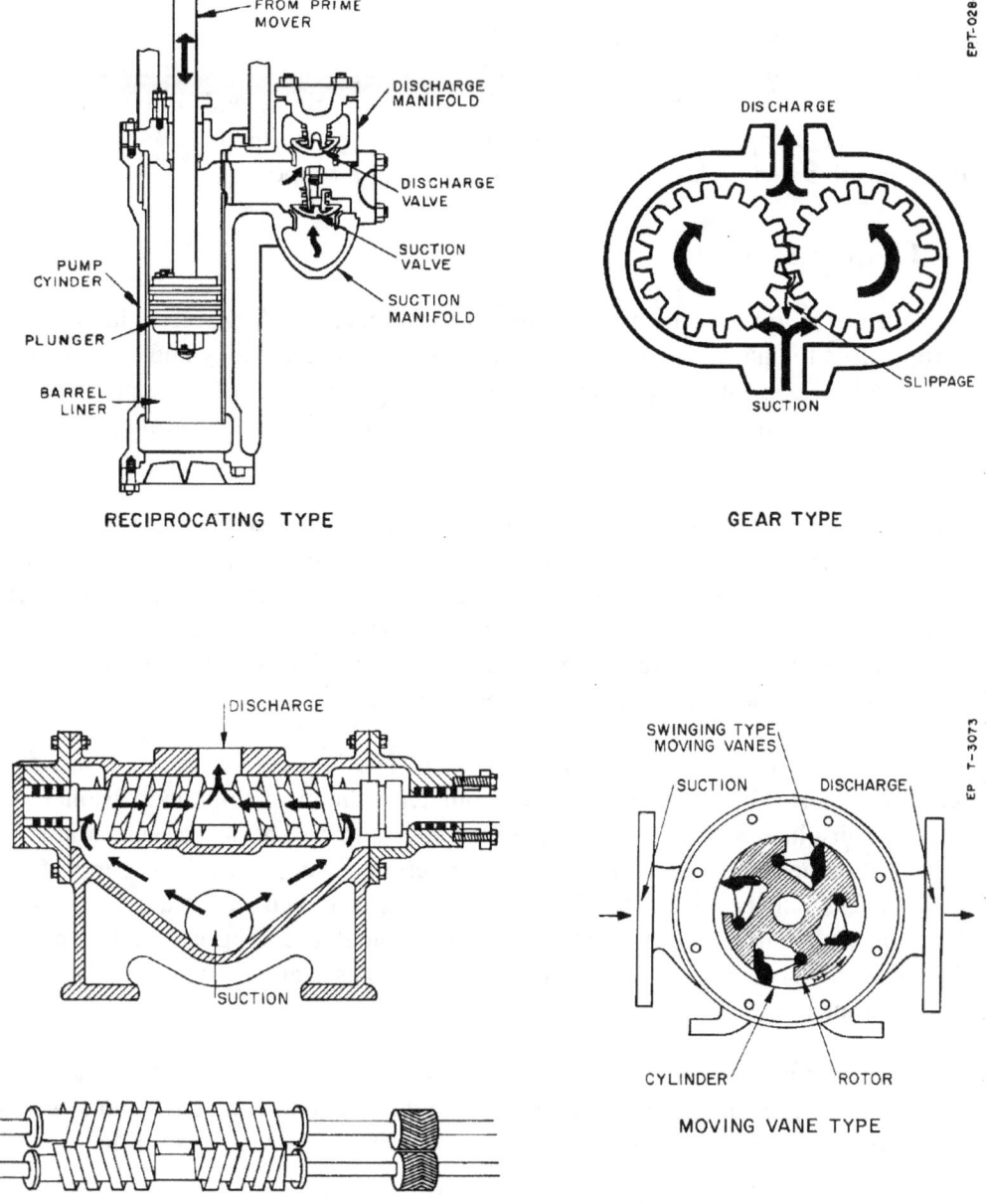

Exhibit 4-24 — Some Positive Displacement Pump Types

A rotary positive displacement pump has a more continuous pumping action. In this pump, the **gears**, **screws**, or **lobes** travel very close to the walls of the pump casing, Exhibit 4-24. The fluid is trapped between the rotating members and the casing and is carried from the suction side to the discharge side. The clearances between the rotating members and the casing are small to minimize fluid slippage back to the relatively low-pressure suction side of the pump. Clearances in positive displacement pumps are critical, and a slight misalignment of the unit, overheating, bad bearings, or a bent shaft can cause the pump to wear out very quickly.

Several precautions must be observed when operating positive displacement pumps:
- Never operate a positive displacement pump with the discharge valve shut, because excessive pressure could build up and cause damage to the pump and downstream piping and components.
- Periodically check the pump relief valves for proper operation and lineup.
- Always keep a positive suction head on the pump to prevent air binding.

Section 4.4 — Condensers

One purpose of a condenser is to lower the backpressure on a turbine, which enables a larger steam pressure and temperature drop across the turbine and thereby increases both the efficiency and capacity of the turbine. Another purpose is to collect condensate for reuse as boiler feedwater. Also, the condenser provides a convenient point in the cycle to introduce makeup water and to collect fresh water from plant low-pressure and high-pressure drain systems.

Heat transfer action in a condenser is hindered by the presence of noncondensible gases that mix with the film of condensate of the tube surface. Tube arrangement must be such that steam passage through the tube nest will not be restricted; steam must retain sufficient velocity to impinge upon the tube surfaces and thereby keep the film of condensate reduced to a minimum. If the sources of air and noncondensibles are numerous, heat transfer will be seriously affected. Fouled tubes, air leakage, and insufficient circulating water are all causes of low condenser vacuum because they inhibit the heat transfer process.

The **condenser vacuum**, measured as a depression below atmospheric pressure, is commonly stated in terms of inches of mercury (inches Hg) because a mercury manometer is the most suitable instrument for measuring that pressure range. The absolute pressure in the condenser equals atmospheric pressure minus vacuum when both are expressed in the same units. Standard atmospheric pressure at sea level supports a mercurial barometer at a height of 29.92 inches of Hg at 32 °F (on what is considered an "ideal" day - on a stormy day this pressure will be lower). At an elevation of 5000 feet the barometer reading is 24.89 inches of Hg at standard conditions. Roughly speaking, atmospheric pressure decreases at the rate of 1 inch of Hg for each 1000 foot increase in altitude.

The absolute pressure of both containers in Exhibit 4-25 is the same although the indicated pressures differ. With this in mind the boiling pressures of both containers would be equal. Hence, vacuum, we can say is relative to our reference of pressures and does not change the boiling point of water unless the absolute pressure is also changed.

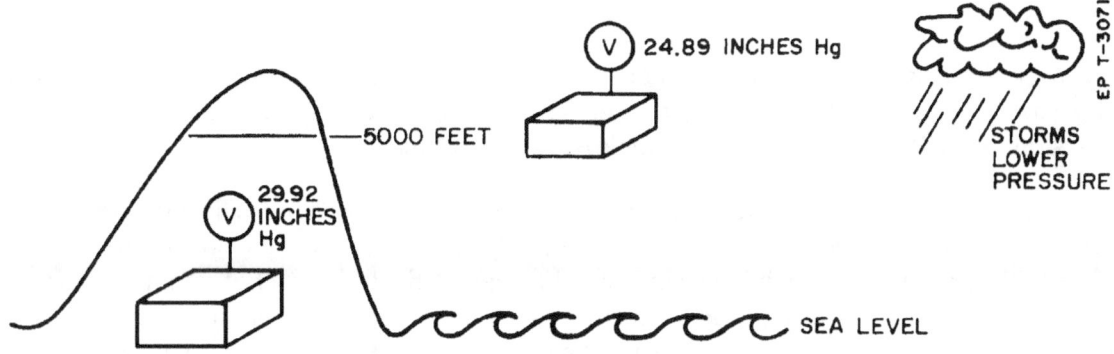

Exhibit 4-25 — Vacuum Indication At Varying Altitude

Section 4.5 — Air Ejectors

The **air ejector** (Exhibit 4-26) is a simple **jet pump**. In most Navy applications, steam is the working medium. The air ejector is used to evacuate a condenser or evaporator of noncondensible gases, thus initially creating and aiding in maintaining a vacuum.

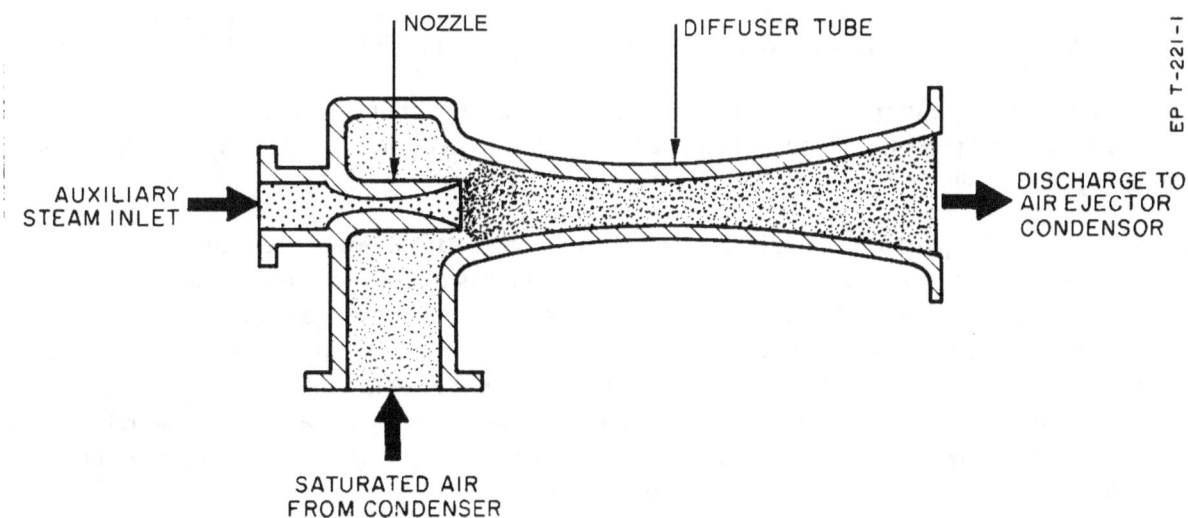

Exhibit 4-26 — Air Ejectors

Auxiliary steam is supplied at about 150 psig to the inlet of the nozzle. This steam expands through the **convergent-divergent nozzle**, where the pressure is reduced and some of the internal energy is converted into the kinetic energy of a fast-moving steam jet issuing from the nozzle. The issuing jet entrains the air-vapor mixture at the low-pressure region surrounding the nozzle exit and accelerates it to the steam's velocity. The high-velocity, air-vapor-entrained mixture then enters a **diffuser tube** where some of its kinetic energy is converted to pressure, so that the pressure at the diffuser outlet is higher than the pressure at the air-vapor diffuser entrance.

The **critical pressure ratio**, the ratio of the throat pressure to the inlet pressure, must equal 0.54 for saturated steam before **sonic velocity** can be reached in the **throat**. Also, sonic velocity must be obtained in the throat before **supersonic velocities** can be reached in the **divergent section**.

Upon starting the air ejector, the pressure in the **mixing chamber** is atmospheric. The chamber and the area to be evacuated are full of air and noncondensible gases. Initially, the velocities in the nozzle are relatively low, or subsonic, due to the pressure in the mixing chamber. When the relatively slow jet of steam is directed through the chamber, it mechanically entrains the noncondensibles and air and carries them into the diffuser. The more gases removed, the lower the pressure in the mixing chamber. As the pressure in the mixing chamber decreases, the velocity of the steam jet increases. Thus, the chamber is evacuated and the steam attains supersonic velocity. These supersonic velocities occur only after the initial evacuation of the condenser.

As the steam leaves the throat of the nozzle, it is directed to the mixing chamber. After initial evacuation of the chamber, the pressure reaches approximately 2 inches Hg absolute pressure. Normal driving steam to an air ejector is 150 psig; the pressure at the throat is 0.54 times the inlet pressure, or approximately 81 psig. The remaining steam pressure serves to force the steam out of the nozzle. The steam sees virtually no restriction, so it expands rapidly into the mixing chamber. The rate of this expansion is greater than the rate of increase in the area of the divergent section. Therefore, it increases its velocity beyond sonic velocity.

As the supersonic jet of steam leaves the nozzle, it goes through the mixing chamber, mechanically entrains the air and noncondensible gases, and carries them to the diffuser. The diffuser gathers the steam and noncondensibles and carries the mixture to a higher pressure by utilizing the velocity of the steam.

The diffuser serves the same purpose as that of a volute on a centrifugal pump: it decreases the velocity of the mixture and increases its pressure; thereby, it enables the ejector to discharge to a relatively higher pressure than is in the mixing chamber.

Up to a 4 inch Hg suction pressure, a single jet is satisfactory; above that, **two-stage ejectors** are required. Two-stage ejectors are used on condensers; the first stage discharges directly into the suction chamber of the second stage, and the second stage discharges into a condenser. The second stage handles all vapor, air, and expanded steam from the first stage and necessarily must be of much larger capacity. This means lower economy, but it saves space and cooling water. The **after-condenser** does not increase the efficiency of the ejector, but does prevent ejector steam from blowing into the atmosphere by condensing it and returning the condensate to the feedwater system. On the dumping condenser ejector, the first-stage ejector exhausts into an **inter-condenser** which reduces the load to the second stage.

Air ejectors are rated on the basis of their **free air handling capacity**. Actually they handle an air-vapor mixture. The calculations relating the actual operating conditions to the free air rating are based on the ejectors being equally able to handle free air or air vapor on a weight basis. For example, at 1.5 inches Hg suction pressure and 80 °F, each pound of air carries with it 1.35 pounds of water vapor. Assuming the leakage of air into the system is 10 lb/hr the ejector will have to handle 23.5 lb/hr of air vapor mixture.

Ejectors are designed for certain suction pressures, and for this reason are quite inefficient and slow in handling large air and vapor loads at low vacuum or when initially evacuating a condenser. The capacity of any stable ejector at a definite suction pressure is constant.

The vacuum in a condenser depends completely on the design capacity of the air ejector assuming design requirements on air leakage, oxygen in the feedwater, and ejector steam supply are met. Once a steady-state condition is met in a turbine with load, circulating water temperature, and condensate removal remaining constant and assuming no air leakage and negligible gases in the steam, the vacuum would remain the same even with the air ejector cut out. A change in any condition could change the vacuum, and it might appear that condensing action was the cause. This is not true as energy cannot be created. Since the above conditions cannot be met, the air ejector is still necessary after full vacuum is attained to remove air and noncondensible gases as fast as they enter the condenser.

The amount of heat absorbed by the cooling water to effect condensation within a surface condenser is equal to the (latent) heat ordinarily required to evaporate water at the same temperature and pressure. The heat unit generally used is the **British Thermal Unit (Btu)** which is the quantity of heat necessary to raise the temperature of 1 lbm of water 1 $^\circ$F. Besides the latent heat that must be absorbed from the steam to condense it, a small amount of heat must be removed from the condensate to yield a subcooled liquid. The amount of heat removed by the cooling water, then, is equal to the latent heat of condensation, 1000 Btu/lbm, plus that heat removed to obtain a 2 to 10 $^\circ$F subcooled liquid, 2 to 10 Btu/lbm.

The subcooling of the condensate is called **condensate depression**, and is necessary if the liquid is to be pumped from the hotwell. An unnecessary amount of subcooling will decrease plant efficiency. For example, if condensate depression is increased 10°F and the reactor is at full power, 1 percent of the reactor power is wasted due to condensate depression.

Section 4.6 — Steam Traps

4.6.1 — Types of Steam Traps

In a steam plant a certain amount of steam condensation must be removed. Large amounts of water in steam pipes result in a **water hammer** (the concussion of moving water against sides of piping). Water carried over with the steam into the power generating equipment may, in a turbine installation, result in damage to the blades. Excessive moisture in steam piping can cause erosion of the inner pipe wall. It is desired to remove the water without loosing any of the steam, which is the thermal energy carrier. The device used to accomplish this is the **steam trap** (or **drain regulator**).

There are three general classes of traps used in the U. S. Navy:
1. **Mechanically operated traps**, the operation of which depends upon a bucket or float mechanism in combination with a suitable valve arrangement.
2. **Impulse type traps**, which depend upon two things for their operation:
   - Hot water flashes to steam when its pressure is lowered below its saturation pressure.
   - The flow characteristics of a liquid through an orifice are different from those of a vapor of the same liquid.
3. **Thermostatically actuated traps**, which depend for their functioning upon the variation of vapor pressure of a volatile liquid or deformation of a bimetallic element with a change in temperature.

4.6.2 — Mechanically Operated Steam Trap

Exhibit 4-27 shows a mechanically operated steam trap. It has a comparatively large capacity and is designed for low pressures and a small temperature range.

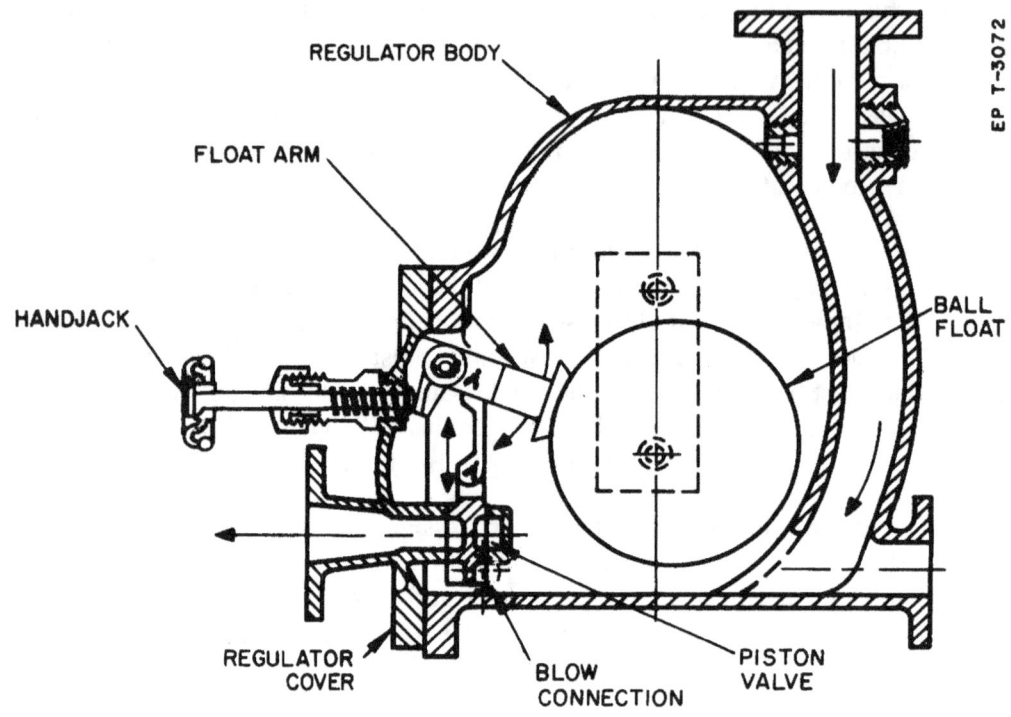

Exhibit 4-27 — Mechanically Operated Steam Trap

The trap inlet is on top of the body, and the condensate passes through the inlet to the inside chamber flooding the chamber and causing the float to rise. This causes the valve arm to lift, the valve to open, and the condensate to discharge as indicated. When the condensate has passed the trap, the float drops, closes the valve, and restricts the passage of steam. This particular type of trap is commonly called a drain regulator.

4.6.3 — Impulse Type Steam Trap

Exhibit 4-28 illustrates a **flash**, or **orifice**, type steam trap that can be used to drain high pressure steam lines. It is often referred to as an impulse trap. When the trap is first cut-in, steam or condensate (or both) enters the trap passing through a strainer before entering the valve body itself. The flow, once established, is continuous. The **control piston** is free to move up and down within the **tapered control cylinder**. As the control piston moves downward, the valve plug closes the seat opening. Since the area of the **control orifice** is slightly greater than the annular space between the cylinder and the **control piston flange**, condensate from the steam line, flows down through the orifice as fast as it flows up through the **annulus**. Since the orifice is connected to an area of lower pressure, the pressure is lowered within the control cylinder, and when it is finally reduced to 85 percent of the inlet pressure, or less, the control piston is forced up by unbalanced pressure below the flange. With the control piston raised within the tapered cylinder the condensate flows freely through the main valve, as shown in Exhibit 4-28.

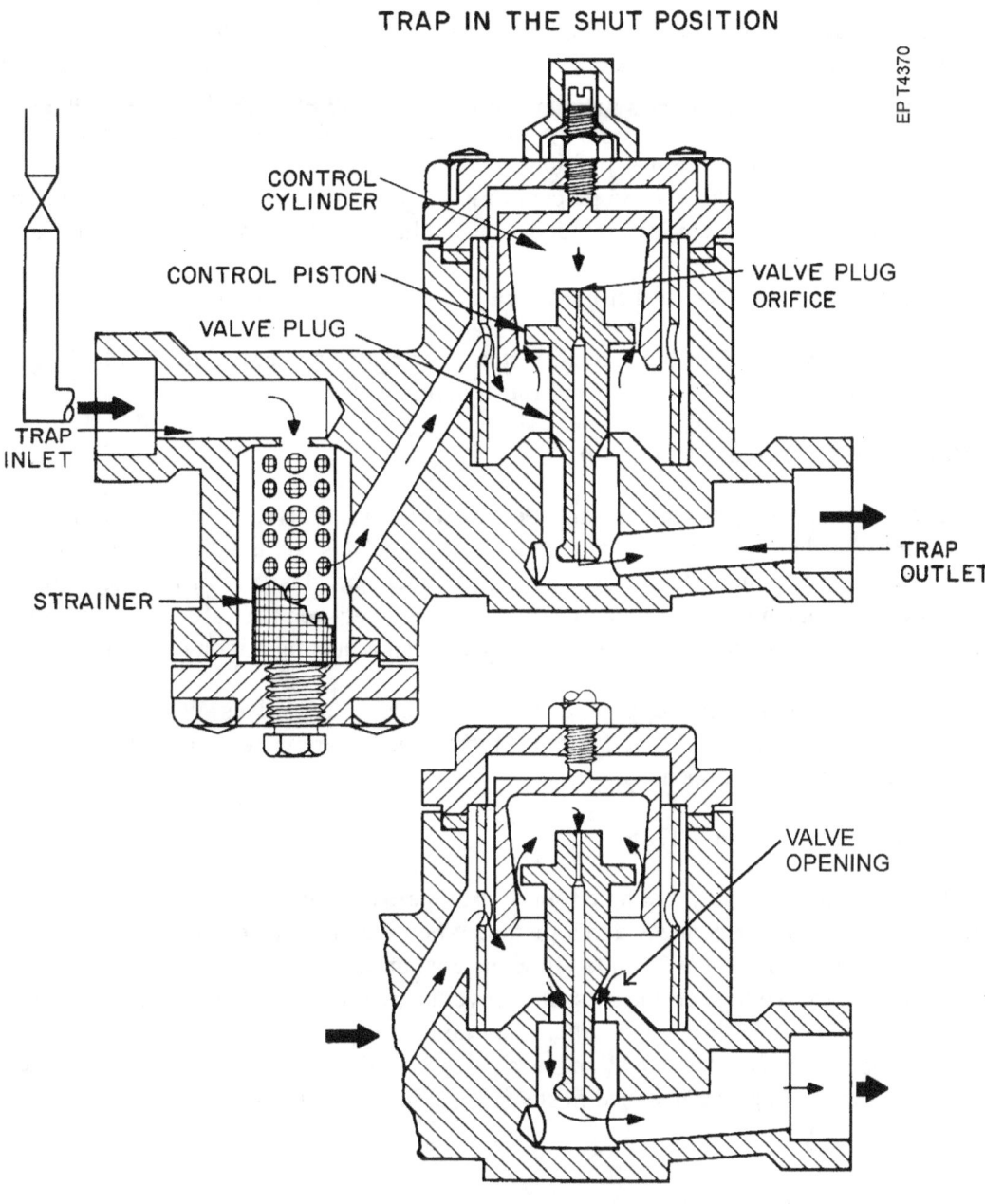

Exhibit 4-28 — Impulse Steam Trap

As the steam trap inlet line warms up, the temperature of the condensate increases approaching the boiling point corresponding to its existing pressure. At normal operating temperatures, the condensate is so hot that part of it will flash into steam, due to the drop in pressure and boiling point as the hot condensate enters the control chamber. The steam at reduced pressure has greater volume than the quantity of condensate from which it was produced; hence, steam chokes the control orifice causing a pressure rise in the control chamber until finally the control piston is forced downward and the valve shuts as shown in the upper illustration of Exhibit 4-28. Steam continues to pass through the trap, going through only the control orifice, into the high-pressure drain line. If condensate tends to enter the trap faster than it can be passed off as steam through the orifice it will back up into the drain line leading to the trap. The trap and a portion of the drain line are not insulated. Hot condensate that has backed up into the drain line cools to a point where it cannot flash into steam in the control chamber. At this point the trap begins to function as previously described under warm-up conditions; that is, when the pressure in the control chamber drops to 85 percent or less, of the inlet pressure, the valve will open, allowing the rapid passage of condensate from inlet to outlet, until either hot condensate or steam enters the trap again.

4.6.4 — Thermostatically Actuated Steam Traps

This category of steam trap includes bimetallic-, piston-, and bellows-operated traps, which are discussed in Paragraph 4.6.4.1 through Paragraph 4.6.4.3 and illustrated in Exhibit 4-29.

4.6.4.1 — Bimetallic-Operated Steam Trap

Steam traps of this type consist primarily of an **integral strainer**, a **valve**, and a **bimetallic element**. The bimetallic element is anchored at one end, and a free-floating valve stem and disc are attached to the other end. The bimetallic elements respond to temperature change to open and close the valve. The temperature at which the valve opens is determined by the amount of free movement between the valve disc and valve seat when the trap is cold. This movement or clearance can be adjusted by an adjusting nut on the valve stem.

The trap is provided with an **integral bypass**, using two double-seated valves. When both valves are open the trap is in service. With both valves closed the trap is bypassed. The trap and bypass are secured with the inlet valve closed and the outlet valve open.

4.6.4.2 — Piston-Operated Steam Trap

Piston-operated steam traps employ a simple **bimetallic-operated** valve as a pilot to open and shut a **free-floating main valve** that is connected to a **piston**. When the pilot valve is open, pressure builds up above the piston and pushes the main valve down to allow the condensate to flow through the trap. Steam entering the trap causes the pilot valve to shut; the pressure above the piston is bled off through a relief port, and the main valve shuts to secure flow through the trap. The trap is also provided with an integral bypass of the type previously discussed. The piston-operated trap has a pressure rating of 600 psig and a flow capacity of 7000 lb/hr. These high-capacity traps are installed in steam separator drain lines.

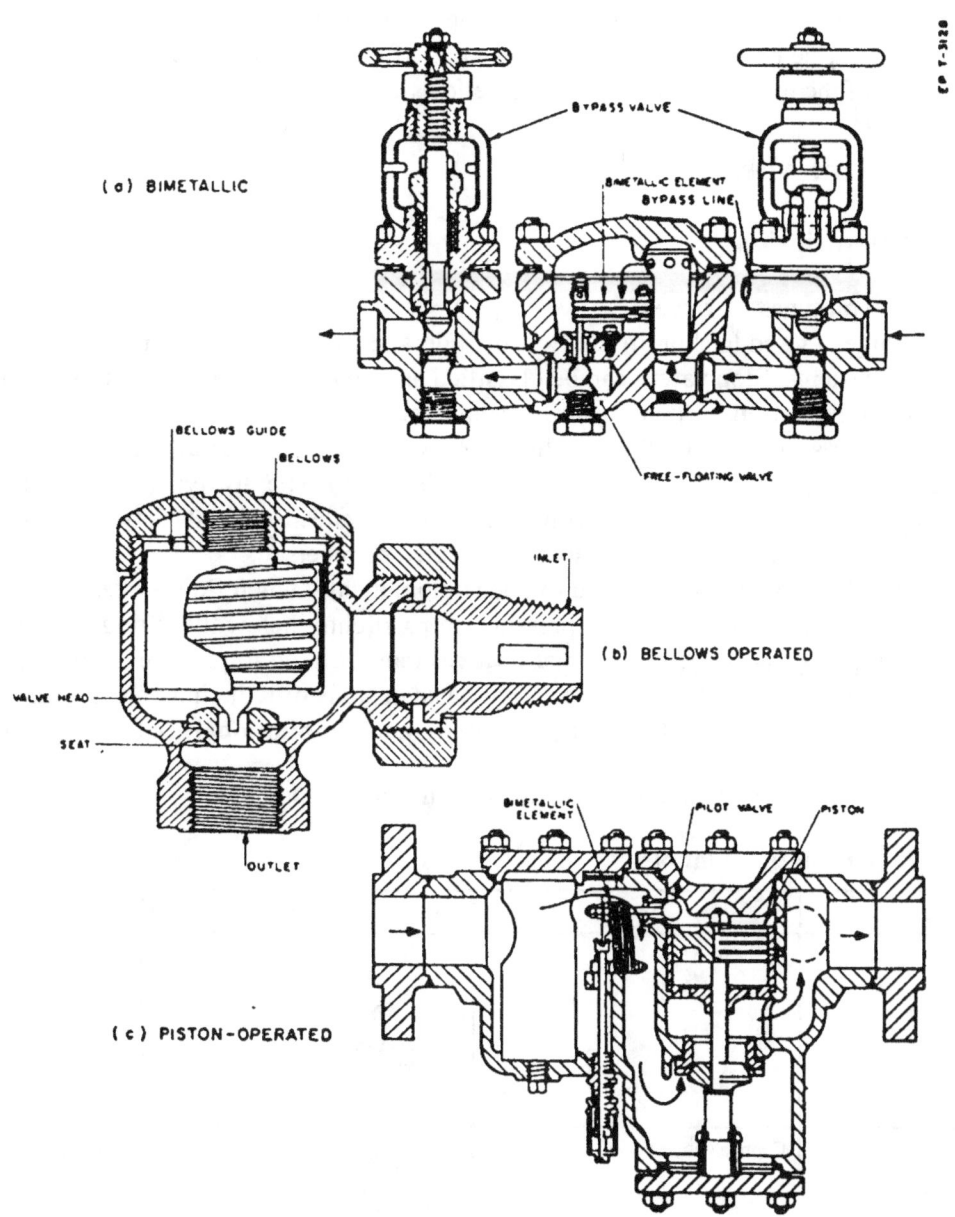

Exhibit 4-29 — Thermostatically Actuated Steam Traps

4.6.4.3 — Bellows-Operated Steam Trap

The bellows-operated steam trap is used mainly for draining heating systems and auxiliary machinery, and is limited to pressures under 100 psi. The flexible, accordion-like, metal **bellows** is partially filled with a volatile liquid and sealed. As long as the trap is cool, the valve is held open by the bellows. If the trap is heated, however, such as when steam enters the inlet, the liquid within the bellows boils, increasing the pressure of the vapor layer above the liquid, and forcing the bellows to expand downward; this action shuts the valve. The valve remains shut until condensate again collects in the trap and is cooled through the uninsulated walls of the trap. The cooled condensate surrounding the bellows causes the vapor within the bellows to condense and to decrease in volume; consequently, the bellows contracts, the valve is lifted from its seat, and the cooled condensate is permitted to flow out of the trap. The capacity of the bellows-operated trap at 15 psid pressure is 1360 lbm/hr.

Section 4.7 — Plant Valves

4.7.1 — General Description

There are several general types of valves.

1. <u>Stop Valves</u> - Used to shut off the flow of fluid and isolate system components or individual parts of fluid systems. Stop valves can be classified as **globe**, **gate**, **ball**, or **butterfly** valves, depending on the configuration of their closure mechanism.

2. <u>Check Valves</u> - Permit the flow of fluid in only one direction. They are operated by the flow of fluid in the piping. A check valve may be of the **swing** type, **lift** type, or **ball** type.

3. <u>Flow Directing Valves</u> - Used to control fluid flow paths. Such valves typically have three or four ports through which fluid flow can be redirected or reversed.

4. <u>Pressure Relief Valves</u> - Designed to open automatically when fluid system pressure exceeds the desired value and close when the pressure drops slightly below the lifting pressure.

5. <u>Pressure Reducing Valves</u> - Automatic valves that are used to provide a constant pressure lower than the supply pressure. The three basic designs of reducing valves are **spring-loaded**, **pneumatic-pressure-controlled (gas loaded)**, and **air-pilot operated diaphragm type**.

Table 4-2 shows schematic symbols for various valve types.

Table 4-2 — Valve Schematic Symbols

Symbol	Name	Symbol	Name
	Globe Valve		Gate Valve, Hydraulically Operated with Position Indication
	Capped Globe Valve		Gate Valve, Locked Open
	Throttle Valve		Ball Valve
	Capped Throttle Valve		Butterfly Valve

Table 4-2 — Valve Schematic Symbols (Cont)

Symbol	Name	Symbol	Name
	Angle Glove Valve		Swing Check Valve
	Glove Valve, Solenoid Operated		Lift Check Valve
	Angle Globe Valve, Hydraulically Operated with Position Indication		Stop Check Valve
	Globe Valve, Locked Open		Angle Relief Valve
	Globe Valve, Locked Shut		Relief Valve, Pilot Operated
	Air Operated Globe Valve, Spring Closing		Pilot Valve

Table 4-2 — Valve Schematic Symbols (Cont)

Symbol	Name	Symbol	Name
	Gate Valve		Three-Way Solenoid Operated Pilot Valve
	Gate Valve, Hydraulically Operated		Pressure Reducing Valve
	Gate Valve, Motor Operated		

Most valve types have the same basic components as shown in Exhibit 4-30. The valve's components are:

1. Body - The body is the lower part of the valve assembly containing the inlet and outlet chambers and the valve seat.

2. Seat - The seat is the fixed surface in the body which makes contact with the disc when the valve is shut.

3. Bonnet - The bonnet is the upper part of the valve assembly containing the gland stuffing box and the yoke. It is usually bolted or screwed to the valve body.

4. Bonnet Stud and Nut or Bolt - This is used to position and hold the bonnet on the body and to compress the gasket to make a tight seal.

5. Disc - The disc is the movable component which mates with the valve seat to form a tight seal. The disc is raised or lowered by rotating the stem to open or shut the valve.

6. Stem and Stem Threads - The stem connects the handwheel to the disc. The stem threads cause the stem and the disc to be raised or lowered as the stem rotates in the stem bushing, which is fixed in the yoke.

7. Packing - The packing is pliable gasket material that is inserted in the gland stuffing box and forms a tight seal against the stem when compressed by the gland. Packing is commonly constructed from graphite, graphite impregnated materials, or Teflon.

8. Packing Gland - The gland fits around the stem and compresses the packing in the gland stuffing box to prevent leakage around the stem. It is commonly in the form of coils, rings, and corrugated ribbon.

9. Packing Gland Retainer - This component is fastened by the gland stud and presses the gland against the packing.

10. Gland Stud and Nut - The gland stud and nut provide the restraining force that holds the gland tightly against the packing to obtain a tight seal around the valve stem. The packing gland retainer

and the gland, in conjunction with the gland studs and gland nuts, prevent internal pressure from blowing the packing out of the stuffing box.

11. <u>Yoke</u> - The yoke is a structural component, integral with the bonnet, which supports the stem bushing, the stem, and the handwheel.

12. <u>Stem Bushing/Yoke Gland</u> - The threaded stem bushing is fixed and engages the threaded section of the stem so that the stem is raised or lowered as it rotates.

13. <u>Handwheel</u> - The handwheel is fastened to the stem and is rotated to open or shut the valve.

14. <u>Backseat</u> - A backseat is usually provided to prevent the stem from being opened too far, to minimize stem leakage past the packing and to permit replacement of the packing without depressurizing the system under emergency conditions. Some valves use backstops rather than backseats.

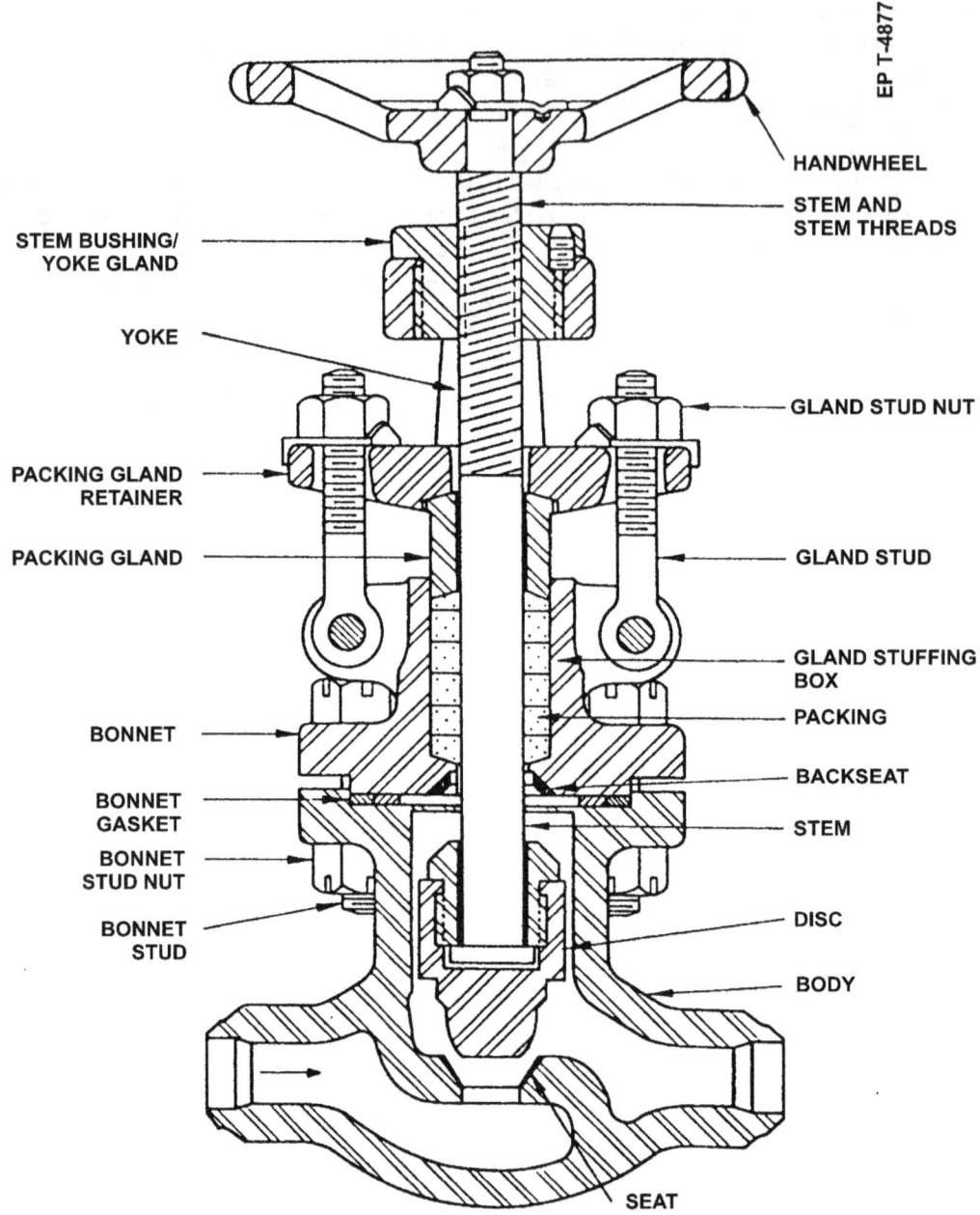

EP T-4877

HANDWHEEL

STEM AND
STEM THREADS

STEM BUSHING/
YOKE GLAND

YOKE

GLAND STUD NUT

PACKING GLAND
RETAINER

PACKING GLAND

GLAND STUD

GLAND STUFFING
BOX

PACKING

BONNET

BACKSEAT

BONNET
GASKET

STEM

BONNET
STUD NUT

DISC

BONNET
STUD

BODY

SEAT

Exhibit 4-30 — Cross Section Of A Typical Secondary Plant Globe Valve

4.7.2 — Pressure Relief Valves

Relief valves are used to protect a system against excessive overpressure conditions. There are two basic types of relief valves that are in general use: **self-actuated** and **pilot-actuated**. The self-actuated relief valve (Exhibit 4-31) is the simpler of the two types. During normal system operation the valve disc is tightly seated by preset spring force and system pressure acts against the spring loaded valve disc. When system pressure increases above the established set pressure, it overcomes the spring force and lifts the disc off its seat, which allows flow through the valve until system pressure decreases again to the set pressure. Self-actuating relief valves are most often used downstream of positive displacement pumps such as charging pumps and hydrostatic test pump setups. Valve discharge is normally to an open onboard drain system or to a holding tank.

The pressure setting of a self-actuated relief valve is established by means of an **adjusting bolt**, which is used to regulate the spring compressive force on the valve disc. If the valve is to be locked shut a **gagging cap-and-screw device** is used to prevent axial movement of the valve stem and retain the valve disc against its seat.

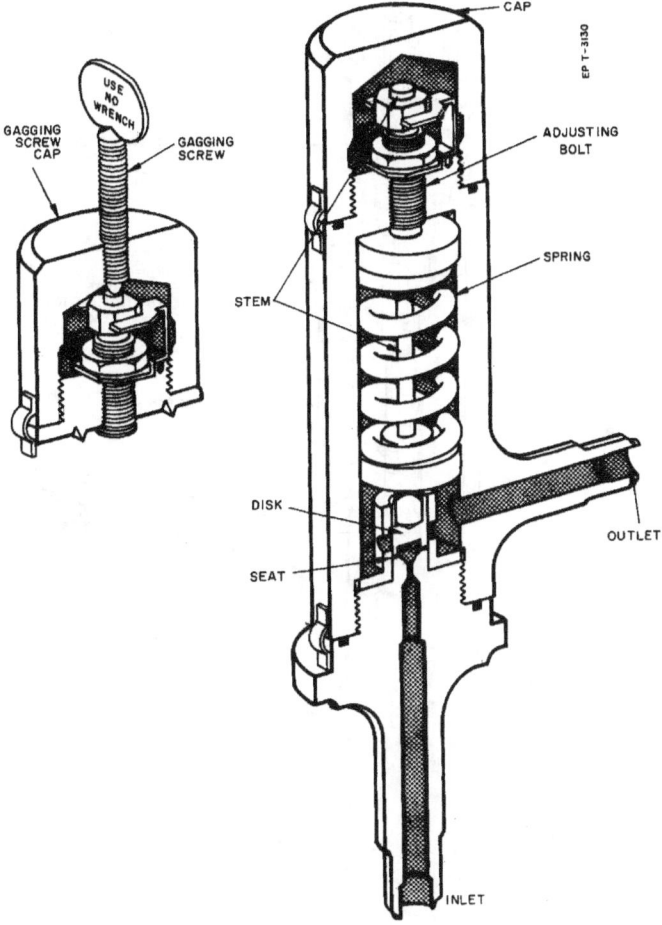

Exhibit 4-31 — Self-Actuated Relief Valve

In pressure relief systems that discharge overboard, the pressure relief valves are subject to seawater leakage back through the overboard discharge piping. Such leakage imposes a backpressure on the disc of the relief valve. Within the pilot-actuated relief valve configuration (Exhibit 4-32) such backpressure acts only on the small area of the bottom of the pilot valve disc; it is opposed by system pressure acting on the relatively larger piston area of the pilot valve, and does not hamper valve operation.

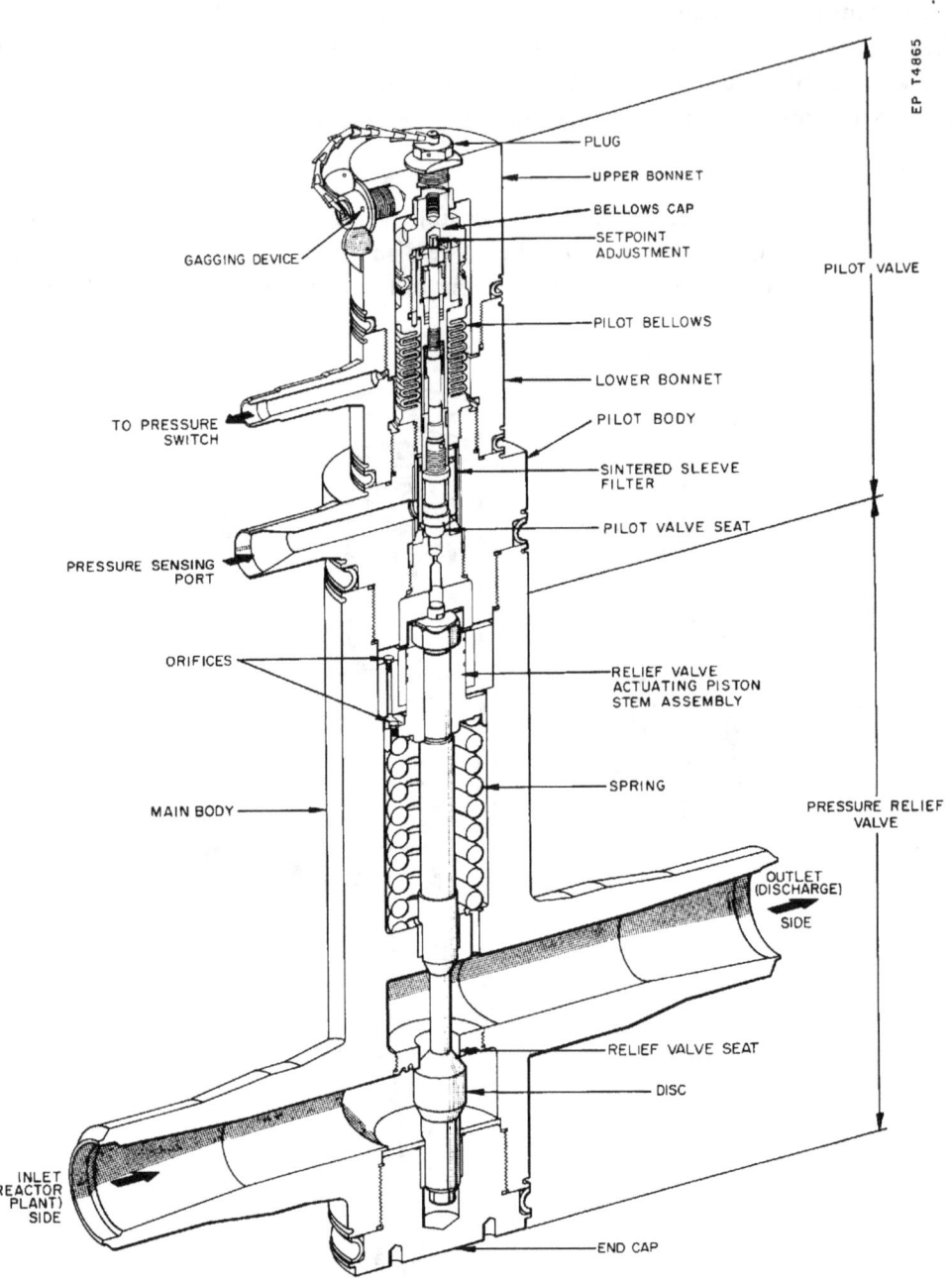

EP T4865

PLUG

UPPER BONNET

BELLOWS CAP

SETPOINT ADJUSTMENT

GAGGING DEVICE

PILOT BELLOWS

LOWER BONNET

PILOT BODY

TO PRESSURE SWITCH

SINTERED SLEEVE FILTER

PILOT VALVE SEAT

PRESSURE SENSING PORT

PILOT VALVE

ORIFICES

RELIEF VALVE ACTUATING PISTON STEM ASSEMBLY

SPRING

MAIN BODY

PRESSURE RELIEF VALVE

OUTLET (DISCHARGE) SIDE

RELIEF VALVE SEAT

DISC

INLET (REACTOR PLANT) SIDE

END CAP

Exhibit 4-32 — Pilot-Actuated Relief Valve

A pilot-actuated relief valve is made up of two sections: a **pilot section** and a **main section**. The main section is a hydraulically operated reverse-seated globe valve. Use of a reverse-seated globe valve allows reactor plant pressure to help keep the valve tightly shut. The valve stem extends through the seat port and connects to an actuating piston in the valve body. A helical compression spring in the valve body holds the valve stem disc in the shut position. A typical pilot-actuated relief valve is shown in Exhibit 4-32. In some installations the pilot valve and the relief valve are individual, separate units that are interconnected by piping that transmits piston actuating pressure from the pilot valve to the top of the relief valve operating piston.

The pilot section valve disc is held shut by reactor plant pressure transmitted through the sensing line and by a compressive force exerted by a spring bellows. The lower end of the bellows is welded to the pilot valve body; the upper end is mechanically attached to the pilot valve stem and is seal welded to the bellows cap. The bellows and bellows cap assembly prevent reactor coolant pressure from entering the pilot valve bonnet. The differential pressure thus established across the bellows actuates the main section of the valve via flow from the pilot section as follows. System pressure, introduced to the bellows via the sensing line, internally pressurizes the bellows and tends to extend the bellows linearly. As system pressure increases and approaches the established pressure relief setting, the extending bellows mechanically unseats the pilot valve disc and allows steam (or water) to flow into the actuating piston chamber of the main valve section. System pressure then acts on the main valve actuating piston, forcing the piston downward and overcoming both the force of the spring and the system pressure acting under the disc of the main valve. This action fully opens the relief valve and permits steam (or water) to discharge and relieve the overpressure condition. Relief valve pressure settings are adjusted by varying the normal position of the pilot valve stem to allow more or less bellows travel.

As system pressure decreases, the pilot bellows contracts and system pressure reseats the pilot disc. When flow from the pilot valve is shut off, pressure in the main valve piston chamber discharges at a controlled rate to the outlet side of the valve through the specially designed piston orifices and via the space around the piston rings. As the piston chamber pressure decreases, the spring force, assisted by system pressure, reseats the main valve. If plant pressure again exceeds the pilot valve setting (minus the setting change due to the temperature sensitivity of some of the valve parts which were heated by flow through the valve) the lifting action is automatically repeated.

Each pilot valve is connected by a **sensing line** to a **pressure switch**. If a leak develops in the pilot bellows, pressure is exerted on an associated pressure switch to initiate an electrical signal that indicates the valve to be malfunctioning. A pilot-operated relief valve may be **gagged** shut by screwing a gagging bolt through the threaded hole in the center of the valve upper bonnet. The gagging bolt bears down on the bellows cap and prevents the bellows and the stem of the pilot valve from rising; thus, gagging only prevents opening of the pilot valve. The main valve, which is reverse-seated, can be opened if backpressure exists that will overcome system pressure and the valve spring tending to hold the valve shut.

This type of valve is used to provide overpressure protection in primary plant pressure relief systems, where both steam and water may be discharged, and as a steam generator safety valve, where steam is the normally discharged medium. Water seals are provided in the inlet line and in the sensing line to steam relief valves. The water seals dampen temperature transients and keep the relief valve cool, thereby minimizing leakage through the valve seat caused by thermal distortion and minimizing changes in relief valve setpoint because of temperature changes. In addition, it minimizes the leakage of gases through the valve seats from the pressurizer steam space.

4.7.3 — Hydraulically Operated Valves

Remotely operated primary system stop valves are typically hydraulically operated gate valves (Exhibit 4-33). Gate valve design provides minimum pressure drop of the fluid flowing through the valve. Valve discs are of the **split-disc type**, and both **parallel face** and **wedge shaped** discs are used. Some discs are fitted with a disc spring such that the discs are loaded toward their respective seating rings in the valve body. Additionally, some disc assemblies incorporate spring loaded latching assemblies which consist of latch arms and latch springs that engage latching lugs on the interior of the valve body to maintain valve position when in the fully open or fully shut position.

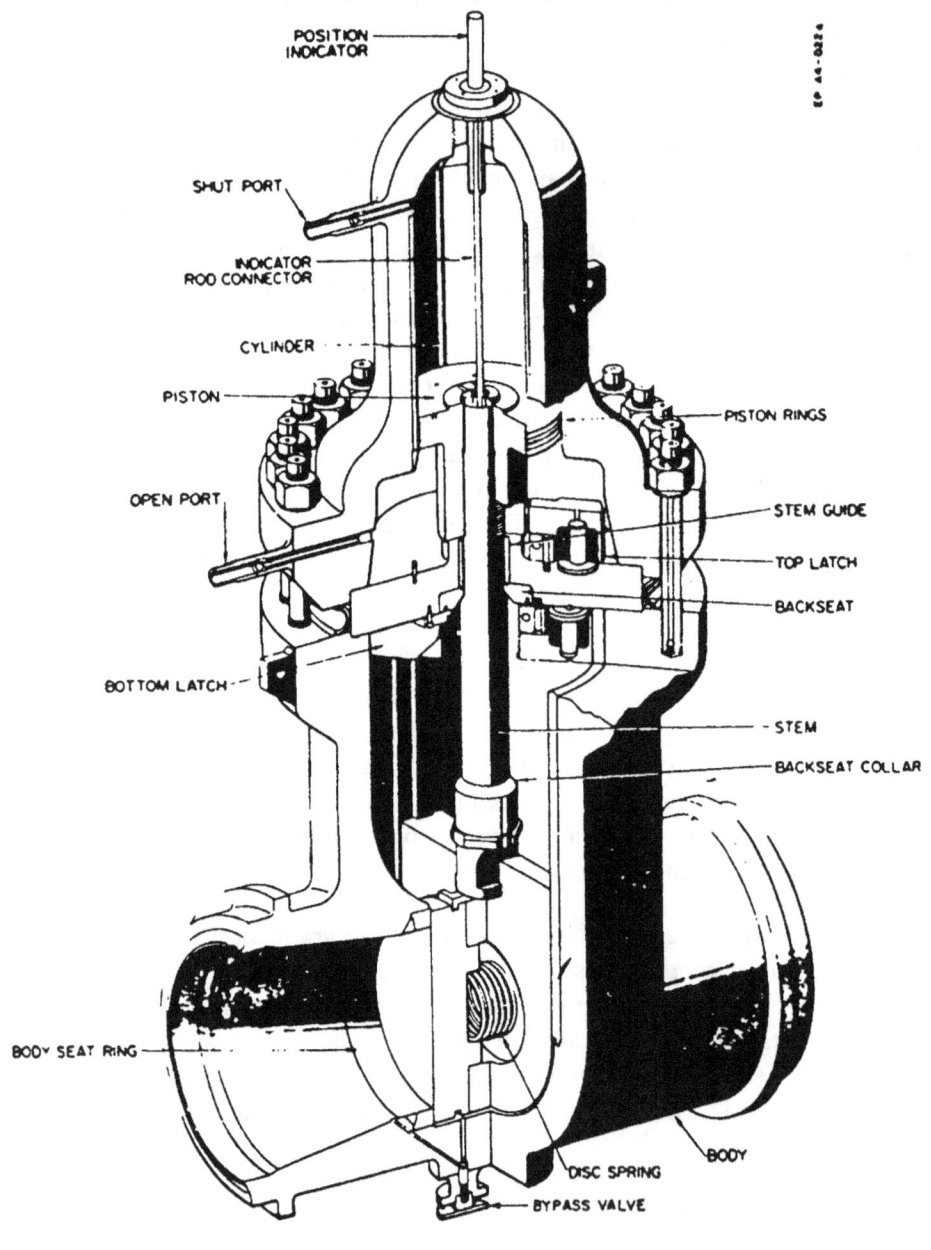

Exhibit 4-33 — Typical Hydraulically Operated Gate Valve

In some designs **bypass channels** are drilled into the valve body, connecting the space between the discs with one side of the valve. When valve operating water is at a higher pressure than system pressure inside the valve body, leakage of this high-pressure water occurs along the valve stem down into the valve body; hence, pressure builds up in the area between the discs of a shut valve such that the discs tend to bind against their seats. The bypass precludes this pressure buildup. When the valve is shut, the bypass acts to preclude depressurization of the closed system of the bypass side of the valve discs when there is a pressure differential across the valve. System pressure transmitted through the bypass pressurizes the space between the discs and forces the opposite disc (low pressure side of the valve) tightly against its seating ring.

When the valve is fully open the valve stem backseats against the backseat collar in the valve body and prevents leakage of primary coolant up along the stem and into the valve operating water; thus, a possible contamination problem outside of the shielded reactor compartment is minimized.

The upper portion of the valve contains the valve operating cylinder, within which the operating piston travels, attached to the top of the valve stem. High pressure water remotely directed to and exhausted from the top and bottom of the cylinder provides for valve piston travel, which opens and shuts the valve.

It is necessary that valve position indication be provided for these valves at remote control stations. Each valve is fitted with electromechanical devices actuated by the valve disc-stem-piston assembly. Movement of the valve stem actuates a **magnetic slug**, the movement of which past a **differential transformer** or **reed switch** assembly produces an electric signal. This signal is subsequently amplified in system circuitry and used to energize valve control panel indicating lights and as input to certain plant protective circuitry. Generally, the early valve designs used simple magnetic core differential transformers, one each mounted on the top and bottom of the valve to indicate fully open and fully shut valve positions, respectively, as the transformer cores (magnetic slugs) are actuated at each extreme of valve stem travel. Later valve designs incorporate single transducers located on the top of the valve, and utilize a single magnetic slug that follows the full stroke of the valve to alternately produce open and shut indicating signals from varying positions of the magnetic slug in an **E-core transformer** or past two sets (open and shut) of reed switches. Valve position indication can also be obtained using electrical limit switches which are mechanically activated by cams or arms that are controlled by motion of the valve stem.

Section 4.8 — Reactor And Propulsion Plant Energy Balances

4.8.1 — Heat Removed From Reactor By Primary Coolant

The rate at which primary coolant removes heat from the reactor ($\dot{Q}_{RX}$) is directly related to the temperature rise of coolant in the reactor (ΔT_{RX}),

Equation 4-12

$$\Delta T_{RX} = T_H - T_c$$

where:

T_H = Reactor Outlet (Hot Leg) Temperature, °F

$$T_C \quad = \text{Reactor Inlet (Cold Leg) Temperature, } ^oF.$$

In steady state, ΔT_{RX} will also match the temperature drop across the steam generator(s). This assumes balanced steam loads if there are more than one steam generator in operation. The rate at which primary coolant removes heat from the reactor is given by

Equation 4-13

$$\dot{Q}_{RX} = \dot{m}_{RX}\, c_{PRI}(T_H - T_c)$$

where:

$$\dot{m}_{RX} \quad = \text{Mass Flow Rate of Coolant through the Reactor, lbm/hr}$$

$$c_{PRI} \quad = \text{Specific Heat Capacity of Primary Coolant, Btu/lbm-}^oF.$$

For a given heat transfer rate, the temperature rise through the reactor is inversely proportional to the flow rate through the reactor.

4.8.2 — Heat Transfer From Primary To Secondary Coolant

The rate of heat transfer from primary to secondary coolant ($\dot{Q}_{S/G}$) depends on the temperature difference between the primary and secondary systems so that

Equation 4-14

$$\dot{Q}_{S/G} = U_{S/G}\, A_{S/G}(T_{AVE} - T_{STM})$$

where:

$$U_{S/G} \quad = \text{Overall Heat Transfer Coefficient, Btu/hr-ft}^2\text{-}^oF$$

$$A_{S/G} \quad = \text{Heat Transfer Area, ft}^2$$

$$T_{AVE} \quad = (T_H + T_C)/2 = \text{Average Coolant Temperature, } ^oF$$

$$T_{STM} \quad = \text{Temperature of Steam in Steam Generator, } ^oF.$$

4.8.3 — Steam Demand

In a nuclear propulsion plant, all power demands are met by drawing high pressure steam from the steam generator to supply secondary systems. Steam demand ($\dot{Q}_{S/D}$) is the rate at which a steam generator supplies energy to secondary systems. It is the rate at which steam flow removes energy from the steam generator minus the rate at which feedwater flow returns energy to the steam generator so that

Equation 4-15

$$\dot{Q}_{S/D} = \dot{m}_{STM} h_{STM} + \dot{m}_{FW} h_{FW}$$

where:

$\dot{m}_{STM}$	= Mass Flow Rate of Steam, lbm/hr
$\dot{m}_{FW}$	= Mass Flow Rate of Feedwater, lbm/hr
h_{STM}	= Specific Steam Enthalpy, Btu/lbm
h_{FW}	= Specific Feedwater Enthalpy, Btu/lbm.

Steam demand is controlled by throttle valves that vary the $\dot{m}_{STM}$. In steady-state,

$$\dot{Q}_{S/G} = \dot{Q}_{S/D}.$$

4.8.4 — Heat Rejected To Seawater

The rate at which secondary coolant transfers heat to seawater in the condensers ($\dot{Q}_{COND}$) is

$$\dot{Q}_{COND} = U_{COND} A_{COND} (T_{COND} - T_{SW,AVE}).$$

where:

U_{COND}	= Condenser Heat Transfer Coefficient, Btu/hr-ft^2-$^\circ$F
A_{COND}	= Condenser Heat Transfer Area, ft^2
T_{COND}	= Condensate Saturation Temperature, $^\circ$F
$T_{SW,AVE}$	= $(T_{SW,IN} + T_{SW,OUT})/2$ = Average Seawater Temperature, $^\circ$F.

The rate at which heat is removed by seawater ($\dot{Q}_{SW}$) is:

$$\dot{Q}_{SW} = \dot{m}_{SW} c_{SW} (T_{SW,OUT} - T_{SW,IN})$$

where:

$\dot{m}_{SW}$	= Mass Flow Rate of Seawater, lbm/hr
c_{SW}	= Specific Heat of Seawater, Btu/$^\circ$F

$T_{SW, IN}$ = Seawater Injection Temperature, ^oF

$T_{SW, OUT}$ = Seawater Temperature Leaving Condenser, ^oF.

In steady-state,

$$\dot{Q}_{COND} = \dot{Q}_{SW}.$$

For additional information, refer to Reference 9.

Chapter 5 — CHEMISTRY REVIEW

More detailed coverage of the topics in this section is presented in Reference 4 and Reference 5 .

Section 5.1 — Elements, Ions, And Compounds

5.1.1 — Atoms And Elements

According to the atomic theory, all matter is composed of atoms existing individually or in combination with other atoms. An **atom** is the smallest particle of an element that keeps the same composition of the element and is electrically neutral. Atoms have dimensions on the order of 10^{-8} cm. They can be treated as distinct particles because they behave as such chemically, but it is now known that atoms themselves are composed of even smaller particles. Atoms are classified chemically by the number of protons in their nucleus. All atoms having the same number of protons in their nuclei have the same chemical behavior and collectively constitute a **chemical element**. The number of protons in the nucleus plays such an important role in identifying the atom it is given a special name, **atomic number, Z**. **Isotopes** are groups of atoms that have the same number of protons (hence, the same element) but have different numbers of neutrons in their nuclei. The **Atomic Mass Number, A**, is the sum of protons and neutrons in the nucleus of a specific isotope of an element. The **atomic weight** of an element is defined as the weighted average of the masses of all its naturally occurring isotopes.

Atomic weights are given in **Atomic Mass Units (AMU)**. These units are a relative weight scale on which the mass of the Carbon-12 isotope is used as a standard and all others are related. Specifically, 1 AMU is defined as 1/12 the mass of the Carbon-12 atom. Since the mass of a proton or neutron is approximately 1 AMU, the mass of any particular atom is approximately equal to its atomic mass number, A. In grams, 1 AMU = 1.66×10^{-24} grams.

Each chemical element has been assigned a special one- or two-letter **symbol**. The first letter is always capitalized no matter where or how the symbol is used. These symbols are used as a type of chemical shorthand when writing chemical formulas.

5.1.2 — Periodicity And The Periodic Table

Over many years of chemical investigation, a remarkable feature of the elements emerged. It was observed that if the elements are arranged in order of their atomic numbers, the chemical properties are repeated somewhat regularly. To a lesser extent, the physical properties are also repeated periodically.

The arrangement of the elements in the form of a table in which elements with similar chemical properties are grouped together is called a **periodic table**, as depicted in Exhibit 5-1. In this table, elements are arranged in order of increasing atomic number in succeeding rows. Each row is called a **period**. Note that some periods are longer than others.

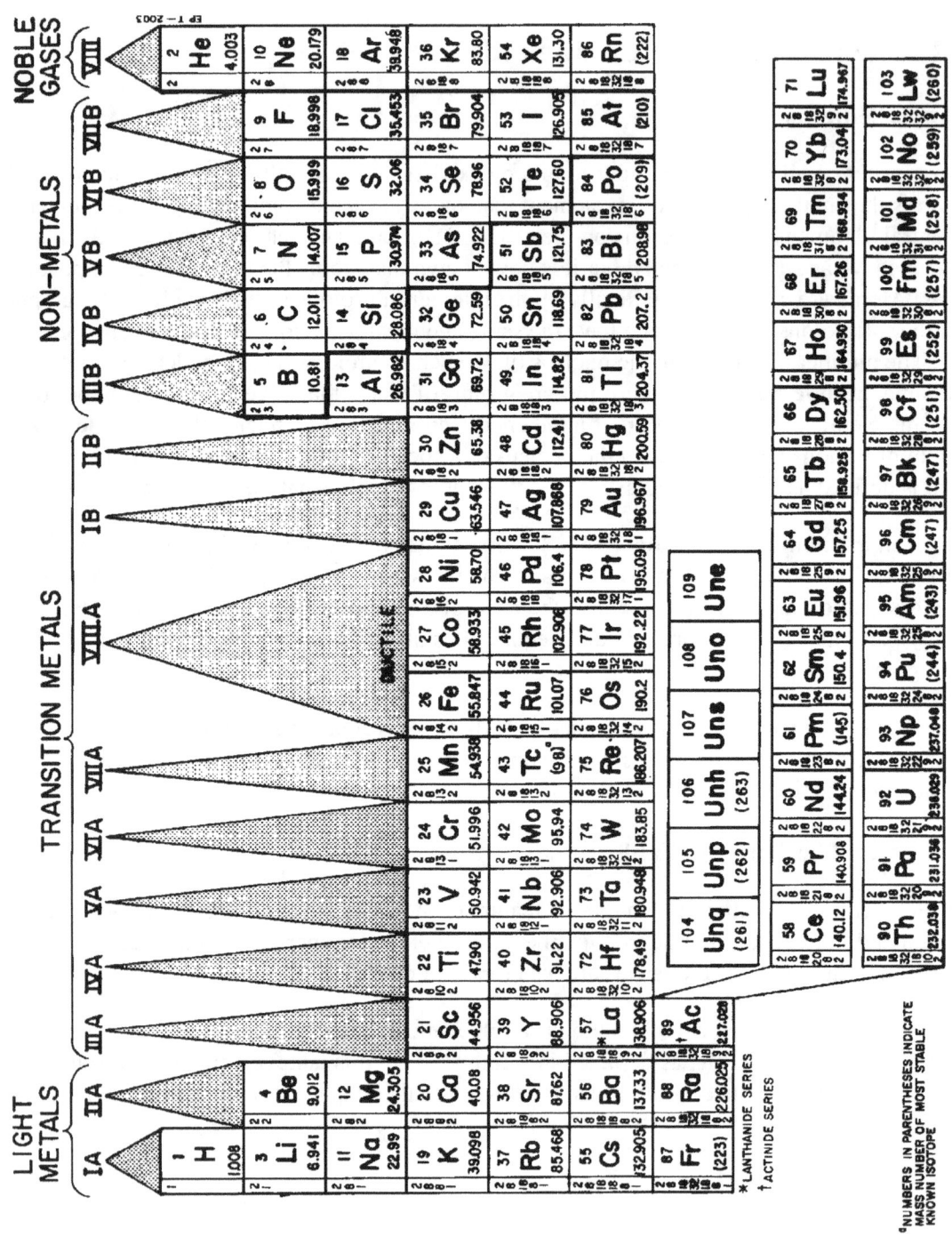

Exhibit 5-1 — The Periodic Table Of The Elements

Elements with generally similar chemical properties appear in vertical columns called **groups**. Each group is designated by a Roman numeral and sometimes a capital letter. It should be noted that some periodic tables have the Noble Gas group, the group on the extreme right hand side, designated as Group O while others designate them as Group VIII. The two left-most groups are very chemically active metals (e.g., Sodium). The groups in the middle contain more structurally useful metals (e.g., Iron, Nickel, Chromium). These metals are capable of being mechanically shaped, and they are good conductors of heat and electricity. The right-most groups are nonmetals (e.g., Nitrogen, Oxygen, and Chlorine). Note that there is a stair-step line separating metals from nonmetals.

5.1.3 — Electron Shells

It is well known that chemical reactions involve only the outermost electrons in atoms. This is why atomic number defines the element. Electrons are in constant motion around the nucleus, possessing both kinetic and potential energy. The total energy of each electron is **quantized**. This means the amount of energy can only be specific values of allowed energies, not a continuum.

These allowed energy levels can be visualized as **shells** of electrons around the nucleus, with forbidden energy levels between different shells. Each shell is further divided in **subshells** with smaller bands of forbidden energy between them. These subshells are also called **orbitals**.

As you move outward from the nucleus, it can be visualized that each shell can hold more electrons because the shells are "bigger" further away from the nucleus. The first shell has one orbital, the second has two orbitals, etc. These orbitals are given letter designations within a single shell. The letters are **s, p, d,** and **f** as you move out from the nucleus. Each orbital has a limit of how many electrons it can hold. The limits are:

Table 5-1 — Electron Limits Within Orbitals

Orbital	Electron Limit
s	2
p	6
d	10
f	14

All electron limits are even numbers. This is because electrons are more stable (lower energy levels) when they are paired. This is due to electrons of opposite spin stabilizing each other.

Filling each orbital gives you the format of the periodic table:
1. Filling the **s** orbital gives you the first two groups on the periodic table.
2. Filling the **p** orbitals gives you the six right-most groups.
3. Filling the **d** orbitals gives you the center groups (i.e., transition metals).
4. Filling the **f** orbitals gives you the series at the bottom of the table (i.e., the lanthanides and actinides).

All elements with the same number of electrons in their outermost shell possess similar chemical properties. Atomic electrons always seek the lowest energy state available. In general, electrons closer to the nucleus have a lower energy state. Note that all the inert (noble) gases have completely filled shells. The inert gas configuration is exceptionally stable energetically since these inert gases are the least reactive of all the elements.

5.1.4 — Molecules, Compounds And Bonding

A **compound** is a distinct substance composed of two or more elements combined chemically in a definite, fixed ratio. A compound can be broken down by chemical means into its component elements. An element cannot be broken down by chemical means.

Just as symbols are used to denote elements, there is a symbolic method of representing compounds. These symbolic representations are called **formulas** and consist of the symbols for the elements which make up the compound (e.g., H_2O, $NaCl$, Fe_3O_4). A **molecule** is the smallest particle in which a compound can be divided without changing its chemical identity.

The forces that hold atoms together in a compound are called **chemical bonds**. Atoms combine to form compounds because by doing so their outermost electrons are rearranged into a more stable (lower energy) state. What then constitutes a stable electron structure? The **inert gas configuration** (i.e. filled or 8 electrons in its outer shell) constitutes a stable electron structure. In general, atoms of other elements gain stability by acquiring the inert gas configuration.

In forming compounds, atoms share or exchange electrons in such a way that each atom in the compound attains the inert gas configuration, **the Octet Rule**. The Octet Rule states:

In forming compounds, atoms share or transfer electrons in such a way that each atom in the compound attains the noble gas configuration. The name, Octet Rule, is derived from the fact that each noble gas, except He, has an octet (8) electrons in its outermost shell. He only has one shell and therefore has only two electrons.

The quantity of electrons shared/exchanged is generally the least number to attain an inert gas configuration (e.g., oxygen will share/exchange two electrons instead of six). The number of electrons shared/exchanged determines the **oxidation number** of an element. Oxidation numbers are numbers assigned to atoms to keep track of electrons exchanged or shared in chemical reactions. For ions, the oxidation number is equal to the sign and magnitude of the ionic charge.

The oxidation numbers of the elements in a compound are what determines the ratio of the elements in a compound. For example, $NaCl$ is not Na_3Cl or $NaCl_3$ because the Na will loose one electron to attain an inert case configuration (oxidation number = +1) and a Cl will gain an electron (oxidation number = -1). Charges cancel with a 1:1 ratio, hence, $NaCl$.

The exact nature of the bond formed between two atoms is highly dependent on the relative attraction of the atoms for electrons. A measure of this relative attraction is **electronegativity**. With only a few exceptions, electronegativity increases from left to right across any row and from bottom to top in any particular column.

Ionic bonding is the complete transfer of one or more electrons from one atom to one or more other atoms. This transfer results in an imbalance of electrostatic charges in these atoms. **Ions** are atoms or groups of atoms which have an excess or deficiency of electrons and which behave as a unit in ordinary chemical reactions. Ions which consist of more than one atom are called **polyatomic ions**. Some of the most common polyatomic ions are:

Table 5-2 — Common Polyatomic Ions

Ammonium	NH_4^+	Phosphate	PO_4^{-3}
Carbonate	CO_3^{-2}	Sulfate	SO_4^{-2}
Chromate	CrO_4^{-2}	Sulfite	SO_3^{-2}

Compounds which are formed by the complete transfer of electrons are **ionic compounds**. The bonding in ionic compounds is caused by the electrostatic attraction between positive and negative ions. Compounds that are most likely to form ionic bonds are those which contain elements with very different electronegativities. In general, ionic compounds result from the combination of distinctly metallic elements with distinctly nonmetallic elements.

Covalent bonding is the sharing of one or more electrons between two atoms. This type of bond is formed when two elements have similar electronegativities. Some common covalent bonds are H_2, O_2, NH_3, CO_2, and H_2O. In solid covalent compounds, the molecules (not the atoms that make-up the molecules) are held together by weak interaction known as **van der Waals forces**. These forces arise from the attraction of the atomic nuclei in one molecule for the electrons of another molecule. Van der Waals forces are roughly 100 times weaker than covalent or ionic bonds, which are approximately equal in strength. Because of the weakness of the van der Waals forces, covalent compounds are liquids or gases at room temperature, while most ionic compounds melt at several hundred degrees Celsius. Examples of van der Waals forces are those between the molecules of liquid nitrogen or liquid oxygen.

Section 5.2 — Water And Solutions

5.2.1 — Properties Of Water

Water possesses high melting and boiling points, as well as a high heat of vaporization, when compared with elements in the same group (i.e., H_2S, H_2Se, and H_2Te). This trend for boiling point is shown in Exhibit 5-2. This extra "cohesiveness" of water molecules is due to **hydrogen bonding**.

Oxygen is more electronegative than other elements in its group since electronegativity decreases as you move down a group. This is due to the decreasing charge density as you move down a group. Oxygen is also much more electronegative than hydrogen. This imbalance in electronegativity causes the sharing of electrons between oxygen and hydrogen to be unequal (i.e., the electrons stay longer in the vicinity of the oxygen).

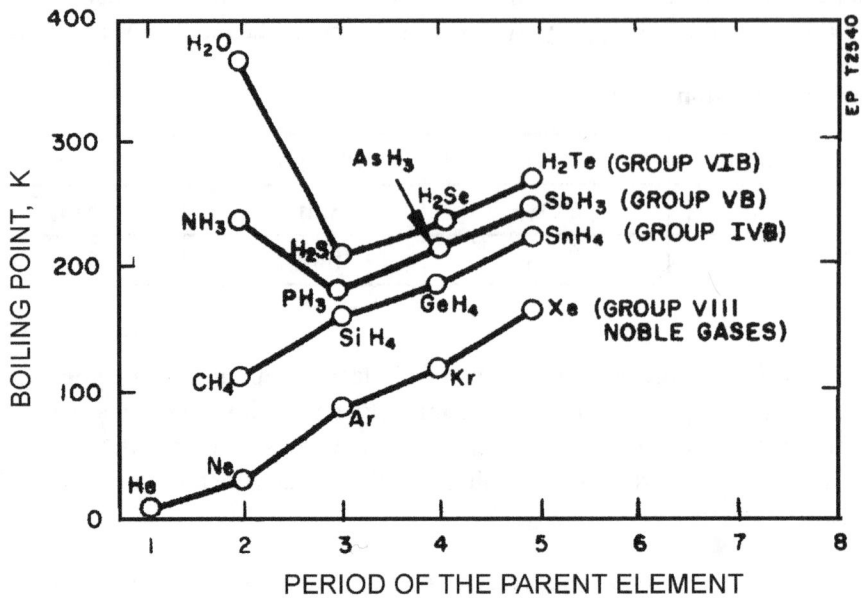

Exhibit 5-2 — Boiling Points Of Noble Gases And Compounds Of Hydrogen With Elements In Groups IVB, VB, VI

This unequal sharing causes the water molecule to have a net negative charge at the oxygen end and a net positive charge at each hydrogen end. A hydrogen bond is formed by the electrostatic attraction between the oxygen end of one water molecule to the hydrogen end of one or more other water molecules. Exhibit 5-3 depicts such an orientation.

5.2.2 — Mixtures And Solutions

A **mixture** is a combination of two or more substances which can be separated by some physical process. A **heterogenous mixture** is a mixture that has a discernable difference in composition between different parts (non-uniform) (e.g. iron filings in sawdust). A **homogeneous mixture** is a mixture that has no discernable difference between various parts (uniform) (e.g. salt water).

Homogeneous mixtures of two or more substances are called **solutions**. The substance that is in the greatest proportion of a solution is called the **solvent**. The substances that are of lesser proportion are called the **solute**. Solutions in which water is the solvent are called **aqueous solutions**

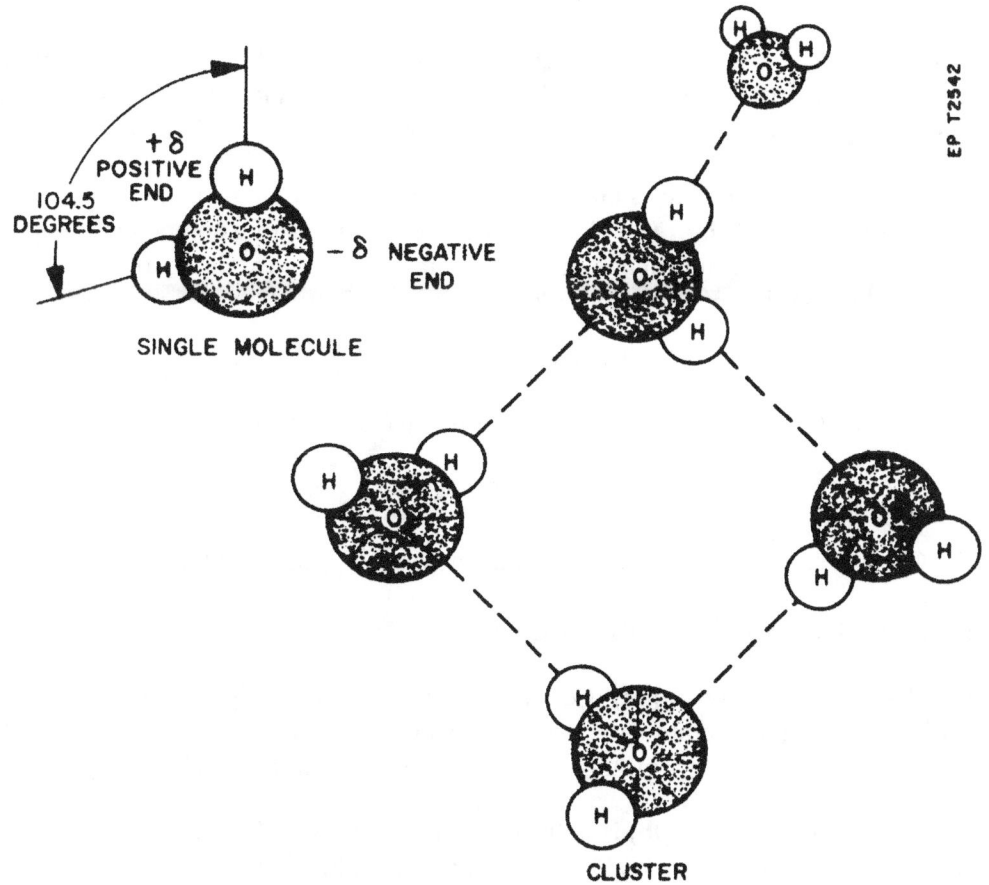

Exhibit 5-3 — Cluster Of Five Water Molecules Held Together By Hydrogen Bonding

5.2.3 — Mole And Molarity

It is known that

$$1\,\text{AMU} = 1.66 \times 10^{-24}\ \text{grams.}$$

Based on this equality it can be shown that

$$6.02 \times 10^{23}\,\text{AMU} = 1\,\text{gram.}$$

The number, 6.02×10^{23}, is called **Avogadro's Number**. A term more frequently used is the **mole**. A mole of any element or compound is Avogadro's Number of units of that element or compound.

Molarity is a system expressing the number of moles of a solute contained in a liter of solution. 1 $\underline{M}$"Compound X" denotes a solution that has one mole of molecule "X" in one liter of the solution. If the solvent is not stated, it is presumed to be water.

5.2.4 — Gram Atomic Weight (GAW) And Gram Formula Weight (GFW)

A mole is a useful quantity because one mole of atoms of an element will be an amount in grams which is numerically equal to the element's atomic weight. The **Gram Atomic Weight (GAW)** is the number of grams of an element numerically equal to its atomic weight. It is the number of grams in a mole of a compound.

The **formula weight** of a compound is the sum of the atomic weights of all the atoms in the formula for that compound. The **Gram Formula Weight (GFW)** is the number of grams of a compound numerically equal to its formula weight, the number of grams in one mole of a compound. For example, to calculate the amount of NaCl that is necessary for one liter of $1\underline{M}$ NaCl solution

$$
\begin{aligned}
\mathrm{NaCl_{GFW}} &= (1 \times \mathrm{Na_{GAW}}) + (1 \times \mathrm{Cl_{GAW}}) \\
&= (1 \times 22.99\,\mathrm{grams}) + (1 \times 35.45\,\mathrm{grams}) \\
&= 58.44\,\mathrm{grams}.
\end{aligned}
$$

58.44 grams of NaCl, with enough water added to make one liter, will be one liter of $1\underline{M}$ NaCl solution.

5.2.5 — Parts Per Million And Parts Per Billion

A concentration of 1 **part per million (ppm)** is defined as one part by weight of solute in one million parts by weight of solution. At room temperature the density of water is 1 gm/ml, or 1000 grams per liter. At this temperature, a 1 ppm solution contains 1 milligram of solute per one kilogram of solution (1 mg/kg).

A concentration of 1 **part per billion (ppb)** is defined as one part by weight of solute in one billion parts by weight of solution. This can also be expressed as 1 microgram of solute per kilogram of solution (1 µg/kg).

The terms ppm and ppb are simply different units of concentration. To demonstrate this, determine what a $1\underline{M}$ NaCl solution will be in ppm:
- From above, $1\underline{M}$ NaCl is 58.44 grams of NaCl diluted to a one liter (1000 ml) solution.
- 1000 ml of water is approximately 1000 grams at room temperature.

Note – The calculation can vary depending on how much error is acceptable. Neglecting differences in solution density and mass contribution of the NaCl will only create a 6% error.

$$
\frac{58.44\,\mathrm{grams\ NaCl}}{1000\,\mathrm{ml\ water}} \times \frac{1000\,\mathrm{ml\ water}}{1\,\mathrm{kg\ water}} \times \frac{1000\,\mathrm{mg}}{1\,\mathrm{gram}}
$$

Multiply through and the result is 58,440 mg/kg or 58,440 ppm NaCl in water.

5.2.6 — Conductivity

Compounds which form ions in aqueous solution (dissociate) and, thus, conduct electricity are called **electrolytes**. All ionic compounds and some covalent compounds (e.g., HCl) are electrolytes. **Non-electrolytes** do not form ions in solution but exist as individual molecules (e.g., sugar). The amount of current which a solution of an electrolyte is able to conduct is directly proportional to the concentration of ions. Since pure water is essentially non-conducting (there is some conductivity due to water dissociation), a quantitative measurement of the ability of an aqueous solution to conduct electricity, **conductivity (L)**, is an indication of the amount of ionic impurities (or solute) in the water.

Electrical resistance is inversely proportional to current flow. For a particular solid conductor of length d and cross sectional area A, the resistance R is given by

Equation 5-1

$$R = \rho d/A.$$

where ρ is a characteristic proportionality constant called **resistivity**. For solutions, this means that resistance is inversely proportional to concentration of electrolyte. It is conceptually simpler to use a quantity which is directly proportional to concentration, such that conductivity (L) is the reciprocal of resistivity or,

Equation 5-2

$$L = 1/\rho = d/RA.$$

If d is in cm and A is in cm^2, conductivity has the units of $(ohm \cdot cm)^{-1}$ or mho/cm. The unit mho, for **conductance**, is derived from a reverse spelling of ohm, the unit for resistance.

5.2.7 — Salinity

Salinity cells are in-line conductivity flow cells which measure the conductivity of water which flows through them. The cell interior has an element that automatically compensates for changes in water temperature. However, instead of being calibrated in conductivity units (μmho/cm), the salinity cells have been calibrated in salinity units. The salinity cells are not selective for chloride, but are calibrated such that if all of the conductivity of the water were caused by seawater salts, the parts per million (ppm) chloride ion concentration in the water would be indicated on the meter. Thus, the salinity cell reading is in units of ppm and is referred to as "X ppm salinity." The following formula is used to relate salinity cell reading to conductivity:

$$\text{Conductivity} (\mu\text{mho/cm}) \text{ corrected to } 25°C (77°F) = \text{Salinity cell reading} (\text{ppm}) \times 4$$

This formula is based upon the conductivity of a seawater solution containing 1 ppm of chloride ion. It can be determined as follows:

Seawater has 19,000 ppm of Cl^- and a total ionic content of 34,500 ppm. The conductivity of a 1 ppm solution of seawater salts is 2.1 μmho/cm. Therefore, the conductivity of a solution of seawater containing 1 ppm of Cl^- would be:

$$\frac{34,500\,\text{ppm}}{19,000\,\text{ppm}} \times 2.15\,\mu\text{mho/cm} = 4\,\mu\text{mho/cm}.$$

Exhibit 5-4 is a simple sketch of a salinity cell. Salinity cells are calibrated by the manufacturer using dilute solutions of seawater of known chloride content. The accuracy of a salinity cell is checked by measuring the conductivity of the same solution with the salinity cell and a flow cell with a portable conductivity bridge. Although the factor calculated above is actually 3.81, it is rounded off to 4. Different manufacturers have used factors from 3.8 to 4.2. The factor of 4 is based on average seawater composition and will give results within 10 percent if the cell is operating properly.

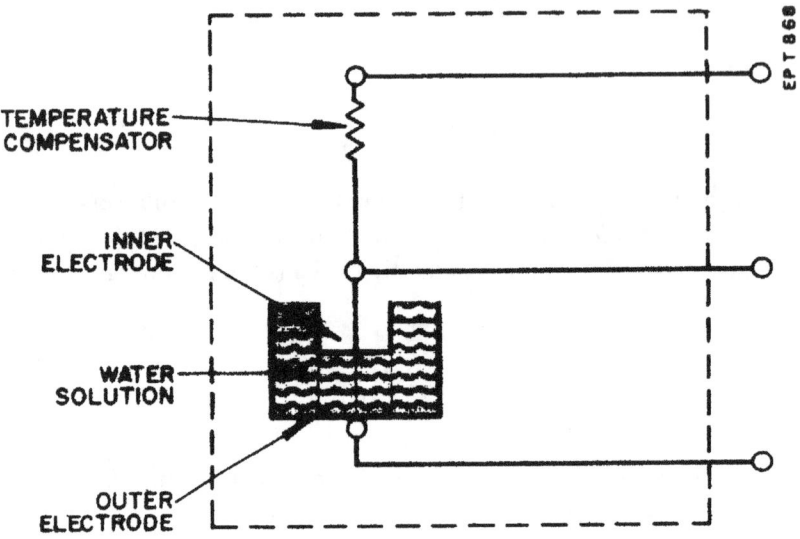

Exhibit 5-4 — Simplified Sketch Of A Salinity Cell

Section 5.3 — Chemical Processes

5.3.1 — Chemical Reactions And Factors That Effect Them

A **chemical reaction** is the conversion of one or more original substances (reactants) into one or more different substances (products). A **chemical equation** is a shorthand method of describing a chemical reaction. For example, the equation

Equation 5-3

$$2H_2 + O_2 \rightarrow 2H_2O$$

can be expressed in words as, two molecules of hydrogen react with one molecule of oxygen to form two molecules of water. Obviously, the equation is faster and easier than the word description.

The speed or rapidity at which a chemical reaction takes place is called the **reaction rate**. Reaction rates vary from infinitely fast (instantaneous) to infinitely slow (imperceptible). Reaction rates are affected by:
- Nature of the Reactants
- Temperature
- Reaction Mechanisms
- Concentration

• Catalysis

Nature of Reactants involves the electronic structure of the reacting particles (atoms, ions or molecules) and the strength of any chemical bonds formed or broken in the course of the reaction. The combination of these factors is a quantity known as the **activation energy** of a reaction (see Exhibit 5-5).

Activation energy is the minimum amount of kinetic energy the particles must have in order to overcome repulsive forces, break bonds, and form new bonds in the products.

Temperature is an indication of the average kinetic energy of the particles in a system (see Exhibit 5-6). The higher the temperature, hence kinetic energy, the more particles in the system will have the necessary activation energy for the reaction to proceed (see Exhibit 5-7).

The **Reaction Mechanism** causes variations in reaction rates since a single step reaction will proceed more quickly than a reaction that requires multiple steps. A "step" in a reaction can be considered to be a collision between two reactant (or intermediate) particles.

Concentration of reactants will also vary the reaction rate. The more reactants in a system, the more likely the chance of collision, and the more likely that the reactants will undergo a reaction. The rate of a chemical reaction is directly proportional to the concentrations of the reactants raised to some power (**Law of Mass Action**).

A **Catalyst** is a substance that when added to a system will alter the reaction mechanism/path so that it can proceed with a lower activation energy.

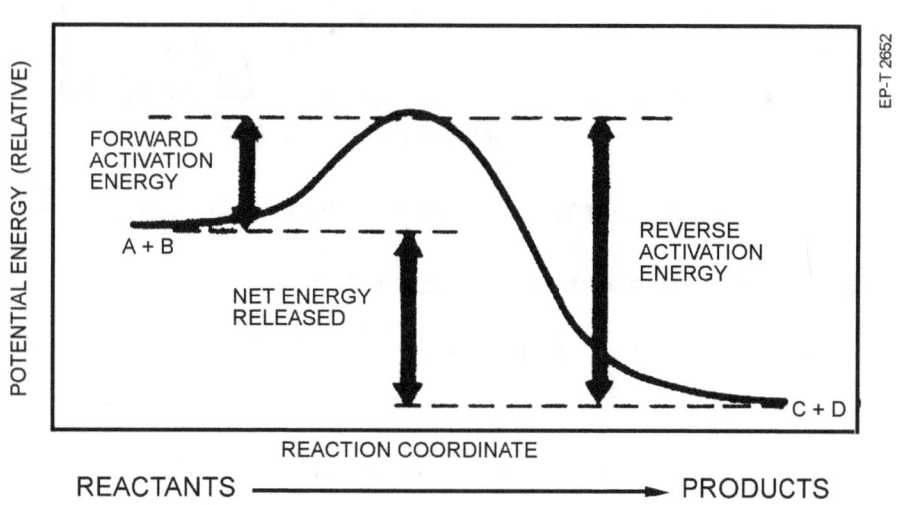

Exhibit 5-5 — Potential Energy Diagram

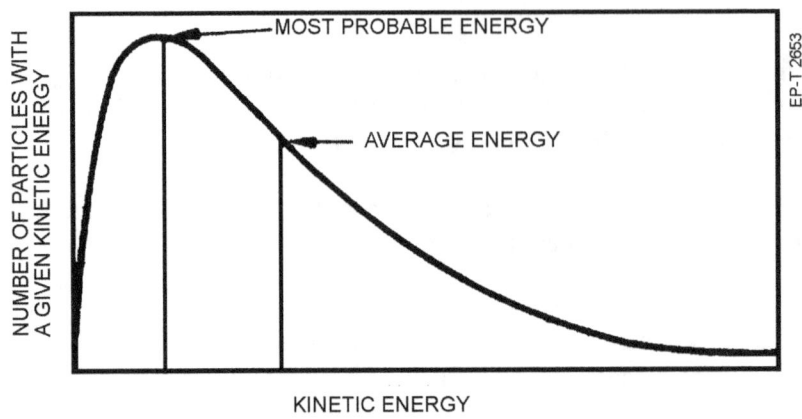

Exhibit 5-6 — Kinetic Energy Distribution (Maxwell-Boltzmann)

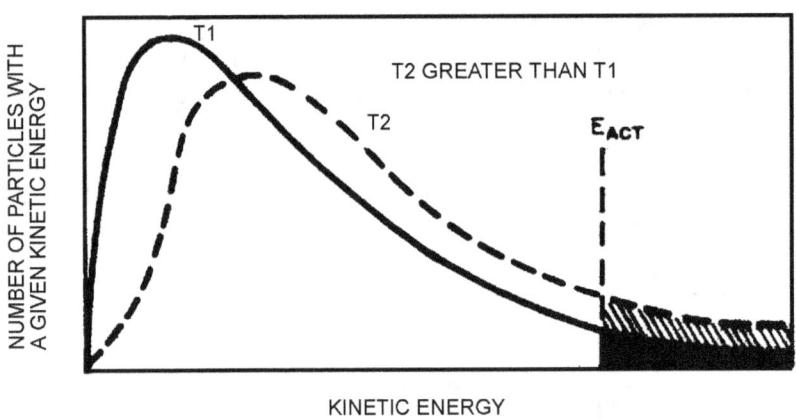

Exhibit 5-7 — Effect Of Temperature On Kinetic Energy Distribution

5.3.2 — Le Chatelier's Principle And Equilibrium Systems

In general terms, a chemical reaction can look like this

Equation 5-4

$$aA + bB \rightarrow cC + dD.$$

The arrow in a single direction implies that the reaction will go "to completion" (i.e., reaction will proceed until one or more reactants are exhausted).

Many reactions do not go to completion, but exist as a system of reactants forming products and products then reacting to reform reactants. This is known as a **reversible** reaction. When the rate of product formation equals the rate of product transition to reactants, then the reaction is considered to be in **equilibrium**. In general terms, a reversible chemical reaction can look like this

Equation 5-5

$$aA + bB \ '' \ cC + dD$$

The double arrows imply the fact that the reaction can proceed in either direction. Occasionally one of the arrows may be larger than the other to denote the direction where the majority species are formed.

All reversible reactions will proceed to an equilibrium condition. The concentrations at this equilibrium condition are the most thermodynamically stable concentrations for that system at that temperature. This point of thermodynamic stability can be designated by a constant known as the **equilibrium constant**, designated by a K_{eq}. The value of the equilibrium constant can be derived from the relation

$$K_{eq} = \frac{[C]^c \, [D]^d}{[A]^a \, [B]^b} \, .$$

The value of K_{eq} will be the constant for a given temperature. If the quantity of one of the reactants or products are externally altered, then the other quantities will adjust to maintain the value of K_{eq}.

This "self-correction" for changes in concentration can be summarized by **Le Chatelier's Principle**, which is

If a system initially at equilibrium is subjected to a disturbance or stress that changes any of the factors which determine the equilibrium state, the system will react in such a way as to minimize the effect of the disturbance or relieve the stress.

Le Chatelier's Principle not only applies to changes in concentrations, but to any of the factors previously discussed as affecting reaction rate.

Section 5.4 — Hydronium Ion And pH

5.4.1 — Acids And Bases

Acids are those compounds which produce hydronium ions (H_3O^+) directly when dissolved in water. Hydrochloric acid (HCl) is an example of an acid.

Bases are those compounds which produce hydroxide ions (OH^-) directly when dissolved in water. Sodium Hydroxide (NaOH) is an example of a base.

Acids and bases will react together by a process called **neutralization**. In this reaction hydronium and hydroxide ions react together to form water and the other ions react together to form a salt. For example,

Equation 5-6

$$HCl + NaOH \rightarrow H_2O + NaCl$$

It should be noted that dissolving acids or bases will generate heat in a solution (due to the acid or base going to a lower energy state). The neutralization process will add even more heat if acids and bases are mixed together. This must be remembered when diluting and mixing solutions.

Acids and bases can be broken into two subclasses, **strong** and **weak**. Strong bases and acids are those that completely disassociate in aqueous solutions (i.e. have a large equilibrium constant). Weak acids and bases do not completely disassociate in aqueous solutions (i.e. have a small equilibrium constant). Examples of strong and weak acids and bases are listed in Table 5-3:

Table 5-3 — Strong And Weak Acids & Bases; Examples

Strong Acids		Weak Acids	
HCl	Hydrochloric Acid	$H_3C_6H_5O_7$	Citric Acid
H_2SO_4	Sulfuric Acid	$HC_2H_3O_2$	Acetic Acid
HNO_3	Nitric Acid	H_2CO_3	Carbonic Acid
Strong Bases		Weak Bases	
NaOH	Sodium Hydroxide	$NH_3(OH)$	Ammonium Hydroxide
$Ca(OH)_2$	Calcium Hydroxide	N_2H_4	Hydrazine
		C_4H_9NO	Morphaline

5.4.2 — Equilibrium Constant For Water

The above reaction had all the hydronium and hydroxide ions reacting to form water. Actually, this is a reversible reaction that exists as an equilibrium in water and all its solutions. The equilibrium equation is

Equation 5-7

$$2H_2O \, '' \, H_3O^+ + OH^-.$$

This leads to an equilibrium constant of

Equation 5-8

$$K_{eq} = \frac{\left[H_3O^+\right]\left[OH^-\right]}{\left[H_2O\right]^2}$$

Since water is the solvent, the changes in the denominator will be insignificant compared to changes in the numerator. The value $[H_2O]^2$ is relatively constant, so it can be multiplied on both sides of the equation and a new term can be defined

Equation 5-9

$$K_w = K_{eq} \left[H_2O\right]^2 = 1 \times 10^{-14}$$

and hence

Equation 5-10

$$K_w = \left[H_3O^+\right]\left[OH^-\right] = 1 \times 10^{-14}.$$

The value K_w will only vary for changes in temperature (since it is an equilibrium constant). Refer to Exhibit 5-8.

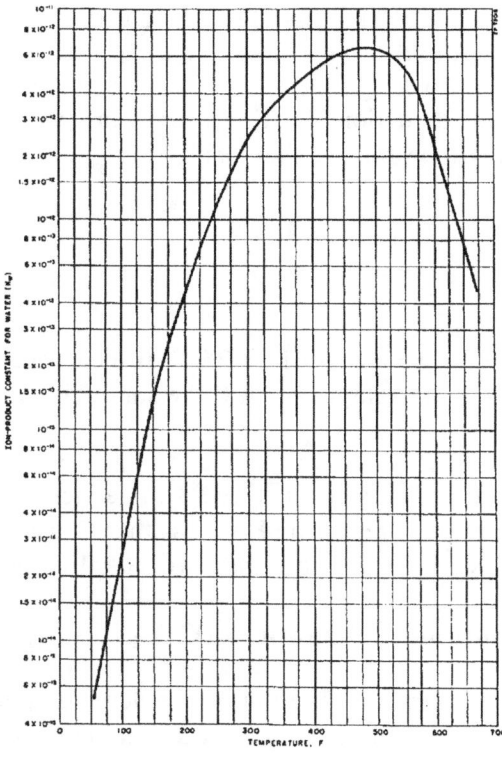

Exhibit 5-8 — K_w For Water, In Molarity Units

5.4.3 — Calculating pH And pOH

The term **pH** historically meant the "power of the hydrogen concentration". This is due to the pH value being equal to the negative common logarithm of the hydronium ion concentration. Mathematically, pH is defined as

Equation 5-11

$$pH = -\log\left[H_3O^+\right]$$

where $[H_3O^+]$ is the molar concentration of the hydronium ion in solution.

An equivalent term for hydroxide ion concentration is **pOH**, which is equal to the common logarithm of the hydronium ion concentration. Based on the relationship:

Equation 5-12

$$K_w = \left[H_3O^+\right]\left[OH^-\right] = 1 \times 10^{-14}$$

it can be easily determined that both $[H_3O^+]$ and $[OH^-]$ will be equal to 1×10^{-7} moles/liter in a pure water solution. This is determined by simply taking the square root of K_w.

It is simple to determine, based on the definition of pH, that for a pure water solution the pH and pOH will both be 7. The sum of the pH and pOH will be 14. The fact that

Equation 5-13

$$pH + pOH = 14$$

is a useful relationship for all values of pH and pOH for 25°C. If the temperature is changed, the value for the sum pH + pOH will also change. For example, at 18°C the sum will be 14.2, and the pH of neutral water would be 7.10.

It is simple to demonstrate mathematically that as the $[H_3O^+]$ value increases for a solution that the pH value of the same solution will decrease. For this reason a pH of less than 7 is considered to be an acidic solution. Therefore, a pH of greater than 7 is considered to be a basic (or **alkaline**) solution.

5.4.4 — Concentration And Dilution Calculations ($C_1V_1=C_2V_2$)

The general approach to solving concentration and dilution problems is application of the principle of **mass balance**. Mass balance is the principle that the amount of mass left in a system is equal to the amount originally in the system plus the amount added minus the amount removed from the system.

By multiplying the concentration of the system by the volume of the system (using similar units), the result will be equivalent to the mass of the solute in the system. If only water is added or removed, the mass of solute will remain the same. If the mass of solute before and after the reaction are set equal to each other, then the following equivalence is derived

Equation 5-14

$$C_1V_1 = C_2V_2$$

5.4.5 — Normality And pH for Strong Acid/Strong Base Reactions

To calculate the final pH of a neutralization reaction the difference must be found between the amount of hydronium or hydroxide ions originally present, then how much of the other is added. The difference between the two values can be considered the amount remaining, and a new pH calculated. This simple method works fine for acids and bases that completely dissociate in solution (i.e., **strong acids** or **strong bases**). In these simple cases the water (K_w) is the only equilibrium system that must be considered. An example of this process would be calculating the original and final pH values for adding a liter of 1$\underline{M}$ HCl solution to two liters of 2$\underline{M}$ NaOH solution.

- The original pH of both solutions would be:

$$\text{Acid Solution: pH} = -\log\left[H_3O^+\right]$$

$$= -\log\left[1\right]$$

$$= 0$$

$$\text{Basic Solution: pOH} = -\log\left[OH^-\right]$$

$$= -\log\left[2\right]$$

$$= -0.3$$

$$\text{pH} + \text{pOH} = 14$$

$$\text{pH} = 14 - (-0.3)$$

$$\text{pH} = 14.3$$

- The new pH of the combined solution will be:

$$\text{Moles of } OH^- : 2 \text{ moles/liter} \times 2 \text{ liters} = 4 \text{ moles}$$

$$\text{Moles of } H_3O^+ : 1 \text{ moles/liter} \times 1 \text{ liter} = 1 \text{ mole}$$

Result of Neutralization: 1 mole of OH⁻ will be neutralized by 1 mole of H_3O^+, which will leave 3 moles of OH⁻.

Result of Dilution: 1 liter + 2 liters = 3 liters.

Resulting Concentration and pH:

$$3 \text{ moles } OH^-/3 \text{ liters} = 1 \text{ moles/liter } OH^-$$

$$\text{pOH} = -\log\left[OH^-\right] = -\log\left[1\right] = 0$$

$$\text{pH} = 14 - \text{pOH} = 14 - 0 = 14$$

Another solution concentration system which can simplify acid/base calculations is **normality**. Normality utilizes the principle of **equivalent weights**. An equivalent weight of acid is that amount of acid which will produce one mole of hydronium ions in solution. For example, the equivalent weight of sulfuric acid (H_2SO_4) is 49.07 grams (GFW/2, since there are two hydrogen atoms per molecule). An equivalent weight of base is that amount of base which will produce one mole of hydroxide ions. For example, the equivalent weight of sodium hydroxide (NaOH) is 40.00 grams (GFW since there is one hydroxide group per molecule).

A 1 normal solution (designated 1$\underline{N}$) contains 1 equivalent weight per liter of solution. The use of normal concentration units greatly simplifies many calculations involving acid-base reactions. This is due to the relationship **one equivalent weight of an acid reacts completely with one equivalent weight of a base**.

An example of a normality neutralization calculation would be the result of mixing a liter of 3$\underline{N}$ H_2SO_4 with a 2 liter solution of 1$\underline{N}$ $Ba(OH)_2$.

$$\text{Number of Equivalents of Acid} = V_a N_a$$

$$= 1\,\text{liter} \times 3\underline{N}$$

$$= 3\,\text{equivalents}$$

$$\text{Number of Equivalents of Base} = V_b N_b$$

$$= 2\,\text{liter} \times 1\underline{N}$$

$$= 2\,\text{equivalents.}$$

There are more acid equivalents than base equivalents, so the resulting solution will be acidic. The 2 base equivalents will neutralize 2 acid equivalents, leaving one acid equivalent.

$$1\,\text{acid equivalent} \times 1\,\text{mole}\;H_3O^+ \big/ \text{equivalent} = 1\,\text{mole}\;H_3O^+$$

$$\text{Resulting volume} = 1\,\text{liter} + 2\,\text{liters} = 3\,\text{liters}$$

$$\text{Final Concentration} = 1\,\text{mole}\;H_3O^+ \big/ 3\,\text{liters}$$

$$= 0.33\,\text{moles/liter}\;H_3O^+$$

$$\text{Final pH} = -\log\!\left[H_3O^+\right] = -\log(0.33) = 0.48$$

Due to simplification of the math, equivalent weights and normality are easier to use in calculations than gram formula weight and molarity.

5.4.6 — Weak Base Or Acid Equilibrium Processes

A weak base (weak acid) is a base (acid) which does not dissociate completely when dissolved in water. An equilibrium system is established between the concentration of the original compound and the concentration of its resulting ions. An excellent example of a weak base is ammonia when dissolved in water.

Equation 5-15

$$NH_3 + H_2O \,''\, NH_4{}^+ + OH^-$$

Equation 5-16

$$K_i = \frac{\left[NH_4{}^+\right]\left[OH^-\right]}{\left[NH_3\right]} = 1.8 \times 10^{-5}$$

Note – Since the equilibrium constant is for an ionization reaction, it is called an ionization constant (K_i).

It should be noted that this equilibrium system causes a $1\underline{M}$ strong base (e.g., NaOH) solution to have a higher pH than a $1\underline{M}$ weak base (e.g., NH_3) solution; pH of 14 for the NaOH solutions as opposed to a pH of 11.93 for NH_3, which is a OH$^-$ concentration two orders of magnitude less.

For more information on stoichiometric and equilibrium calculations, refer to Reference 4.

Section 5.5 — Gas Laws

5.5.1 — Ideal Gas Laws

The ideal gas law is

Equation 5-17

$$PV = nRT$$

where:

P	= Absolute Pressure
V	= Volume
n	= Moles of Gas
R	= Ideal Gas Constant
T	= Absolute Temperature

Absolute temperature is measured in either **Kelvin** or **Rankine** where:

$$K = 273 + {}^0C$$

$$R = 460 + {}^0F$$

Equation 5-17 may be solved for a constant value, such as:

Equation 5-18

$$\frac{PV}{T} = nR$$

If no mass is lost or gained, nR will be constant. The relationship of PV/T will be constant for all changes in P, V or T. It is noteworthy that 1 mole of an ideal gas at STP (25°C and 1 atm) has a volume of 22.4 liters. Changes between two states can be solved by the equation

Equation 5-19

$$\left(\frac{PV}{T}\right)_1 = \left(\frac{PV}{T}\right)_2$$

This relationship can be shown graphically. Exhibit 5-9 (a) shows how pressure varies with volume while temperature remains constant. Exhibit 5-9 (b) shows how pressure varies with temperature while the volume remains constant.

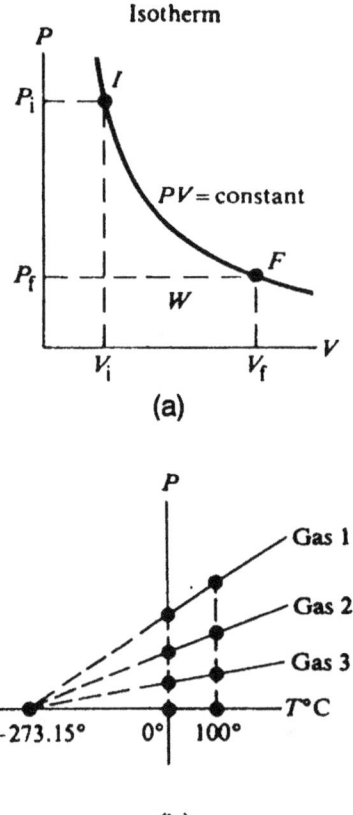

(a)

(b)

Exhibit 5-9 — Ideal Gas Law Relationships

5.5.2 — Henry's Law

The effect of pressure on the solubility of a gas in water is described quantitatively by Henry's law. For a pressure, P, of a pure gas in direct contact with water at a temperature, T, the maximum concentration (solubility) of gas in the water is given by:

Equation 5-20

$$C = H(T)P$$

where H(T) is the Henry's constant for the particular gas and that temperature, T, of the solution. If P is in atmospheres, atm, and C in cc/kg, then H(T) has the units cc/(kg • atm). In the case of a mixture of gases in contact with water, the partial pressure (p) of each component of the mixture represents the concentration of component i.

Equation 5-21

$$C_i = H_i(T)_{P_i}$$

In effect, Henry's law says that the solubility of a gas in water is directly proportional to the partial pressure of the gas above the surface of the water. For a given amount of gas in a pressurized water system, if the pressure of the water is greater than the partial pressure necessary to maintain the gas concentration in the solution, all the gas in the system will be dissolved and there will be no gas space. If the pressure is then reduced to less than the partial pressure corresponding to the concentration of gas in solution, bubbles of gas may start to form in the water. A common example of this phenomenon is the foaming (gas coming out of solution) which occurs when the cap is removed (sudden decrease in pressure) from a bottle of carbonated beverage.

Section 5.6 — Corrosion

5.6.1 — Corrosion Concerns

Corrosion is the destructive attack of a metal by chemical or electrochemical reactions with its environment. Every metal corrodes under certain conditions. Because of the difficulty of developing general principles of corrosion for all metals, and also for reasons of practical interest, we will concentrate on the chemical aspects of iron and iron alloys. Refer to Reference 1 or Reference 2 for information on specific alloys.

5.6.2 — Electrochemistry

Since most corrosion processes are electrochemical in nature, a brief discussion of electrochemistry is in order. An electrochemical cell consists of two electrodes connected by either metallic contact or electrolytic solutions to form a closed circuit for the corrosion current.

Oxidation is the process where the metal loses electrons. It is called oxidation because when a metal reacts with oxygen to form an oxide it is always observed to lose its electrons to the oxygen. The electrode where oxidation occurs is the **anode**. The term anode is from the fact that **anions** (negative ions) in the solution will migrate to that electrode. An example of an oxidation reaction is:

Equation 5-22

$$Fe \rightarrow Fe^{2+} + 2e^-.$$

Reduction is the process where an atom or molecule gains electrons. It is called reduction because when an atom or molecule gains electrons its overall charge is reduced. The electrode where reduction occurs is the **cathode**. The term cathode is from the fact that **cations** (positive ions) in the solution will migrate to that electrode. An example of a reduction reaction is:

Equation 5-23

$$O_2 + 2H_2O + 4e^- \rightarrow 4OH^-.$$

Every electrochemical cell must have an anodic reaction (oxidation) to supply electrons to support a cathodic reaction (reduction). This coupling of a reduction and oxidation reaction is called a **redox** reaction. The site of the anodic reaction is where corrosion is occurring. The reduction of oxygen, hydronium ions or water can consume the electrons formed by the anodic reaction.

The driving force for the reaction is the difference in oxidation potential between the anodic reaction and the cathodic reaction. For example, the reaction in Equation 5-22 has a standard oxidation potential of +0.45V, while the reaction in Equation 5-23 has a standard reduction potential of +0.40V. By combining the oxidation and reduction half reactions, the overall redox reaction can be determined

$$2Fe \rightarrow 2Fe^{2+} + 4e^- \qquad E^0 = +0.45V$$

$$O_2 + 2H_2O + 4e^- \rightarrow 4OH^- \qquad E^0 = +0.40V$$

$$\overline{2Fe + O_2 + 2H_2O \rightarrow 2Fe(OH)_2 \qquad E^0 = +0.85V.}$$

The standard potential for the reaction is for conditions where all chemical concentrations are 1 mole/liter, a temperature of 25°C, and a pressure of 1 atmosphere. Variations in concentration, temperature, pressure or competing reactions will change the potential of the reaction. The potential of every half reaction is based on the thermodynamic potential for the reaction to occur. For additional information, refer to Reference 4.

5.6.3 — Types Of Corrosion

General Corrosion is the uniform attack of a metal surface by its environment. That is, the corrosion rate is uniform over the surface of the metal. The corrosion is due to areas of the metal surface switching between being anodic or cathodic. No area is preferentially anodic. Examples of general corrosion are the rusting of untreated iron and the tarnishing of silverware.

The effect of pH upon corrosion depends upon the metal or alloy involved and the specific conditions to which it is exposed. Factors such as temperature and the presence of oxygen can significantly alter the dependence of corrosion rate upon pH. It is instructive to consider two examples.

If the corrosion rate of iron exposed to oxygenated water at room temperature is plotted versus pH, the results shown in Exhibit 5-10(a) are obtained. For very low pH, the corrosion rate is large because it is limited only by the rate at which atomic hydrogen (H), adsorbed onto cathodic sites as a result of reduction of hydronium ion (H_3O^+), can combine to form molecular hydrogen and be released into the solution. As pH increases, the amount of hydronium ion available for reduction decreases and the rate of corrosion becomes dependent instead upon the rate at which oxygen can (1) diffuse to the metal surface, and (2) be replaced in solution by absorption of O_2 from air. Neither of these processes is appreciably affected by pH. As the bulk solution pH increases beyond pH = 7, appreciable hydrolysis of ferric ion takes place, and high concentrations of $Fe(OH)_3$ (aq) and $Fe(OH)_4^-$ ions are produced adjacent to the surface of the metal. For bulk solution pH between 7 and about 10, this ionic layer shields the surface of the metal from the bulk pH change but allows free passage of oxygen. The corrosion rate is, therefore, still determined by oxygen diffusion in this range. As the solution pH exceeds 10, however, the shielding action of the ionic layer is insufficient to prevent formation of Fe_2O_3. Increasing pH results in the formation of a passivating Fe_2O_3 oxide layer, with a consequent decrease in corrosion rate to a very small value as pH approaches ~14. Although not shown in the figure, the corrosion rate subsequently rises as pH increases beyond 14 and the passivating Fe_2O_3 film dissolves to form $Fe(OH)_4^-$ ions, allowing relatively free diffusion of oxygen to the metal surface.

At elevated temperatures and in the absence of dissolved oxygen, the behavior of corrosion rate versus pH is somewhat different, as illustrated by Exhibit 5-10(b). The corrosion rate is large at very low values of pH for the reason given in the preceding paragraph. As pH increases, the corrosion rate decreases because the hydronium ion available to be reduced to H_2 decreases. The corrosion rate does not level off at the plateau region seen in Exhibit 5-10(a) because no oxygen is available to assume the reduction reaction role; rather, it continues to decrease slowly as bulk pH increases due to formation of a passivating Fe_3O_4 film, unless some other oxidizing agent such as Cu^{2+} is available. This film is in equilibrium with OH^-, $Fe(OH)^+$, $Fe(OH)_2$ and $Fe(OH)_3^-$ in the solution adjacent to the oxide layer. The minimum solubility of Fe_3O_4 occurs in the vicinity of pH ~10. The corrosion rate is seen to decrease to a minimum at about that value, and then increase. The rapid increase in corrosion rate as pH exceeds ~13 occurs because Fe_3O_4 becomes increasingly soluble above that point, going into solution primarily as $Fe(OH)_4^-$ ions.

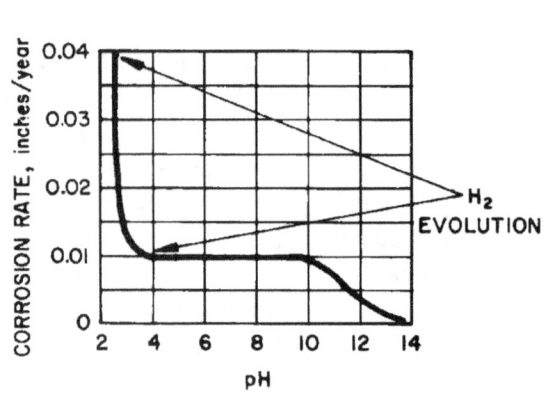

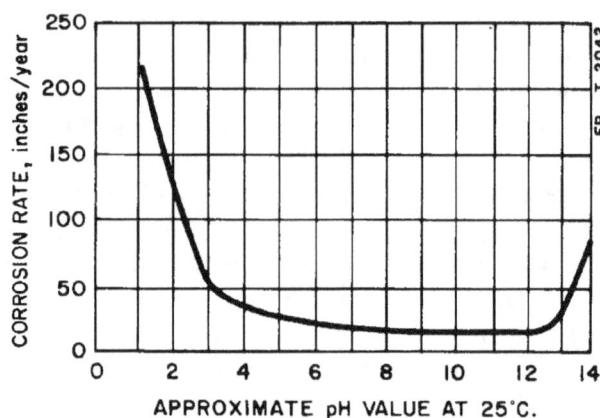

(a) EFFECT OF pH ON THE CORROSION OF IRON EXPOSED TO AERATED WATER AT ROOM TEMPERATURE

(b) RELATIVE RATE OF ATTACK ON IRON BY 590°F WATER OF VARYING pH.

Exhibit 5-10 — Effect Of pH On The Corrosion Rate Of Iron In Water

In nuclear reactor plants, it is desired to minimize the corrosion of iron and steels exposed both to high-temperature reactor coolant and to high temperature boiler water. This is accomplished by keeping dissolved oxygen to a minimum, and by maintaining the pH in the range where the passivating oxide film is least soluble. Thus, both reactor coolant and boiler water are normally maintained in an alkaline condition.

Galvanic Corrosion results when two dissimilar metals are connected and exposed to a water environment. One metal becomes cathodic and the other anodic, setting up a galvanic cell. For example, if copper and steel are directly connected in an aqueous environment, the steel will preferentially corrode as the anode of the cell as depicted in Exhibit 5-11. The smaller the anodic area relative to the cathodic area, the sooner the anode will fail.

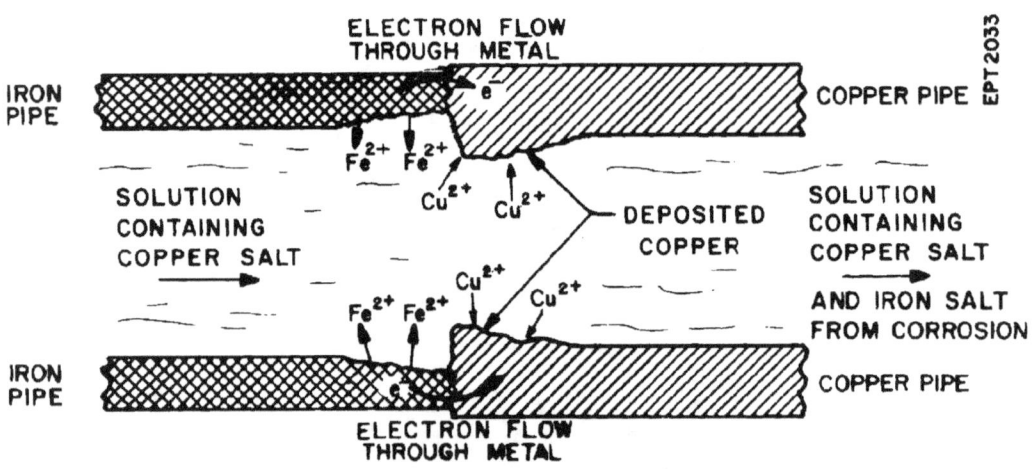

GALVANIC CORROSION ABOUT IRON COPPER JUNCTION IN A PIPE

Exhibit 5-11 — Galvanic Corrosion At Iron-Copper Pipe Junction

Galvanic corrosion is commonly prevented by avoiding direct metal coupling of dissimilar metals. The degree of dissimilarity is dependent on the environmental conditions where the system will exist. A **galvanic series** lists metals in order of their activity level for corrosion. Metals close together on the list will exhibit a low level of galvanic corrosion. Metals further apart on the list will corrode, with the more active metal as the anode.

A method called cathodic protection is often used to retard or eliminate galvanic corrosion. In this method, a second metal is added to the metal to be protected. This metal has a reduction potential even greater than that of the metal which is to be protected. The most active metal tends to corrode in place of the protected metal.

For example, the corrosion of a steel structure in an acid environment involves the following electrochemical equations:

$$Fe \rightarrow Fe^{2+} + 2e^-$$

$$2H + 2e^- \rightarrow H_2$$

$$Zn \rightarrow Zn^{2+} + 2e^-$$

If electronics are supplied to the steel, metal dissolution (corrosion) will be suppressed and the rate of hydrogen evolution increased. Thus if electrons are continually supplied to the steel structure, corrosion will be suppressed. The electrons are supplied through the corrosion (oxidation) of the Zinc as shown in the above equation.

Galvanic corrosion can also be reduced by bimetallic or electrical insulating connections

A **Concentration Cell** is developed when a single metal is exposed to two different concentrations of water solutions. The concentration that can best support the anodic reaction will be the portion of the metal which is corroded. An example would be crevice corrosion induced by a difference in oxygen concentration inside and outside the crevice. The crevice area is soon depleted of oxygen, while surfaces outside the crevice have the oxygen continually replenished. This will cause the oxygenated surfaces to be preferentially cathodic and the crevice preferentially anodic as depicted in Exhibit 5-12.

Pitting Corrosion is a type of corrosion in which the corrosion rate in a localized area is significantly greater than the general corrosion rate, usually due to concentration differences in the metal's environment. The pit or crack formed may penetrate the entire thickness of a section of metal before any surface corrosion is visible to the naked eye. Corrosion reactions within the pit can also produce an environment that will accelerate corrosion within the pit after initiation. Exhibit 5-13 is a schematic of surface pitting.

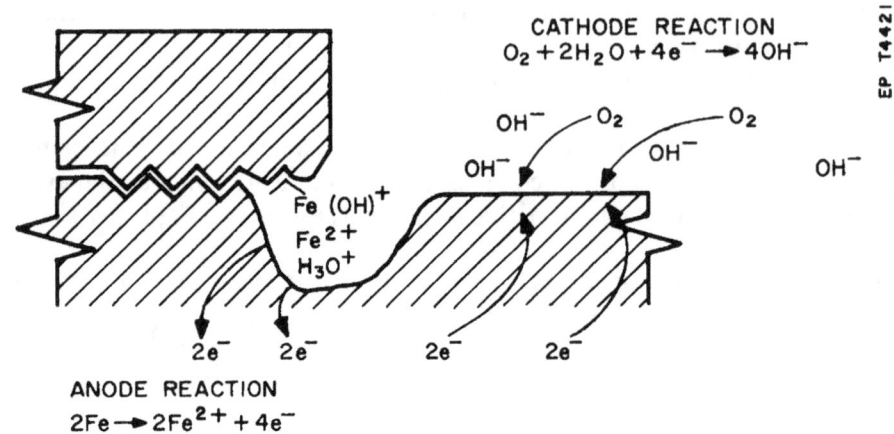

Exhibit 5-12 — Crevice Corrosion

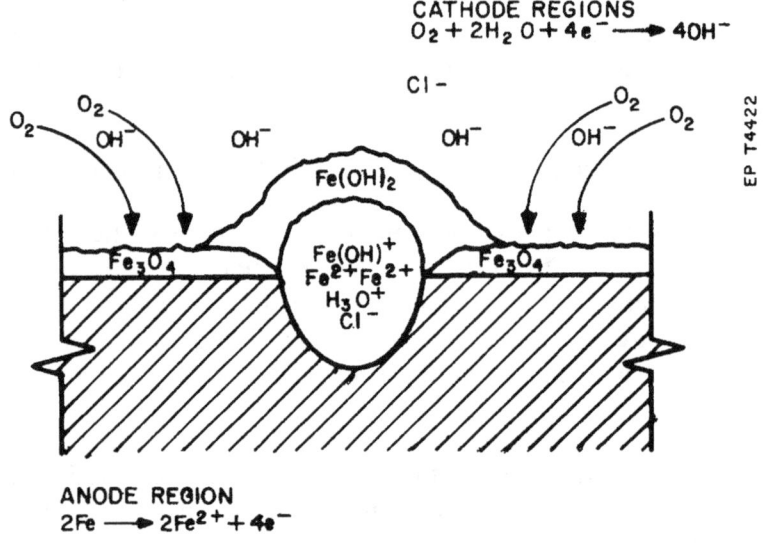

Exhibit 5-13 — Surface Pitting

Stress Corrosion Cracking (SCC) is a type of corrosion which results in cracks in a metal when the metal is in a stressed condition and exposed to a specific environment. An example would be stressed stainless steel exposed to chloride ions and oxygen. The likelihood of SCC occurring in this environment is depicted in Exhibit 5-14. Stresses may be due to heat treatments or residual stresses from cold working or processing. The conditions necessary for SCC to occur is highly variable between various metals and alloys. Common concerns for SCC in boilers are related to either chloride contamination or caustic (highly basic) conditions. SCC will preferentially occur in restricted flow areas beneath sludge piles on boiler tube sheets.

Phosphate Wastage is corrosion resulting from formation of soluble phosphate complexes in environments containing excessively high phosphate ion concentration. It is a problem in boilers using phosphate treatment as a means of corrosion control. Under localized flow restriction and concentrating conditions, the solubilities of the disodium phosphate and trisodium phosphate (treatment chemicals) are exceeded. When the phosphate salts precipitate out of solution the environment becomes acidic due to the loss of alkalinity. While common in boilers with Alloy 600 tubes, it has also been observed on Type 304 stainless steel tubes in the Shippingport S1B steam generator.

5.6.4 — Passivity

Passivity is the property of a metal being more resistant to corrosion than would be expected from its oxidation potential due to the formation of a thin protective oxide layer on the surface of the metal. Certain metals tend to develop a tightly adherent oxide layer on their surface (e.g., Chromium, Zirconium). The oxide layer inhibits reactions of the base metal with the surrounding environment. Stainless steel is an iron alloy with ≥ 12 w/o chromium. Exhibit 5-15 depicts the general corrosion rate of steel as chromium content is increased.

Damage to the oxide layer can be caused by the presence of Cl^- and F^- ions, by changes in the environment, and by stresses developed within the film as a result of temperature changes or transitions in the oxide layer crystal structure.

Corrosion can be limited by controlling impurities in the water (e.g. Cl^-, O_2), using similar metals or sacrificial anodes where possible, keeping chemical concentrations and pH in their control bands, and pretreating (operating in a high temperature, high pH environment until a passive layer forms).

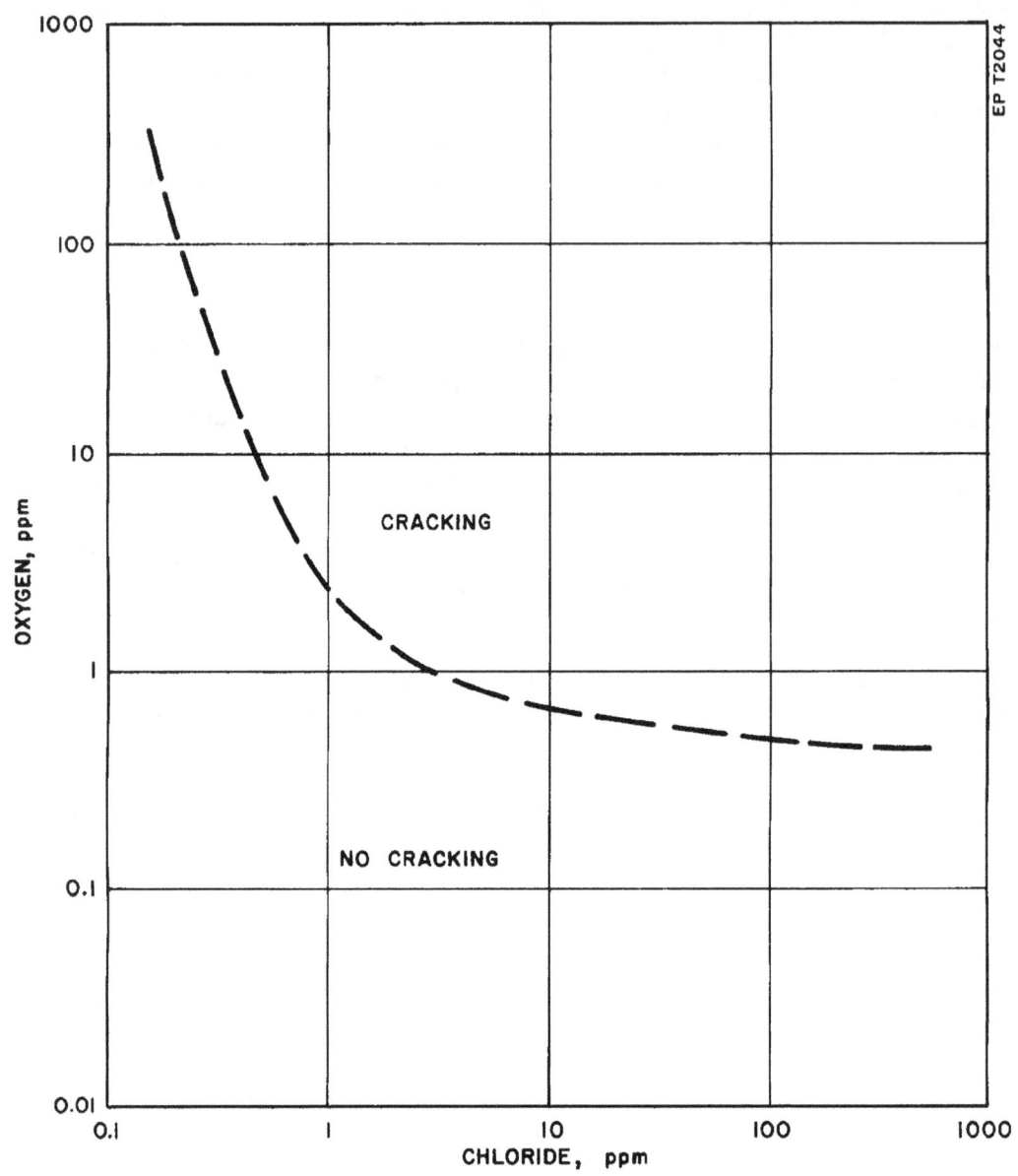

EP T2044

Exhibit 5-14 — Regions Of Chloride And Oxygen Content Where Austenitic Stainless Steels (18-8) Undergo SCC (Temperature Range, 470-500 °F)

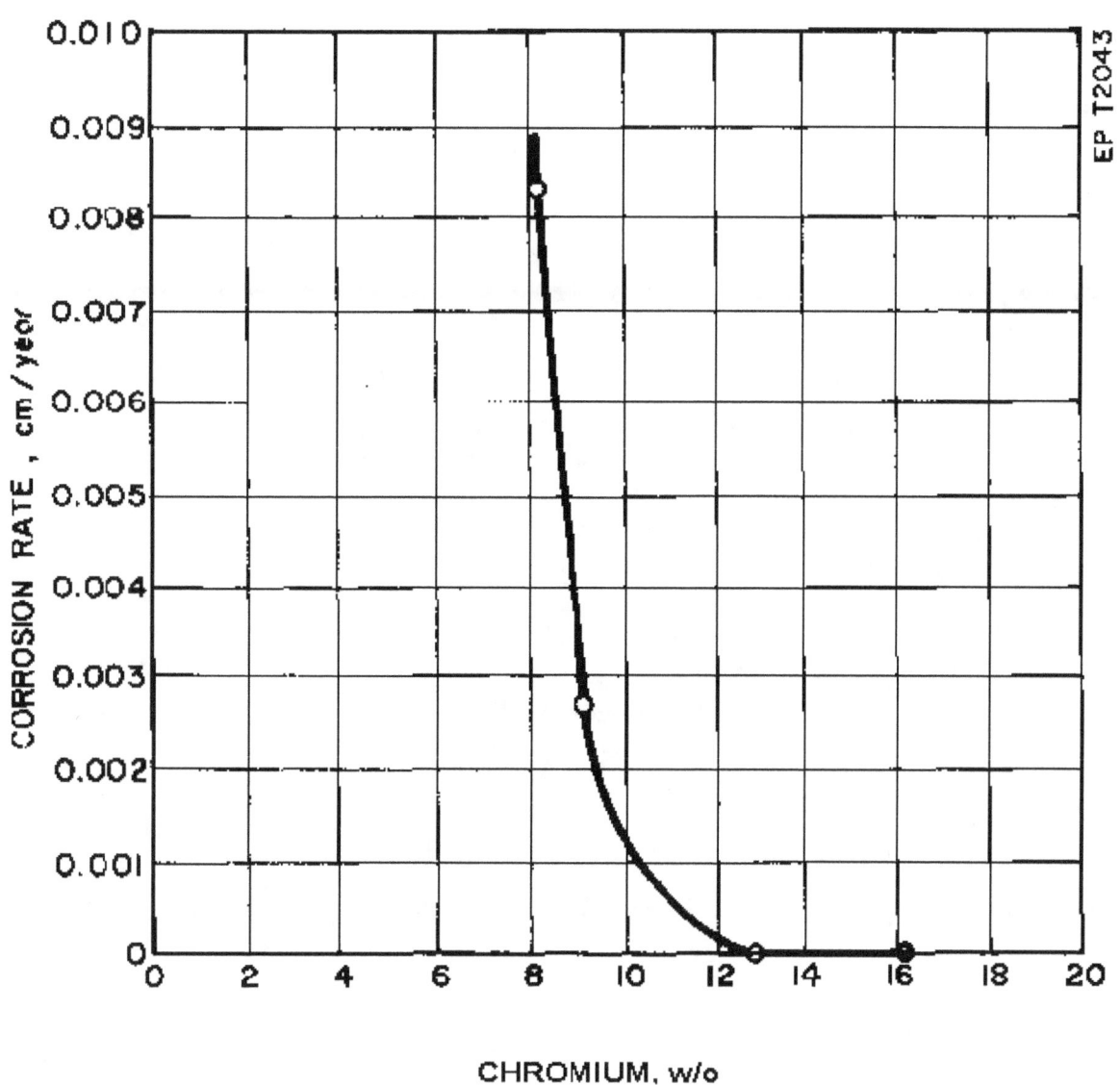

Exhibit 5-15 — Effect Of Cr Content On Corrosion Of Steel

(Intentionally Blank)

Chapter 6 — MATERIALS REVIEW

Section 6.1 — Structure Of Metals

6.1.1 — Crystal Structures

Solid materials can be divided into two general classes, **crystalline** and **amorphous**. All metals have a crystalline structure. A crystalline substance is one in which the constituent atoms, ions, or molecules form a repetitive three dimensional array and thus the crystalline substance exhibits long-range order. Salt is an example of a crystalline substance. In contrast, the constituent atoms of an amorphous substance (a substance without definite form) only has short-range order. Glass is an example of an amorphous substance.

The predominant feature of a crystal is the regularity of its structure. This regular arrangement of atoms or molecules is a **crystal lattice**. In a crystal lattice, the smallest group of atoms having all the characteristics of the whole crystal is called the **unit cell**. The simplest unit cell is the cubic, depicted in Exhibit 6-1. Salt crystals have a cubic structure.

Most engineering metals have unit cells that are body-centered cubic (BCC), face-centered cubic (FCC), hexagonal closed-packed (HCP) or body-centered tetragonal (BCT).

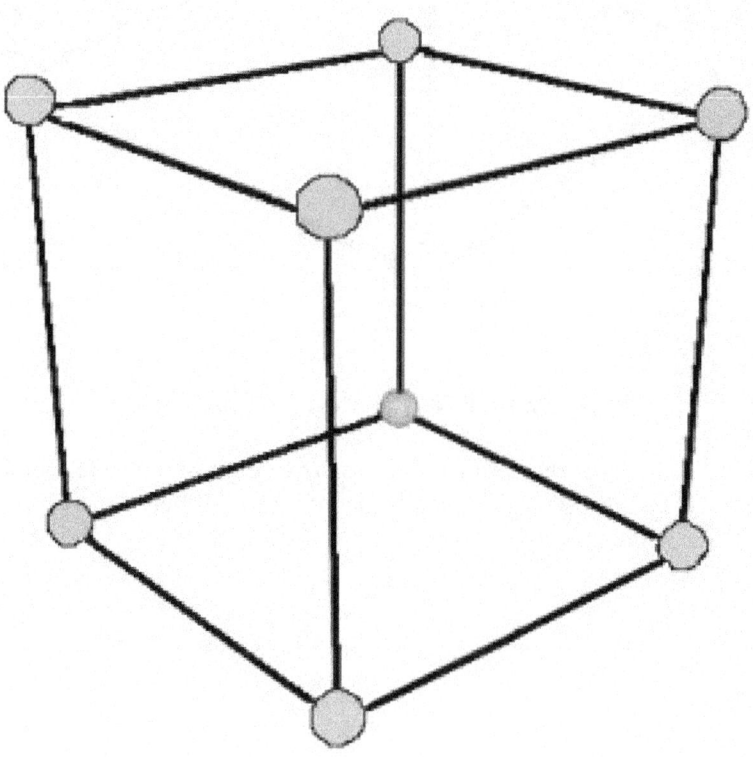

Exhibit 6-1 — Simple Cubic

The **body-centered cubic (BCC)** shown in Exhibit 6-2 consists of eight corner atoms and one atom in the center of the cube. Each corner atom is shared by seven other adjacent unit cells.

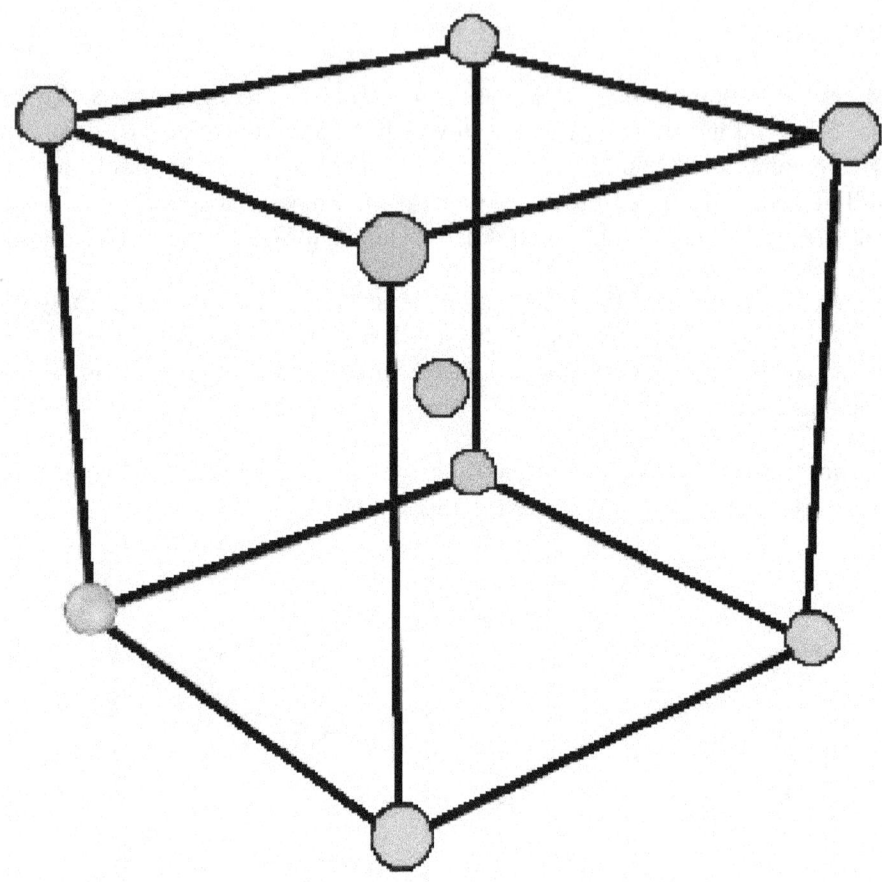

Exhibit 6-2 — Body-Centered Cubic (BCC)

The **face-centered cubic (FCC)** shown in Exhibit 6-3 consists of eight corner atoms and one atom on each of its six faces. Each face atom is equally shared by it's adjacent unit cell face.

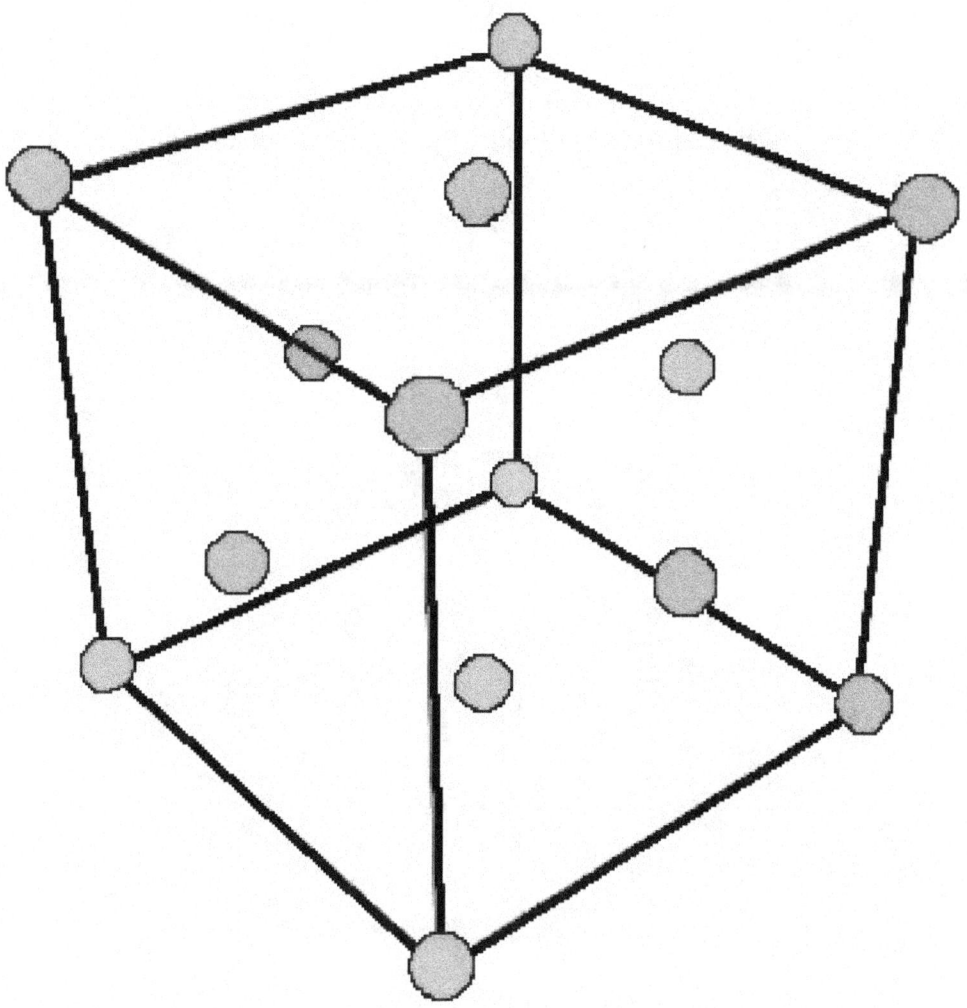

Exhibit 6-3 — Face-Centered Cubic (FCC)

The **hexagonal closed-packed (HCP)** unit cell, shown in Exhibit 6-4, is bound by two hexagons at the bottom and at the top. There is an atom centered on each hexagonal face as well as three atoms forming an equilateral triangle between, and parallel to, the hexagonal faces.

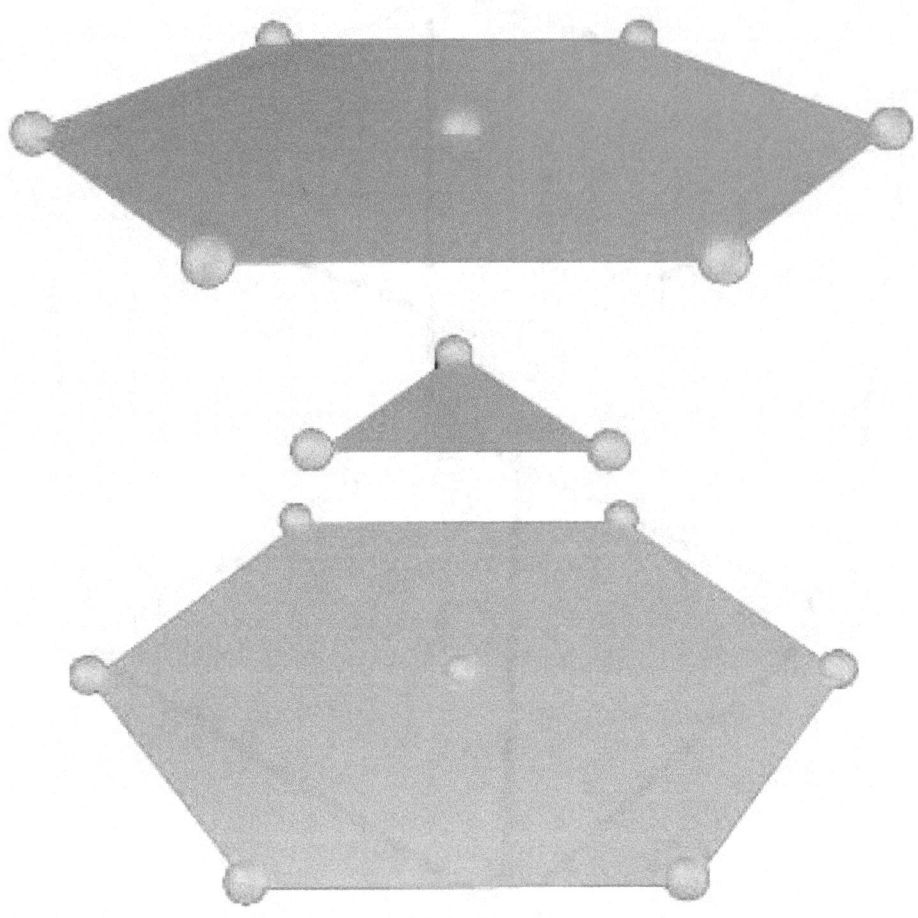

Exhibit 6-4 — Hexagonal Closed-Packed (HCP)

The **body-centered tetragonal (BCT)**, in Exhibit 6-5, is similar to the body centered cubic, but is elongated along one axis.

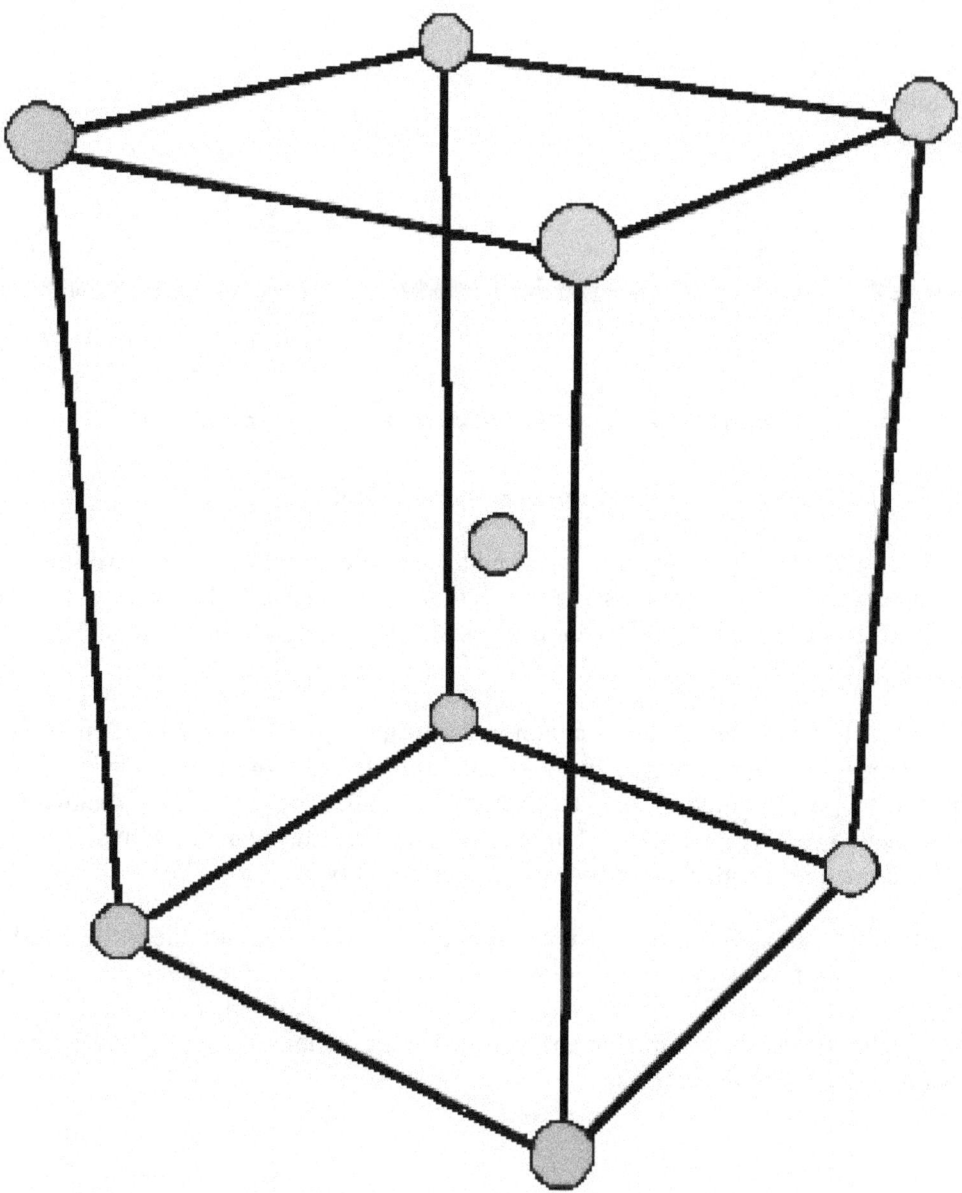

Exhibit 6-5 — Body-Centered Tetragonal (BCT)

6.1.2 — Crystal Imperfections

Real crystals are not perfect. Mechanical strength is profoundly affected by imperfections. If iron could be made free of crystalline imperfections, it could support stresses up to 1.8×10^6 psi in tension. The actual strength of iron is $\sim 10^4$ psi. Imperfections also have some advantages. For example, in the absence of defects, materials would not exhibit ductility.

Point defects are imperfections involving single atoms. There are three obvious possibilities for defects (depicted in Exhibit 6-6),

- vacancy
- substitutional
- interstitial.

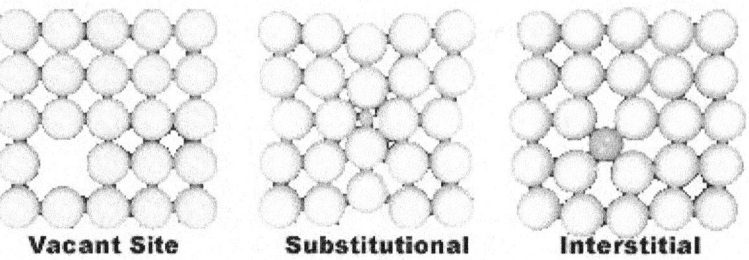

Vacant Site   **Substitutional**   **Interstitial**

Exhibit 6-6 — Point Defects

For a **vacancy defect**, an atom is simply missing from a lattice position. For **substitutional defects**, a different element's atom is substituted for an original element's atom. For **interstitial defects** an additional atom is in a position that is not a lattice position (between positions). This is an atom at a site that is not occupied in a perfect crystal.

Point defects in crystals occur during solidification in the formation of crystals, by thermal agitation, by cold working, and by exposure to high energy neutrons. **Line defects** are formed by a partial plane of atoms. At the edge of this partial plane of atoms there is a line defect which is called an **edge dislocation**. The atomic arrangement results in a compressive stress along the sides of the extra plane of atoms and tensile stress below the extra plane of atoms. In metals many dislocation lines are curved.

A dislocation is mobile and moves in a direction perpendicular to its length under an applied force. As this process occurs, blocks of metal **slip** past one another and the metal deforms plastically, as depicted in Exhibit 6-7. At elevated temperatures edge dislocations may also move vertically by a process known as **climb**. Such movement is controlled by the rate of movement of atoms and vacancies and can result in the slow deformation (creep) of a metal under very low stresses.

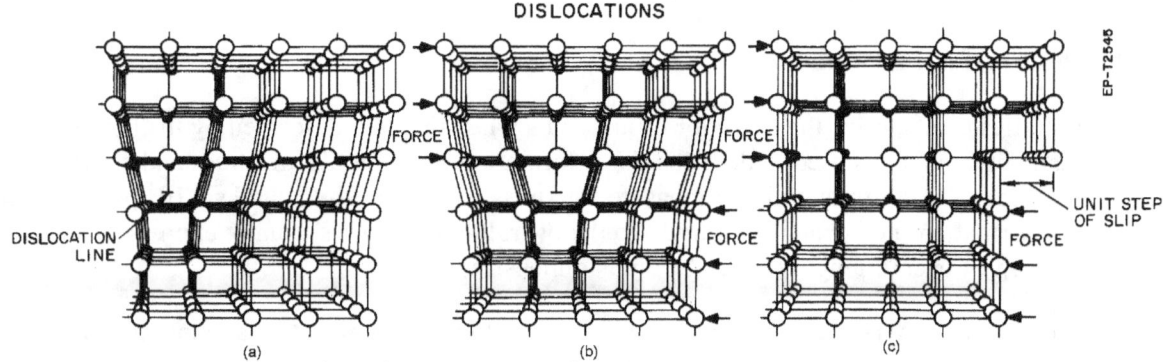

Exhibit 6-7 — Movement Of A Single Edge Dislocation Through A Crystal Resulting In Plastic Deformation

The dislocation is the most important of all the lattice defects. Billions of these imperfections arise during solidification of each cubic inch of metal, and they are intimately connected with nearly all mechanical deformation phenomena. Most methods of strengthening metals produce this effect by **pinning** edge dislocations and impeding slip.

6.1.3 — Grain Boundaries

When a molten metal solidifies, thousands of microscopic, randomly oriented crystals are formed. These crystals are seldom aligned relative to each other. Therefore, local regions of mismatch develop as adjacent crystals grow together. These regions of mismatch, where the crystalline structure is distorted, form the surfaces (**grain boundaries**) of separate crystals or grains.

The internal surfaces formed by grain boundaries strongly influence many material properties. Grain boundaries act as barriers to dislocation movement and therefore affect a metal's strength. A metal with fine grains has more resistance to deformation and thus is stronger than the same metal with large grains.

Grain boundaries are high energy disordered regions which, compared to regions within grains, contain more defects. Hence, polycrystalline metals are stronger than their single crystal counterparts. Impurity atoms, which tend to segregate at grain boundaries, can embrittle a metal and enhance its failure by intergranular corrosive attack. Diffusion is more rapid along grain boundaries especially for small interstitial atoms, such as hydrogen, which embrittles metals.

At elevated temperatures phase transformations nucleate at grain boundaries. In addition, the phenomena known as **grain growth** occurs because the higher energy grain boundaries move outward and grow at the expense of grains with lower energy grain boundaries. At high temperatures grain boundary segregation of impurities diminishes, reducing the tendency of a metal to embrittlement and corrosive attack at grain boundaries.

Section 6.2 — Mechanical Properties Of Metals

6.2.1 — Methods Of Measurement

Mechanical properties characterize the behavior of materials when they are deformed by the application of force. The fundamental variables used to describe the deformation of materials are stress and strain. **Stress** is the force per unit area. The common units of measure for stress are psi (lbf/in^2) and ksi (10^3 psi). **Strain** is the ratio of the deformation (i.e., change in length) to the overall length and is usually expressed in inches of deformation per inch of overall length (in/in). Strain is dimensionless.

Mechanical properties of materials are measured by a variety of mechanical tests that simulate conditions materials must withstand while in service. The most common of these tests are:

1. Tension Test – This test measures the tensile load-carrying ability and amount of deformation a material can withstand before rupture.
2. Hardness Test – This test measures the ability of a material to resist indentation or penetration.
3. Charpy V-Notch Test and Fracture Toughness Test – These tests measure the ability of a material to resist fracturing in the presence of a notch or crack.
4. Creep Test – This test measures the amount of continuous plastic deformation of a tensile specimen over a period of time.
5. Fatigue Test – This test measures the ability of a material to resist breaking when subjected to a fluctuating or alternating load.

6.2.2 — Tension Test

The tension test is the most widely used and one of the most informative of all mechanical tests. In this test, a circular or rectangular test specimen is subjected to a tensile load applied axially to the specimen's ends. In the usual tension test, a constant strain rate test (elongation), the testing machine pulls the ends of the specimen apart at a constant, slow rate while the elongation and resisting force of the specimen are continuously measured. Tested and untested specimens are shown in Exhibit 6-8. From these measurements, the testing machine plots the load versus the elongation. The resulting curve is readily interpreted in terms of stress and strain. The stress value used, termed the **engineering stress**, is calculated using the original cross-sectional area, even though the actual area continuously decreases throughout the test due to necking. The strain value used, termed the **engineering strain**, is defined as the increase in length divided by the original length.

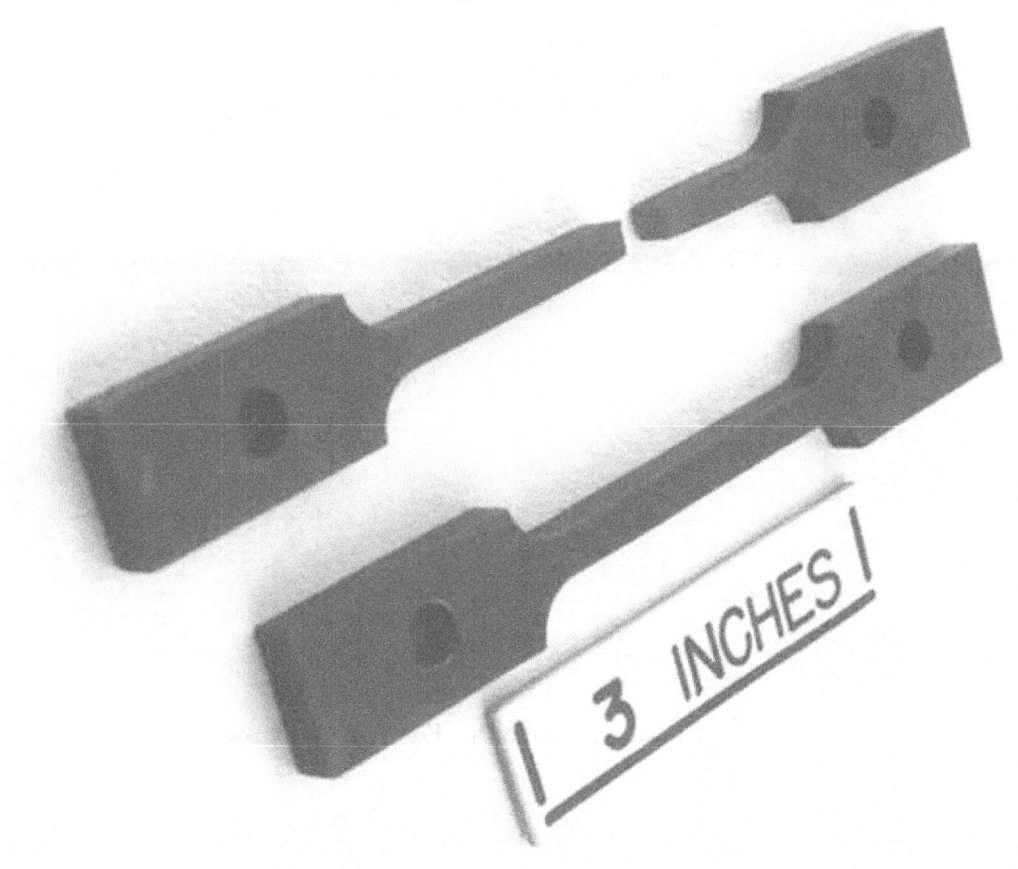

Exhibit 6-8 — Tension Test Specimens

The results of tension tests are **stress-strain curves**. The shape of a stress-strain curve depends on the composition of the material tested, the test temperatures, the strain rate, the heat treatments used to prepare the material, and the test specimen's prior history of plastic deformation, particularly cold working. Exhibit 6-9 illustrates a typical stress-strain curve for a low carbon, low strength structural steel. Because this material exhibits a large amount of straining before the specimen breaks (Point B), it is called a **ductile** material. Another typical stress-strain curve is illustrated in Exhibit 6-10. This material breaks (Point B) after a very small amount of strain and is considered a **brittle** material.

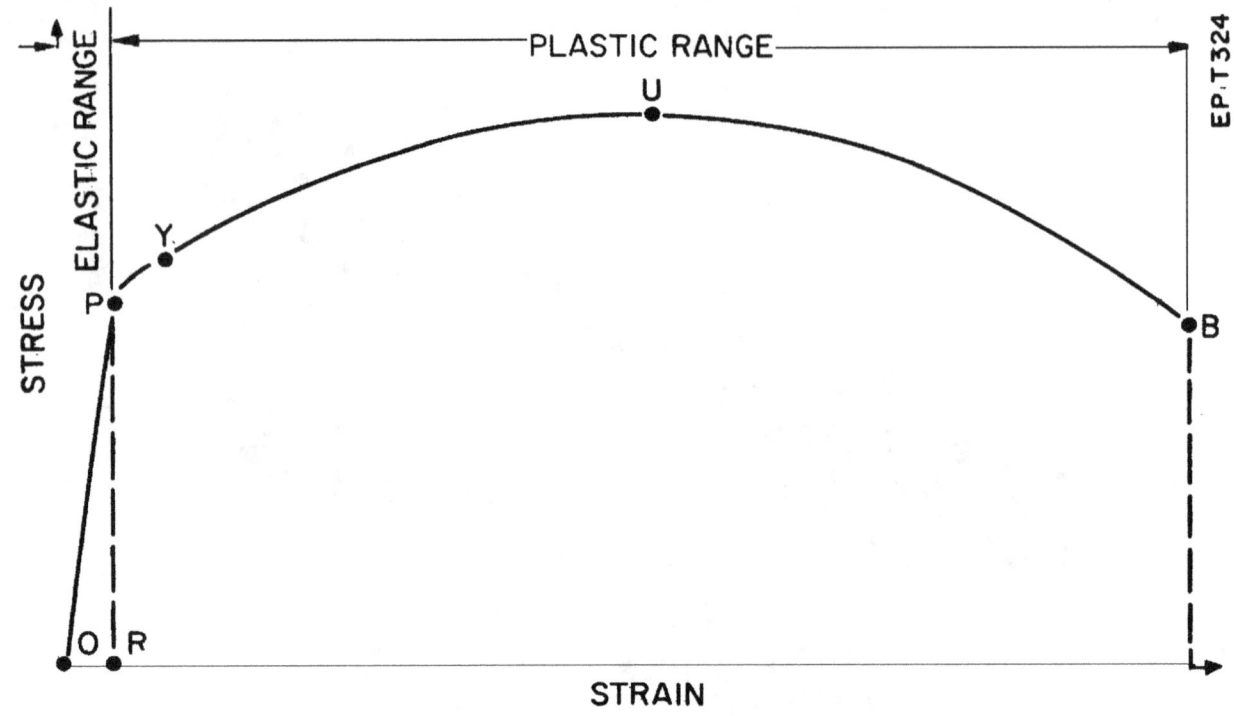

Exhibit 6-9 — Stress-Strain Curve For A Ductile Material

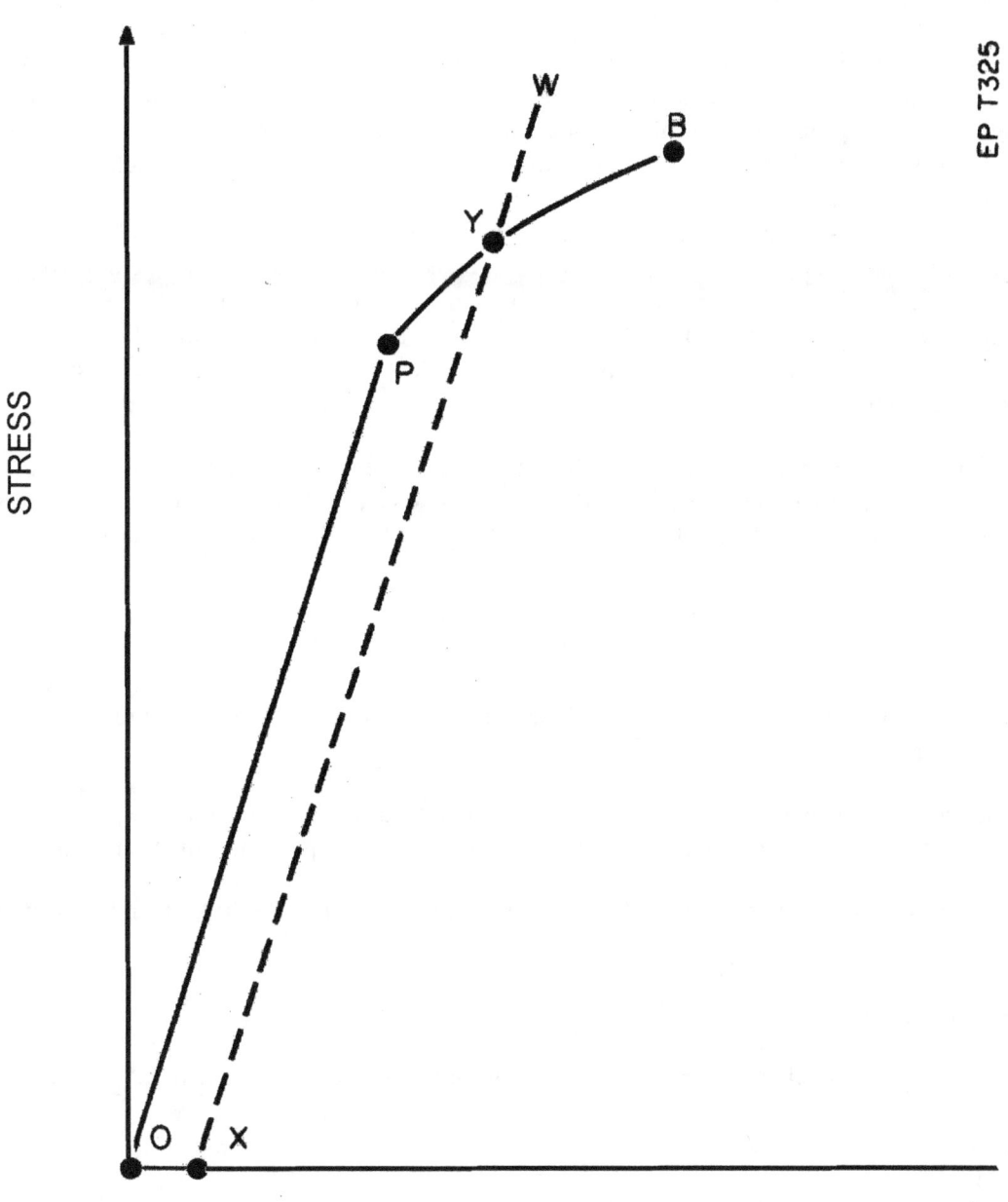

Exhibit 6-10 — Stress-Strain Curve For A Brittle Material

Characteristic points on a stress-strain curve are:

1.   <u>Proportional Limit (P)</u> – The highest tensile stress where stress is proportional to strain is the proportional limit.

2.   Elastic Limit – The highest tensile stress a specimen can sustain and still return to its original dimensions when the stress is removed. If a stress greater than the elastic limit is applied, permanent plastic deformation occurs within the material, and the specimen does not return to its original size when the stress is removed. Since the elastic limit has a value only slightly higher than the proportional limit for most metals, the plastic limit is approximately represented by Point P. Rubber, however, behaves nonlinearly and has an elastic limit much higher than the proportional limit.

3.   Yield Point (Y) – Point on stress-strain curve when plastic straining is easily observed. The usual convention is to define the yield point by the intersection of the deformation curve with a straight line parallel to the proportional/elastic portion and offset a 0.2% on the strain axis.

4.   Ultimate Tensile Strength (U) – The maximum stress developed in the material, based upon the original cross-sectional area of the specimen. In a brittle material, the ultimate tensile strength and breaking point will be the same.

The slope of the stress-strain curve up to the proportional limit is a measure of the material's stiffness in the elastic region. This slope is called the **Young's modulus of elasticity (E)**. For example, a tensile stress of 30,000 psi stretches steel about 0.001 inch for each inch of length. Thus, the modulus of elasticity for steel is

$$E = \frac{30,000 \text{ psi}}{0.001 \text{ in/in}} = 30 \times 10^6$$

Nearly all steels have approximately this modulus of elasticity at room temperature. The modulus of elasticity drops to 27×10^6 psi at 400°F.

Ductility is the mechanical property of a material that measures its ability to deform in the plastic range. The two commonly used measurements of ductility are percent elongation and percent reduction of area.

Metals are broadly classified as low strength (25-50 ksi), medium strength (51-100 ksi), and high strength (>100 ksi).

6.2.3 — Hardness

The simplest measure of the mechanical properties of a material is its hardness. In general, high hardness materials have
- more strength
- less ductility
- more susceptibility to brittle fracture.

Exhibit 6-11 illustrates the relationship between hardness (Brinell) and tensile strength for a typical low alloy steel.

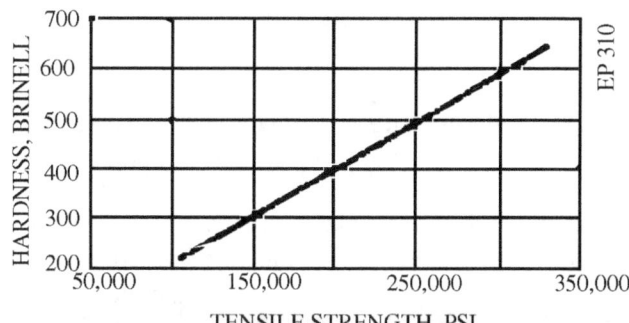

Exhibit 6-11 — Relationship Between Hardness And Tensile Strength For Typical Low-Alloy Steel

The hardness of a material is commonly measured by its ability to resist indentation or penetration. Common hardness tests are:

1. Brinell Test – A hydraulic press forces a small, very hard ball into the test material forming a dent. **The Brinell Hardness Number (BHN)** is the ratio of the applied load to the area of the surface of indentation. The indenter is relatively large (10 mm diameter) compared to the material homogeneities so that the local hardness variations are averaged out. BHNs range from approximately 50 for soft metals to 700 for very hard metals.

2. Rockwell Test – Uses a smaller indenter than the Brinell Test. The Rockwell testing machine applies a load to the indenter and then releases the load, giving a dial reading which is inversely related to the permanent depth of the indentation. This test has a variety of loads and indenters for different ranges of hardness and materials. Results for different loads or indenters are denoted by letters (e.g. Rockwell A, R_A).

3. Knoop Microhardness Test – A pyramid-shaped diamond is pressed into the test material using a small fixed load for a specific time. The hardness is determined by measuring the size of the tiny indentation with a microscope. This test can measure the hardness of very small particles in materials or thin sheets.

6.2.4 — Toughness

Toughness characterizes a material's ability to absorb energy without breaking. It is the opposite of **brittleness**. The manner in which absorbed energy is dissipated is different for breaks in tough and brittle materials. A tough material generally has higher ductility due to dislocation motion along slip planes and the ability of dislocations to overcome barriers by changing their direction of motion along secondary slip planes at an angle to the original slip plane. Dislocation motion requires energy and thus dissipates absorbed energy over a large volume. In a brittle material, a dislocation can move only a very short distance along a primary slip plane before it is stopped; it cannot take off in another direction along a secondary slip plane. As a result, absorbed energy is concentrated at the source and may be sufficient to break interatomic bonds. This will cause the material to **fracture**.

When a metal of high toughness breaks, the broken pieces show considerable deformation. Their irregular and fibrous surfaces result from dislocations slipping in many intersecting planes before a ductile rupture occurs. In contrast, a brittle fracture exhibits little or no plastic straining around the break because dislocation motion along intersecting planes cannot occur in brittle material. The smooth appearance of the broken surfaces indicates an abrupt transgranular cleavage.

One measure of toughness is the area under the stress-strain curve. This area is the work per unit volume done on the material by the tension test machine. It is the product of strain and stress.

Real structures are not smooth test specimens, but have geometric irregularities or flaws that concentrate stresses and may grow to critical size under load. Most materials fracture below their yield strength when loaded in the presence of a flaw or discontinuity. The resistance of a material to fracture under these conditions is **notch toughness**. Methods that test for notch toughness are:

1. Charpy V-Notch Test - Employs a square bar (55 mm × 10 mm × 10 mm) with a 2 mm deep notch on one side. The test apparatus (see Exhibit 6-13) has a swinging pendulum that is released from a specified height, striking the specimen at the bottom of the pendulum swing. The test specimen is fractured, absorbing energy (slowing down) the pendulum such that it does not swing up as high as it would on a free swing (i.e., no specimen in holder). A calibrated scale on the apparatus converts the height of the pendulum swing to the energy absorbed by the breaking specimen in ft-lbf. Low alloy steels exhibit a **ductile-to-brittle transition** with decreasing temperature (see Exhibit 6-14). It should be noted that tests are usually performed in triplicate because of scatter in the test results. There are several criterion for determining the transition temperature between ductile and brittle cleavage.

Exhibit 6-12 — Charpy Specimens

2. Fracture Toughness Test - A specimen used for this type of test is illustrated in Exhibit 6-15. Specimens range in size from 0.2 inch up to several inches thick, with other dimensions scaled up proportionally. A fatigue crack is grown in each specimen prior to testing by applying a low cyclic load. This simulates the sharp edged cracks encountered in service better than machined

slots or notches. The specimen is loaded, as indicated in Exhibit 6-15, using a tension testing machine as shown in Exhibit 6-16. The fracture toughness (K_{IC}) is determined from the load required to break the specimen and the dimensions of the specimen. Breaking takes place along the plane of the crack. Exhibit 6-17 shows both tested and untested fracture toughness specimens.

3.   Drop Weight Test - The drop-weight test is basically a "go, no-go" test which determines the temperature below which a material exhibits essentially no plastic deformation in the presence of a sharp notch and impact loading conditions. This temperature is called the **nil ductility transition temperature (NDTT)**. The specimen is a flat plate with a bead of brittle weld-deposited material at the center. A notch is cut partially through the weld bead as a crack starter. The welded side of the specimen is placed face down on the test fixture, as shown in Exhibit 6-18, and the center of the specimen is struck with a falling weight. The stop in the center of the fixture limits the amount of deflection when the drop-weight strikes the specimen. Under impact, the slight flexure of the specimen opens a crack at the root of the weld notch. The test is repeated at 10°F intervals. If the temperature is low enough, the crack propagates into the specimen material causing it to fracture with virtually no plastic deformation. The NDTT is the maximum temperature at which specimen breakage occurs. The test measures a material's resistance to propagation of a crack leading to brittle fracture.

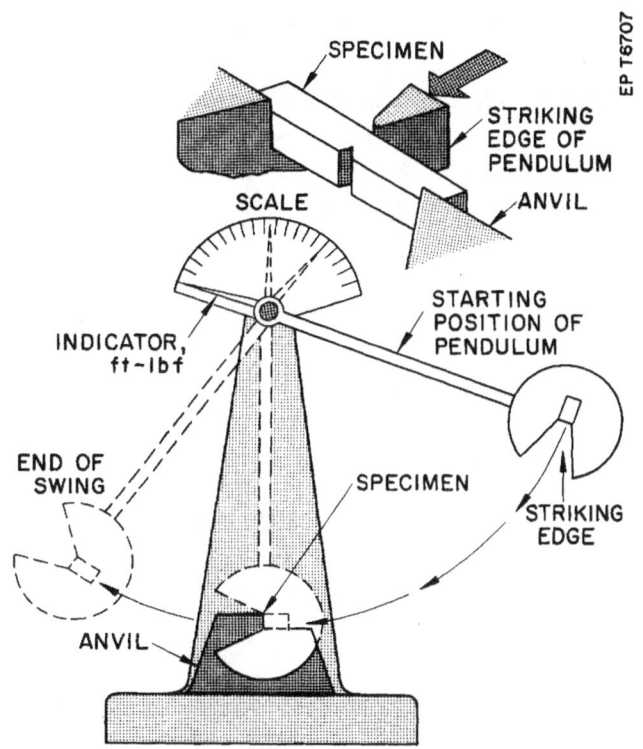

Exhibit 6-13 — Charpy Testing Apparatus (Schematic)

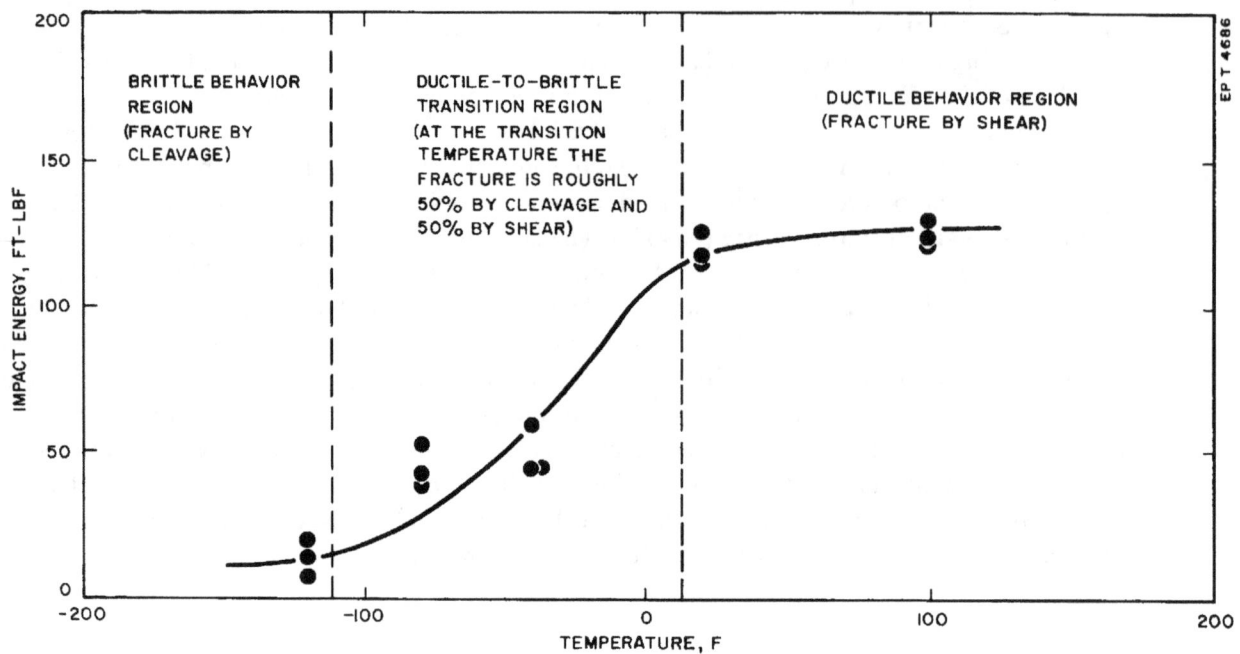

Exhibit 6-14 — Charpy V-Notch Transition Temperature Curve For A Low Alloy Pressure Vessel Steel

Exhibit 6-15 — Test Specimen Used To Measure Fracture Toughness

Exhibit 6-16 — Fracture Toughness Test Stand

Exhibit 6-17 — Fracture Toughness Specimens

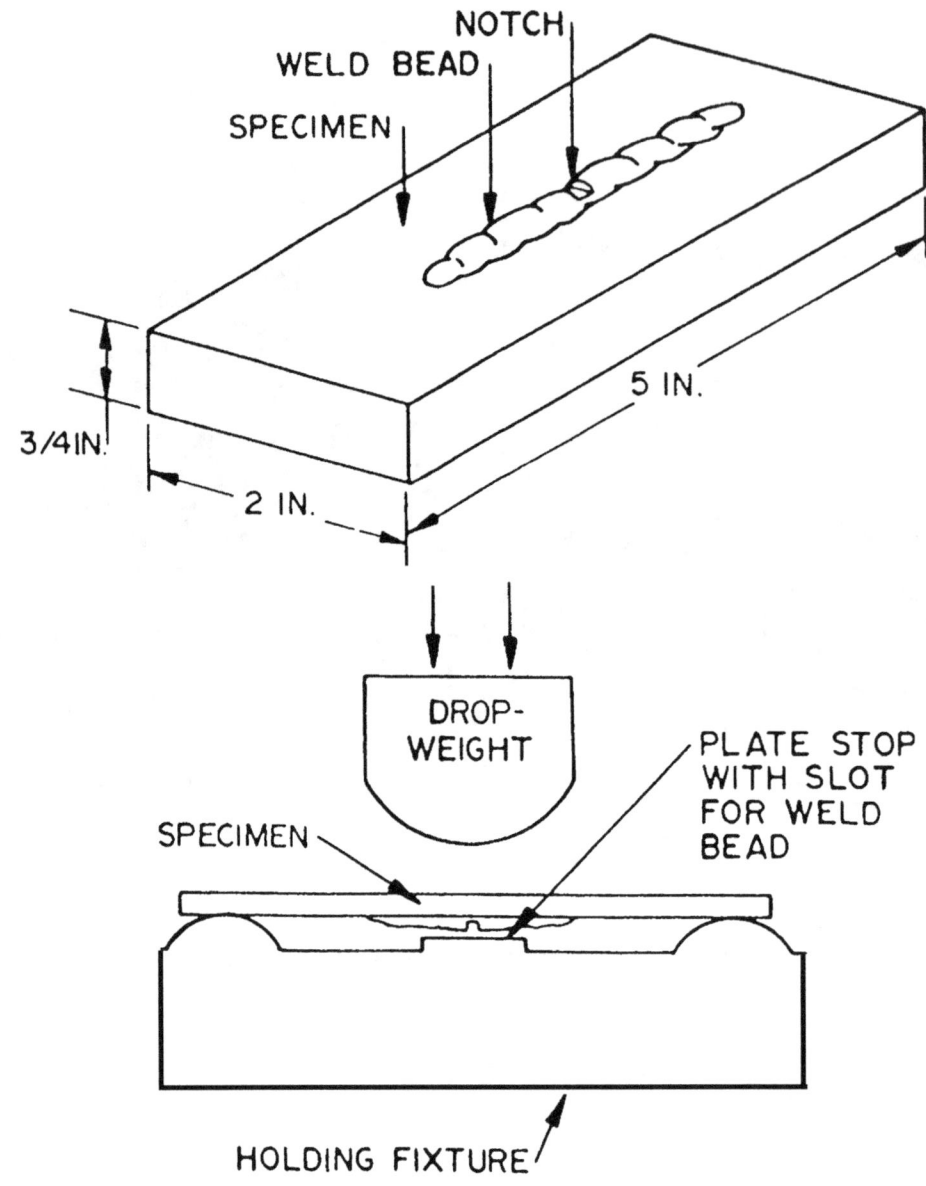

Exhibit 6-18 — Drop-Weight Test

Section 6.3 — Material Failure

6.3.1 — General

Failure is any event which satisfies one of the following conditions: the part becomes completely inoperable; the part remains operable, but it is no longer able to perform its intended function satisfactorily; or the part has suffered serious deterioration that has made it unreliable or unsafe for continued use, thus necessitating its immediate removal from service or replacement. It is good engineering practice to provide a margin between

normal operating loads and those loads expected to cause failure. This type of margin is a **design factor** (sometimes called design margin, safety margin, or factor of safety). The size of the design factor depends on:
- how well actual equipment loadings are known
- what uncertainty exists in our knowledge of material properties
- how precise is our stress analysis
- what manufacturing deviations can exist in the actual component
- what material deterioration can be expected in service
- what experience exists with similar equipment
- how important are cost and weight.

6.3.2 — Ductile Failure

A ductile material will sustain considerable plastic deformation prior to cracking or rupturing. The general characteristics of a ductile rupture are:
- there is extensive distortion which increases progressively as the overload condition is approached, and
- no fragments are formed.

In many structures involving bolted closures, extensive leakage would occur prior to rupture, giving advance warning of ultimate failure. Ductile failure is prevented by assuring that the maximum expected stress is some fraction of the minimum specified yield strength or ultimate strength.

6.3.3 — Brittle Fracture

Pure **brittle fracture** is the type of breaking associated with ceramics or glass. There is essentially no plastic deformation, failure is generally quite sudden, and frequently fragments are formed. Examinations of brittle fractures which have occurred in metal structures have revealed three items that they have in common:
- pre-existing cracks/flaws
- tensile stress
- low fracture toughness.

Brittle fracture of steel occurs at low temperatures. As temperatures are varied it is noticed that there is an abrupt transition between ductile and brittle failure. The temperature of this transition will vary between various alloys and heat treatments.

The most accurate method of determining this transition temperature is with the drop weight test (Paragraph 6.2.4, Item 3). Because of the weld bead preparation and the specimen's larger size, the drop weight test is more cumbersome than the Charpy V-notch test. The Charpy V-notch test is relatively easy to perform, the specimen is conveniently small, and there exists a great amount of data for a wide range of materials. The small size facilitates placing specimens in test reactors to allow estimation of radiation effects.

The correlation between the drop-weight test and the Charpy V-notch test is the **Charpy fix energy (CFE)**. The NDTT is determined by the drop-weight test. Referring to the curve in Exhibit 6-14, an energy level can be determined for a specific temperature. That energy for the NDTT is the CFE. For future correlations, the temperature that matches the CFE is the **nil ductility temperature - Charpy (NDT$_{Charpy}$)**.

A series of specimens of the same alloy is now irradiated (i.e., exposed to neutron flux) to simulate maximum expected service conditions. The Charpy V-notch test results for the irradiated samples provide a curve that is shifted to the right of the unirradiated sample. By reading the CFE on the irradiated curve, we find a new NDT_{Charpy} for the irradiated material. To account for variations in the data collected, experimental error, and conservatism, 60°F is added to the NDT_{Charpy} to give what is defined as the **reference transition temperature (RTT)**. This is illustrated in Exhibit 6-19. Below RTT the stress on the material must be strictly controlled to prevent brittle fracture.

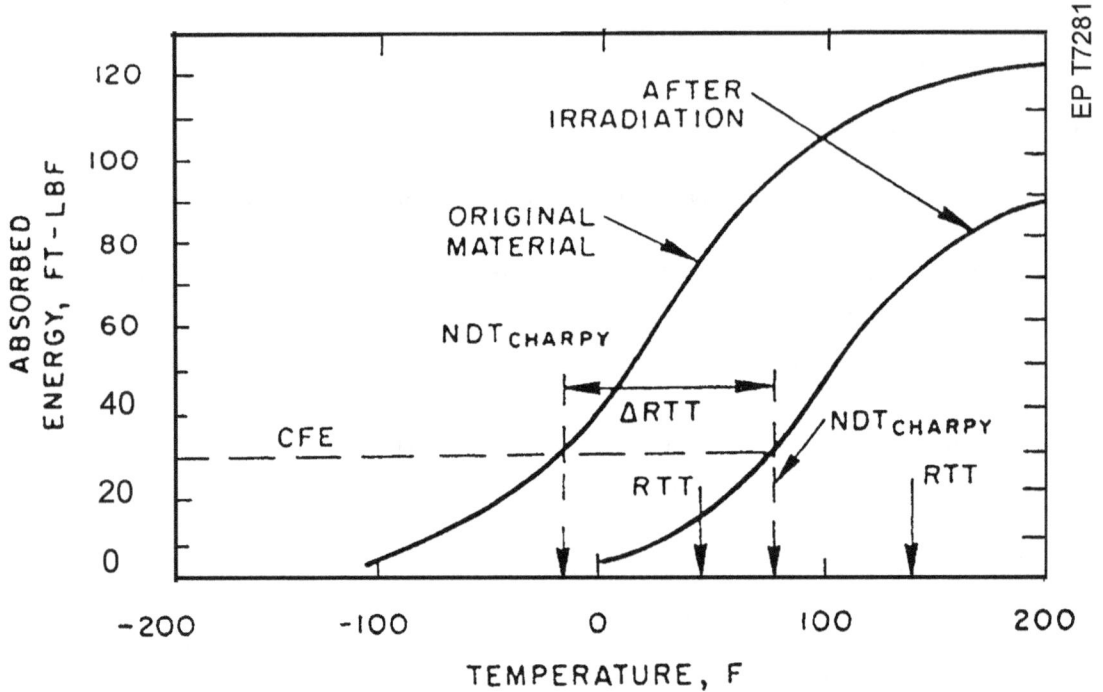

Exhibit 6-19 — Charpy Transition Curves Showing Typical Shift In RTT Due To Radiation Exposure

A fracture mechanics approach is necessary to determine allowable stresses below RTT. **Fracture mechanics** is the study of the behavior of materials containing cracks. At temperatures below RTT, the reduction in load carrying capacity of a structure is closely related to the size of the crack in the structure.

Application of fracture mechanics theory involves establishing design values of the fracture toughness of the material involved, performing detailed stress analyses of the structure, and determining conservative upper bounds for a potential crack's size, shape, and orientation. These values are then used to calculate the maximum plant pressure and heat-up and cooldown rates which can be allowed while preserving an adequate brittle fracture design factor. Finally, allowable pressures and corresponding metal temperatures are related to primary coolant pressures and temperatures which are available to operators through instrumentation.

The **stress intensity factor (K_I)** is a calculated load parameter proportional to the stress magnitude near the tip of a sharp crack in a structural member under tensile load. Mathematically, K_I is defined as

Equation 6-1

$$K_i = C \bullet \sigma_G \bullet \sqrt{a}$$

where:

C		= Dimensionless coefficient dependent upon the component and crack geometry and method of loading.
σ_G		= Gross Stress (ksi)
a		= Crack size or depth (inches).

When the **gross stress (σ_G)** is equal to the **fracture stress (σ_F)**, i.e., the stress at which brittle fracture occurs, then the stress intensity factor (K_I) equals the **critical stress intensity (K_{IC})**. K_{IC} is also known as the fracture toughness. **Fracture toughness** is a material property which is a measure of a material's ability to withstand tensile stress in the presence of a flaw without crack growth by brittle fracture. Mathematically, fracture toughness is defined to be:

Equation 6-2

$$K_{IC} = C \bullet \sigma_F \bullet \sqrt{a}$$

K_{IC} is a measurable value. It is determined using the fracture toughness test (Paragraph 6.2.4). Test results can determine limiting values of K_{IC} over a range of temperatures.

Allowable stress (σ_a) is the maximum value of gross stress (σ_G) which may be permitted without exceeding stress limits (including design factor) imposed by brittle fracture theory. It is defined as

Equation 6-3

$$\sigma_a = \frac{K_{IC}}{DF \bullet C \bullet \sqrt{a}}$$

where:

DF		= Minimum Specified Design Factor.

The allowable stress in a design is the sum of **residual**, **thermal**, **reaction** and **pressure** stresses, as shown in Equation 6-4.

Equation 6-4

$$\sigma_{total} = \sigma_P + \sigma_R + \sigma_T + \sigma_{reaction}$$

Residual and reaction stresses are unaffected by operator action. Thermal and pressure stresses can be controlled by the operator. Thermal stresses are limited by fatigue limits for heat-up and cooldown rates.

Pressure stresses are controlled by controlling system pressure. If thermal stresses are controlled within specific limits for heat-up and cooldown rates, then **allowable pressure stress (σ_P)** is the only parameter to be controlled by the operator. By methods beyond the scope of this manual, a **pressure-stress conversion factor (PSCF)** can be determined such that

Equation 6-5

$$\sigma_P = \sigma_{total} - \sigma_R - \sigma_T - \sigma_{reaction}$$

Equation 6-6

$$P_a = \frac{\sigma_P}{PSCF}$$

where:

$\qquad$ **P_a** $\qquad$ = allowable system pressure (psi).

With corrections for instrument errors, limits on pressure for specific plant temperatures can now be determined for all plant components susceptible to brittle fracture. The most limiting component limits, combined with additional design factors, are combined to form a series of pressure-temperature limits known as **brittle fracture protection limits (BFPL)**. Refer to Reference 2 for a detailed discussion.

6.3.4 — Creep

Creep, the slow, continuous, permanent (plastic) deformation of a metal under a steady load can occur at stress levels less than the material's yield stress. Creep rate is proportional to load (i.e., creep rate decreases as the load decreases). Creep can be accelerated by fast neutron irradiation. Creep behavior is important to core design, as discussed in Reference 1 .

In a standard creep test, a specimen similar to a tension load specimen is subjected to a constant axial load in an oven that maintains a uniform test temperature. A typical test setup is shown in Exhibit 6-20. The results consist of measurements of specimen elongation with time. From this elongation data, the known tensile load, and specimen dimensions, stress and strain are determined. Exhibit 6-21 shows both tested and untested creep test specimens.

Exhibit 6-20 — Creep Test Stand

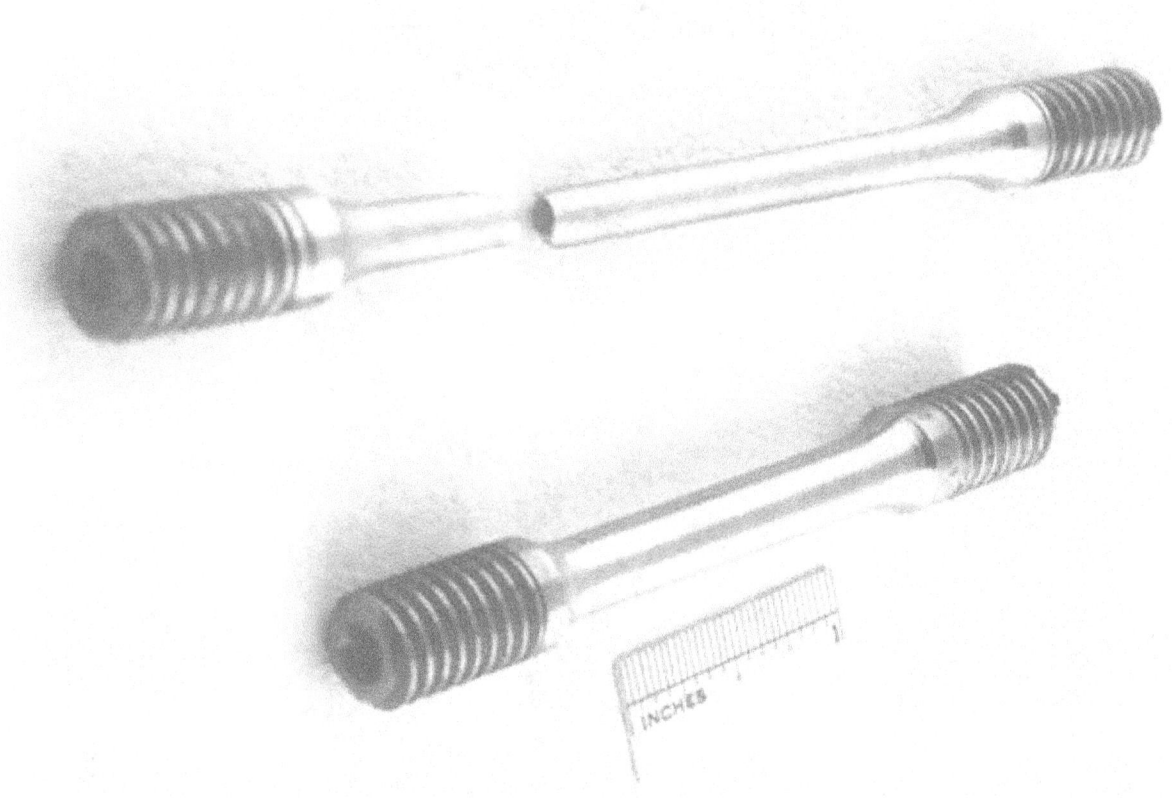

Exhibit 6-21 — Creep Test Specimens

Exhibit 6-22 is an example of a typical creep curve of total strain versus time. The initial strain (ϵ_0) occurs immediately when the load is applied. The curve generally exhibits three characteristic regions:

1. <u>Primary Creep</u> - In primary creep, the creep rate decreases due to initial strain hardening
2. <u>Secondary Creep</u> - Secondary creep is characterized by a relatively constant creep rate. It represents a kind of equilibrium situation in which the tendency to further strain harden is offset by annealing effects.
3. <u>Tertiary Creep</u> - Creep is accelerated due to necking and/or grain structure changes within the material (accumulation of voids at grain boundaries and grain boundary sliding). In a constant stress test, i.e., load reduced to maintain a constant stress as the specimen necks, tertiary creep may not occur.

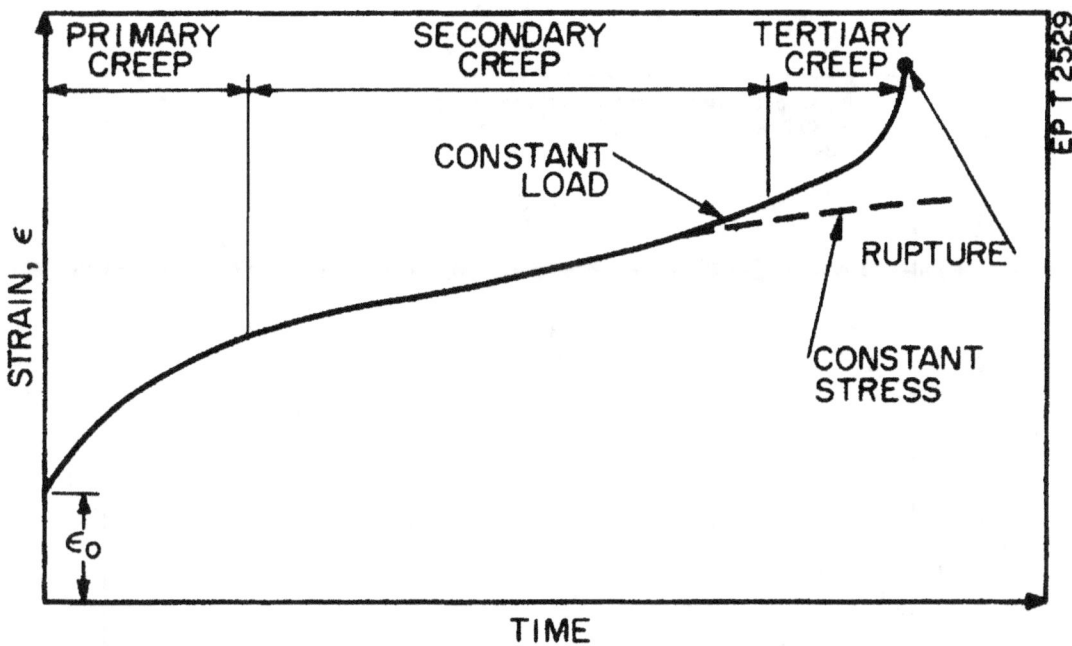

Exhibit 6-22 — Curve Of Three Stages Of Creep

6.3.5 — Fatigue

Fatigue of metals is the behavior of metals under the action of cyclic stress and strain, as opposed to steady stress and strain. Fatigue can be dangerous because a structure can be so weakened by a fatigue crack that complete fracture through the remaining cross-section occurs suddenly. Three basic factors related to stress are necessary to cause a fatigue failure:
* a tensile stress of sufficient magnitude
* a sufficient variation (range) of stress
* a sufficient number of cycles (fluctuations).

Fatigue failure is the failure mode of unflawed materials.

The fatigue life capability of a material can be determined by a series of fatigue tests run at different strain ranges. The resulting data is summarized by a fatigue curve, a plot of **stress amplitude (S)** versus the **cycles to failure (N)**, sometimes called an **S-N curve**. Some materials, including low alloy and plain carbon steels, have fatigue curves that flatten out at very high cycles ($>10^6$). The stress value at which this flattening out occurs is called the **endurance limit**.

Another type of fatigue test is the fatigue-crack-growth test. This test is conducted by subjecting a fatigue-cracked specimen (see Exhibit 6-15) to constant-amplitude cyclic-load fluctuations. The fatigue-crack length (a) is then measured and plotted against the number of load cycles (N) for various stress intensities (σ). The resulting plot is shown in Exhibit 6-23. A more useful presentation of fatigue-crack-growth data is to plot the rate of fatigue-crack growth per cycle of load versus the stress intensity factor (ΔK_I) on a log-log plot as shown in Exhibit 6-24.

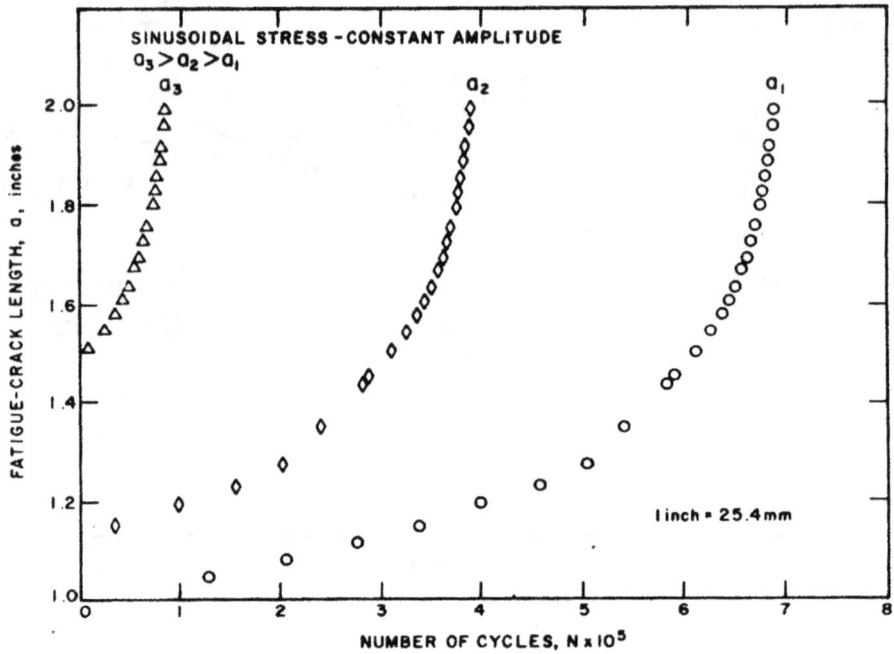

Exhibit 6-23 — a-N Curve

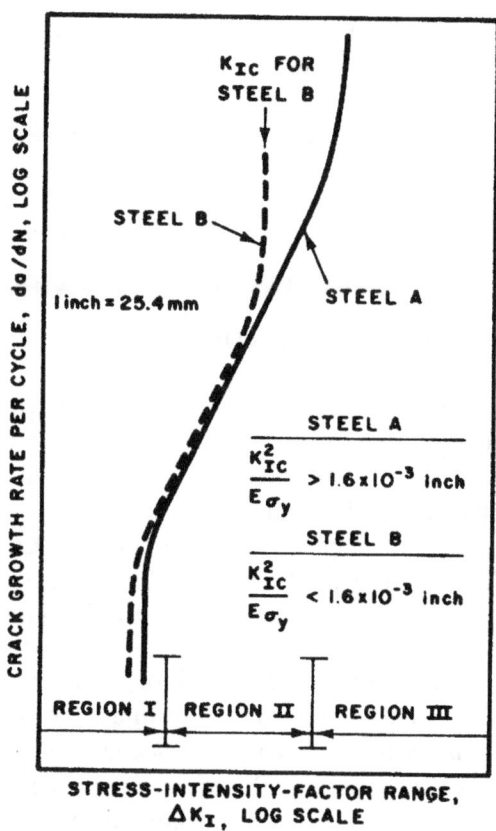

Exhibit 6-24 — Rate Of Crack-Growth Versus Stress Intensity Factor

The ability of a metal component to withstand many stress cycles depends on many factors in addition to the inherent strength of the material. Some factors, such as corrosive environments and temperature, are related to service conditions. Others, such as stress concentrations (e.g., notches, flaws) and metallurgical structural flaws can result from poor design, metallurgical, fabrication or handling practices. They must all be considered in reactor plant design. In addition, designers use a conservative design fatigue curve which is lower than the best fit of the experimental data. This conservative approach accounts not only for scatter in the test data but also for fabrication uncertainties.

Section 6.4 — Effects Of Irradiation On Metals

6.4.1 — Neutron Irradiation

Radiation exposure may alter the properties of metals used in reactor plant components by changing the arrangement of atoms in the metal's crystalline lattice. All radiation-induced changes in material properties are referred to as radiation damage, although not all of the resulting property changes are harmful. In fact, radiation can sometimes result in increased strength and fatigue life.

Of all the radiation emitted from a nuclear power plant, neutron irradiation is of primary concern. Gamma radiation will only generate heat in metals. Alpha, beta and fission product particles have too small of a range to be of a concern outside the core region (refer to Reference 1 for more details).

The initial kinetic energies of neutrons produced by fission have an average value of 2 MeV, but can exceed 10 MeV. Since neutrons are electrically neutral, they interact to a negligible extent with atomic orbital electrons and can travel well beyond the reactor core. A neutron's kinetic energy is dissipated by nuclear reactions (scattering or absorption) with atoms along its path, displacing atoms from their normal lattice sites and causing radiation damage in metal.

6.4.2 — Effects Due To Neutron Scattering Reactions

Most of the radiation damage caused by neutrons results from the scattering of fast neutrons. In scattering collisions with metallic atoms, the neutron retains most of its energy and moves on, having been deflected from its initial direction of travel. If the neutron's kinetic energy is less than some critical value the collision simply increases the struck atom's vibrational energy. If the neutron's kinetic energy exceeds this critical value the atom is ejected from its lattice position. This displaced atom, a **primary knock-on** atom, may strike other atoms with sufficient energy to dislodge them from their normal lattice sites, thus producing secondary, and higher order knock-on atoms. The critical energy to displace an atom is generally considered to be about 25 eV.

When an atom is knocked out of its normal lattice site, a void, or vacancy is created in the lattice structure. The moving atom then comes to rest some distance from its origin, either wedged between other atoms in normal lattice positions as an interstitial or in another lattice vacancy. The mobility of interstitial atoms at normal reactor operating temperatures is appreciable; therefore, many vacancies and interstitials recombine by diffusion or thermal activation. This process, called **microannealing**, reduces the damage to the lattice structure.

6.4.3 — The Effect Of Neutron Irradiation On The Mechanical Properties Of Metals

The effects of neutron irradiation on metals are similar to those of cold working. Neutron induced lattice defects impede slip, thereby increasing strength and decreasing ductility. Neutron irradiation also affects the brittle-to-ductile transition exhibited by body-centered cubic metals, such as low alloy ferritic steels.

In general, neutron exposure rate affects creep in metals in somewhat the same manner as elevated temperature. The concentration of vacancies during neutron radiation is greater than in the absence of a neutron flux by an amount established by the equilibrium between vacancy production rate and vacancy-interstitial recombination. Since an increased vacancy concentration promotes diffusion dependent phenomena, property changes associated with the increasing vacancy concentration produced by increasing temperature are also produced by exposure to high neutron fluxes. At elevated temperatures, the thermally produced vacancy concentration is so large that normally encountered levels of neutron radiation have a negligible effect. Thus, the greater the rate of neutron exposure, the greater the rate of creep deformation of a loaded component.

Refer to Reference 2 for additional effects and more information specific to reactor plant materials.

Section 6.5 — Iron And Steel Alloys

6.5.1 — Allotropes Of Iron

Iron changes crystal structure (allotropic changes) from one type of unit cell to another at fixed transition temperatures. Below 1670°F (910°C) iron is body-centered cubic (α iron), between 1670°F (910°C) and 2552°F (1394°C) it has a face-centered cubic (γ iron) structure. Above 2552°F (1394°C) the structure returns to body-centered cubic (δ iron). It remains a body-centered cubic (δ iron) up to 2800°F (1538°C), where it will melt. See Exhibit 6-25.

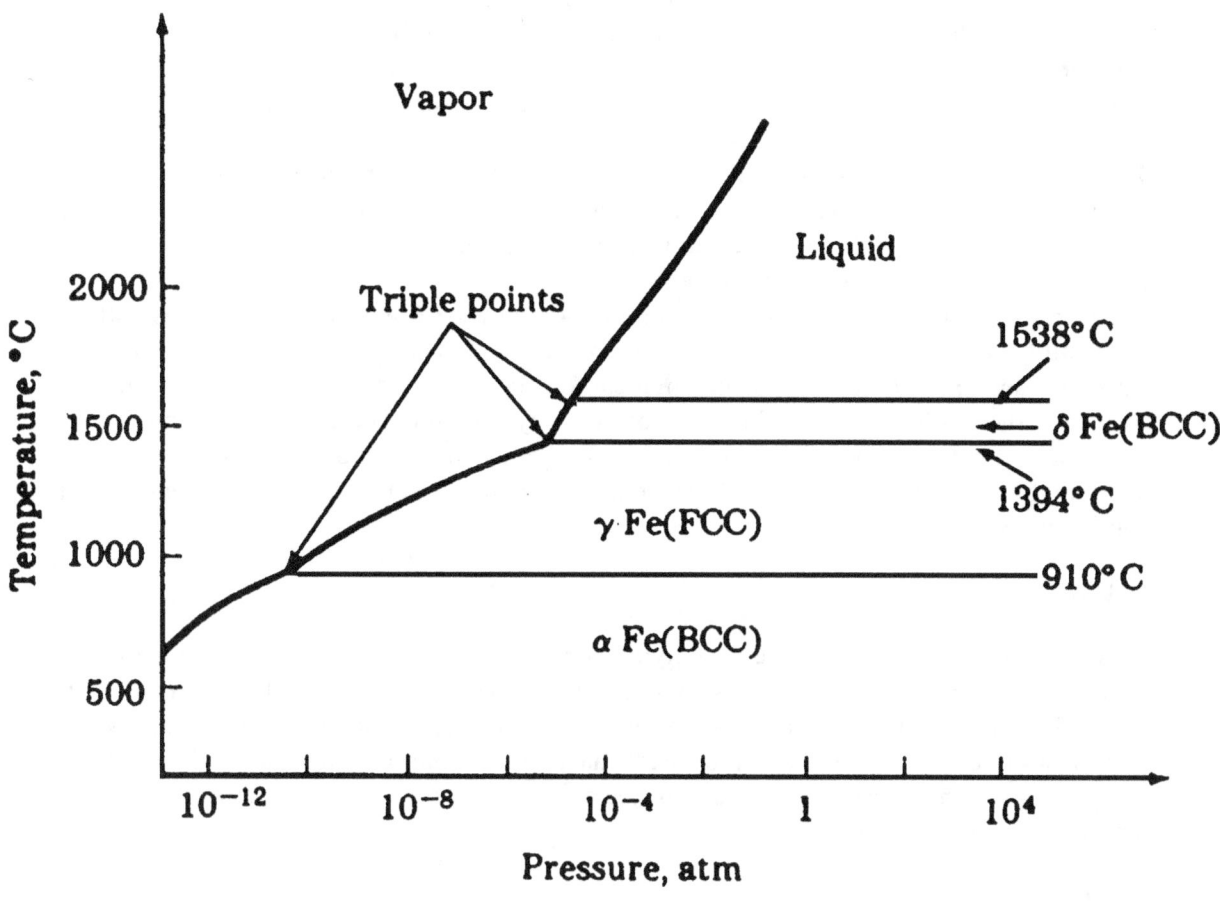

Exhibit 6-25 — Allotropic Changes Of Iron

6.5.2 — Classification Of Steels

Steels constitute a large percentage of structural materials. These iron based alloys can be divided into three broad categories:
*   carbon steels
*   low-alloy steels
*   high alloy steels, such as stainless steels.

6.5.2.1 — Carbon Steels

Carbon steels contain small amounts of manganese (up to 0.9 w/o) and silicon (up to 0.3 w/o) whose principle function is to combine with oxygen and sulfur (commonly occurring impurities in steel) to reduce their harmful effects on the fracture toughness and ductile-to-brittle transition temperature. Carbon steels contain varying amounts of carbon, depending on their intended use. The final carbon content is the major determinant of the strength of these steels and their response to heat treatment. These alloys together with the unalloyed cast irons are the lowest in cost.

1. <u>Low Carbon Steels</u> - These steels contain up to approximately 0.2 w/o carbon. Because of the low carbon content it is not possible to increase hardness or strength by heat treatment. The strength of these steels is slightly greater than that of iron, but because of their strength properties, they are used for low strength boilerplates, structural supports, deck plates, and a number of general purpose applications. These steels are generally tough and ductile, and are easily formed, machined, and welded.

2. <u>Medium Carbon Steels</u> - These steels are stronger than the low carbon steels and can be strengthened further by heat treatment. They range in carbon content form approximately 0.2 w/o up to 0.45 w/o and are of higher strength than the low carbon steels. The medium carbon steels are used for piping systems in the secondary plant.

3. <u>High Carbon Steels</u> - These steels contain from 0.45 to over 1.0 w/o carbon. Hardness, strength, and ductility vary greatly over this range of carbon content. These steels find very limited use in power plant applications because the high carbon content reduces the ductility and makes fabrication such as forming, welding, and forging very difficult. These steels respond well to heat treatment but are susceptible to cracking during quenching and do not harden throughout except in thin sections. They are used for cutting tools, drills, and similar applications requiring high hardness.

6.5.2.2 — Low-Alloy Steels

Alloying elements such as chromium, manganese, molybdenum, and nickel improve strength, fracture toughness, corrosion, wear resistance, physical, and magnetic properties. Low-alloy steels have a carbon range of approximately 0.15 - 0.40 w/o. They contain up to a total of 5 w/o alloying elements which improve the hardening characteristics of carbon steels. Low-alloy steels have high strength, increased toughness, and greater resistance to abrasion and corrosion than carbon steels. Low-alloy steels, which are readily welded, are used to construct most of pressure vessels.

6.5.2.3 — High-Alloy Steels

These steels contain over 5 w/o alloying elements. One group of high-alloy steels (stainless steels) contain a large percentage of alloying elements to increase resistance to corrosion and improve high temperature strength. The two most important alloying elements in these steels are chromium and nickel. **Stainless steels** are used in many applications because of their excellent corrosion and oxidation resistance at elevated temperatures. Their corrosion resistance is due to a thin, tightly adherent chromium oxide film that protects the steel from most corroding media. This property is not evident in chromium steels until the chromium content is 12 w/o or more. For this reason, stainless steels contain at least 12 w/o chromium in addition to other alloying elements.

Some stainless steels can be heat treated to high strength while others cannot. The most common stainless steels are the Fe-18Cr-8Ni-0.08C (18-8) and the Fe-12Cr-0.10C alloys. The 18-8 stainless steel known as type

304 cannot be strengthened by heat treatment. However, on rapid cooling, Fe-12Cr-0.10C alloys (type 403 and type 410 stainless steels) form martensite which can be tempered to high strength.

(Intentionally Blank)

Chapter 7 — CALCULATIONS AND THUMBRULES

Section 7.1 — Conversion Factors

7.1.1 — Length

1 foot	= 0.3048 meters
1 inch	= 2.54 centimeters
1 meter	= 100 centimeters
	= 39.37 inches
	= 3.28084 feet
1 micron	= 1.0×10^{-6} meter
1 mile	= 5280 feet
	= 1.609 kilometers
1 nautical mile	= 1852 meters
	= 6076 feet
	= 1.151 miles

7.1.2 — Area

1 square foot	= 144 square inches
	= 0.0929 square meters
1 square inch	= 6.4516 square centimeters

7.1.3 — Volume

1 cubic foot	= 1728 cubic inches
	= 7.4805 gallons
	= 0.028317 cubic meters
	= 28,317 cubic centimeters, cc
1 gallon	= 128 fluid ounces
	= 231 cubic inches

	= 3785 cubic centimeters, cc
	= 3.7854 liters
1 cubic inch	= 16.387 cubic centimeters, cc
1 cubic cent.	= 1 milliliter

7.1.4 — Time

1 hour	= 3600 seconds
1 year	= 365.2422 days
	= 52.117 weeks

7.1.5 — Volume Flow Rate

1 cubic foot/second	= 448.83 gallons/minute, gpm
	= 1699.3 liters/minute
1 gallon/minute	= 3785 liters/minute
	= 0.13368 cubic foot/minute

7.1.6 — Mass

1 kilogram	= 2.2046 pound-mass, lbm
1 ounce (mass)	= 28.350 grams
1 pound-mass	= 16 ounce (mass)
	= 0.4536 kilograms
1 ton (short)	= 2000 pound-mass, lbm
1 ton (metric)	= 2204.62 pound-mass, lbm
	= 1000 kilograms
	= 1.1023 tons (short)

7.1.7 — Force

1 Newton	= 100,000 dynes
	= 0.2248 pound-force, lbf

| **1 pound-force** | = 4.448 Newtons |

7.1.8 — Pressure

1 atmosphere	= 760 mm Hg @ 0 °C
	= 101,325 Newtons/square meter
	= 33.899 feet of water @ 39.2 °F
	= 14.696 pounds-force/square inch, psi
1 bar	= 10^6 Newtons/square meter
	= 14.891 pounds-force/square inch, psi
	= 1.01325 atmospheres
1 mm Hg @ 0 °C	= 0.019337 psi
	= 2989 Newtons/square meter
	= 0.4335 psi
	= 0.0295 atmosphere
1 Pascal	= 1 Newton/square meter
1 psi	= 6894.8 Newtons/square meter
	= 51.715 mm Hg @ 0 °C
	= 0.068046 atmosphere

7.1.9 — Energy Or Work

1 Btu	= 1055.06 Joules
	= 778.173 ft-lbf
1 Electron Volt, eV	= 1.602×10^{-19} Joules
	= 1.602×10^{-12} ergs
1 ft-lbf	= 1.355816 Joules
1 kilowatt-hour	= 3412.13 Btu

7.1.10 — Power

1 Btu/hour	= 0.293 Watt
1 ft-lbf/second	= 1.356 Watts
1 horsepower	= 550 ft-lbf/second
	= 0.7457 kilowatt
	= 0.7068 Btu/sec
	= 42.4 Btu/min
	= 2544 Btu/hr
1 kilowatt	= 3412 Btu/hr
	= 1.341 horsepower
1 Watt	= 1 Joule/second

7.1.11 — Temperature

°F	$= 1.8(°C) + 32$
°K	$= 273.15 + °C$
°R	$= 1.8(°K)$
	$= 459.67 + °F$

7.1.12 — Physical Constants

Density of water at 39°F	= 8.345 lbm/gallon
	= 62.43 lbm/cubic foot
Density of water at 60°F	= 8.338 lbm/gallon
	= 62.37 lbm/cubic foot
Density of seawater at 60°F	= 64.0 lbm/cubic foot
Density of mercury at 0°F	= 13.5955 grams/milliliter

$$= 848.74 \text{ lbm/cubic foot}$$

**Standard Acceleration
of Gravity** $= 9.80665 \text{ meters/second}^2$

$$= 32.174 \text{ feet/second}^2$$

Section 7.2 — Mathematics

This section is not suitable as a review of mathematics. It is a compilation of various mathematical formulas and principles which may be used for solving problems related to the principles described in this manual. For more comprehensive information on mathematics, refer to Reference 8. For more advanced information (i.e., vector algebra or calculus) refer to Reference 10.

7.2.1 — Algebra

1.   Laws of Exponents

$$a^x \bullet a^y = a^{x+y} \quad (ab)^x = a^x \bullet b^x$$

$$(a^x)^y = a^{xy} \quad a^0 = 1 \text{ if } a \neq 0$$

$$a^{x/y} = \sqrt[y]{a^x} \qquad a^{1/y} = \sqrt[y]{a}$$

2.   Complex Numbers

$$i = \sqrt{-1} \qquad\qquad i^2 = -1$$

$$a + bi = c + di, \text{ if and only if, } a = c \text{ and } b = d$$

$$(a + bi) + (c + di) = (a + c) + (b + d)i$$

$$(a + bi)(c + di) = (ac - bd) + (ad + bc)i$$

$$\frac{a + bi}{c + di} = \frac{(a + bi) + (c - di)}{(c + di)(c - di)} = \frac{ac + bd}{c^2 + d^2} + \frac{(bc - ad)i}{c^2 + d^2}$$

Based on Exhibit 7-1, complex numbers can be represented as a radius and an angle such that:

$$a + bj = r\angle\theta$$

where:

$$r = \sqrt{a^2 + b^2} \quad \text{and} \quad \theta = \tan^{-1}(b/a)$$

(This format is known as polar coordinate system or phasor notation.)

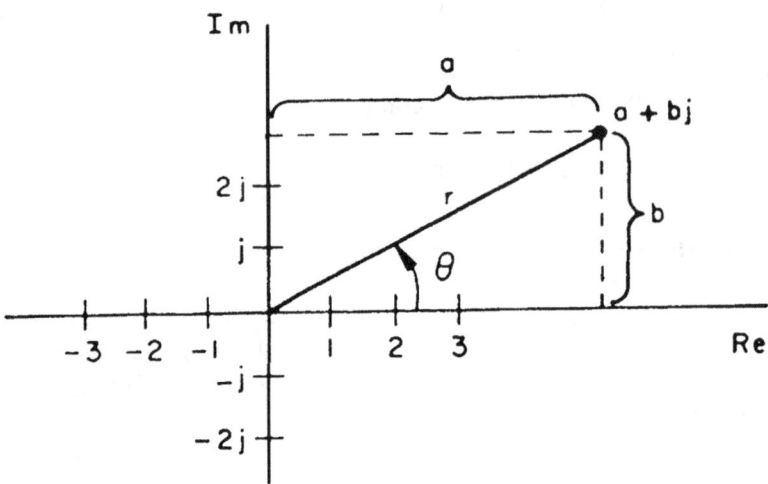

Exhibit 7-1 — Alternative Complex Number Format

3.   Law of Logarithms

If M, N, & b are positive numbers and $b \neq 1$:

$\log_b MN = \log_b M + \log_b N$

$\log_b M/N = \log_b M - \log_b N$

$\log_b M^P = P \bullet \log_b M$

$\log_b (1/M) = -\log_b M$

$\log_b b = 1$

$\log_b 1 = 0$

$\log_b (b^N) = N$

$\log_b M = \ln M = 2.3026 \log_{10} M$

Note – When the term "log" is used without the base (i.e., b) designated, then it is assumed to be base 10.

4.   Ratio and Proportion

If $a : b = c : d$ or $\dfrac{a}{b} = \dfrac{c}{d}$, then $ad = bc$ and $\dfrac{a}{c} = \dfrac{b}{d}$.

5.   Constant Factor of Proportionality (or Variation), k
*   If y varies directly as x, or y is proportional to x, then y = kx.
*   If y varies inversely as x, or y is inversely proportional to x, then y = k/x.

- If y varies jointly as x and z, then y = kxz.
- If y varies directly as x and inversely as z, then y = kx/z.

6.   Quadratic Equations

If $ax^2 + bx + c = 0$, $a \neq 0$, and a, b, & c are real numbers, then

$$x = \frac{-b \pm \sqrt{b^2 - 4ac}}{2a}$$

7.2.2 — Elementary Geometry

1.   Triangle:

Area = bh/2 where **b** denotes the base and **h** the height/altitude.

2.   Rectangle:

Area = ab where **a** and **b** denote the lengths of the sides.

3.   Trapezoid:

Area = $\frac{1}{2}$h(a + b) where **a** and **b** are the sides and **h** is the height/altitude.

4.   Circle:

Circumference = 2πr where π is $\cong$ 3.1416 and **r** is the radius.

Area = πr^2.

5.   Cube:

Volume = a^3

Diameter $= a\sqrt{3}$

Total Surface Area = $6a^2$ where **a** is the length of a side.

6.   Rectangular Parallelepiped:

Volume = abd

Diameter = $= \sqrt{a^2 + b^2 + c^2}$

Total Surface Area = 2(ab + bc + cd) where **a**, **b**, and **c** are the lengths of the sides and **d** is the length of the diagonal.

7.   Cylinder:

Volume = πhr^2

Total Surface Area = 2πr(r + h) where **r** is the radius of the end and **h** is the height.

8.   Sphere:

   Surface Area = $4\pi r^2$

   Volume = $(4/3)\pi r^2$.

7.2.3 — Trigonometry

Table 7-1 — Trigonometry Based On A Right Triangle, Exhibit 7-2

Sine θ	= sin θ	= y/r
Cosine θ	= cos θ	= x/r
Tangent θ	= tan θ	= y/x
Cotangent θ	= cot θ	= x/y
Secant θ	= sec θ	= r/x
Cosecant θ	= csc θ	= r/y

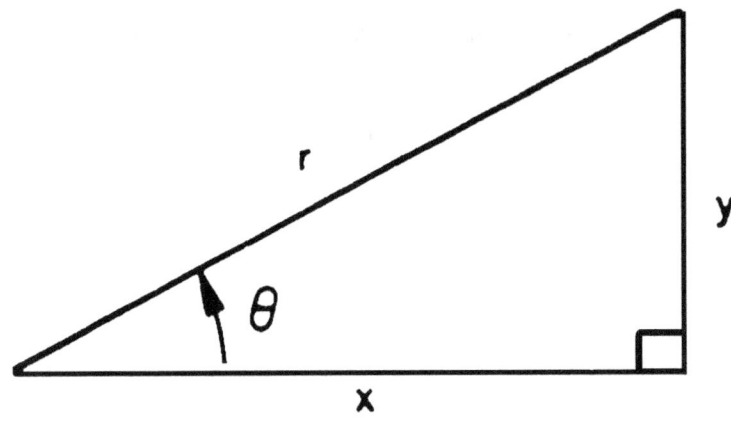

Exhibit 7-2 — Right Triangle

Section 7.3 — Radiological Controls

This section provides basic formulas and thumbrules for performing radiological controls calculations. It is not meant to be a review of radiological fundamentals or controls. For a detailed information refer to Reference 7.

7.3.1 — Radioactive Decay

The activity or activity level is the decay or disintegration rate of a radioactive sample. Activity will follow the radioactive decay law

$$A = \lambda N = A_0 e^{-\lambda t}$$

where:

A	= Activity, Ci
λ	= Decay Constant or $(\ln 2)/t_{1/2}$, sec^{-1}
N	= Number of Radioactive Atoms
A$_0$	= Activity at time t = 0, Ci
t	= time, sec
t$_{1/2}$	= Half-Life of Radioactive Atoms, sec.

7.3.2 — Quality Factors

The **quality factors (QF)** for various types of radiation are simply proportionality factors that relate the biological effects, as measured in rem, to the actual radiation energy absorbed, as expressed in rads; that is, REM = RAD × QF. A **rad** is the amount of radiation that will deposit 100 ergs of energy per gram of material. The **rem** is that amount of radiation which will cause damage to the tissue of our bodies equivalent to the damage that would be caused by absorbing 100 ergs of gamma radiation per gram of body tissue. Quality factors for various types of radiation are detailed in the following table:

Table 7-2 — Quality Factors For Various Types Of Radiation

Type of Radiation	Factor
Beta or Gamma	1
Neutron (thermal)	2
Neutron (fast) or Protons	10
Alpha	20

Most radiation survey instruments measure gamma radiation in roentgen. A **roentgen** is that amount of gamma radiation which will produce one electrostatic unit of charge (of either sign) in 1 cubic centimeter of dry air at standard conditions (0°C and 1 atm pressure). For gamma radiation, it can be assumed for practical purposes that roentgen = rem = rad.

7.3.3 — Radiological Control Thumbrules For Dose Assessment

It should be noted that many of these thumbrules are stated for ^{60}Co, and must be used with care, if at all, when other nuclides are involved.

1. <u>Waterborne Activity</u>

Swallowing 1 μCi of ^{60}Co results in a committed dose to the GI tract of approximately 80 mrem, and a committed effective dose of approximately 10 mrem, most accumulated in the first year.

2.   Airborne Activity

a.   Breathing air containing 1×10^{-9} μCi/ml of ^{60}Co for one 40-hour week results in a committed dose to the lungs of approximately 60 mrem, and a committed effective dose of approximately 7 mrem, of which about 1/3 is accumulated in the first year.

b.   A 1 μCi source of ^{60}Co retained in the lungs one day after breathing results in a committed dose to the lungs of approximately 6 mrem, and a committed effective dose of approximately 700 mrem, of which about 1/3 is accumulated in the first year.

c.   Breathing air containing 1×10^{-7} μCi/ml of ^{60}Co for one hour results in a committed dose to the lungs of approximately 150 mrem, and a committed effective dose of approximately 18 mrem, of which about 1/3 is accumulated in the first year.

d.   Breathing air containing radioactive short-lived fission-product particulates at a level of 1×10^{-6} μCi/ml continuously for one week (168 hours) results in a committed dose to the lungs of approximately 100 mrem, and a committed effective dose of approximately 20 mrem, most of which is accumulated in the same week.

e.   Continuous exposure to airborne radioactivity inert gas or short-lived fission-product particulates at a level of 1×10^{-6} μCi/ml results in an external whole-body dose of about 10 mrem/week. This rule is applicable for a finite volume such as a room in a building.

f.   The following thumbrules estimate the percent of particulate activity, released under various conditions, that initially becomes airborne. Environmental factors may modify these estimates:

0.01%	= COLD, LIQUID SPILL
0.1%	= COLD, LIQUID SPRAY
	DRY SPILL
	SLOW STEAM LEAK
	RESIN SPILL
1%	= HOT, LIQUID SPRAY
	DUST SPILL
10%	= FIRE

3.   Surface Contamination

a.   10,000 μμCi/100 cm^2 of ^{60}Co will produce a **beta**-gamma survey meter reading (AN/PDR-27 or equivalent with beta shield open) equivalent to a dose rate of approximately 0.1 mrem/hr above background close to the contaminated surface.

b.   100,000 μμCi/100 cm^2 of ^{60}Co will produce a **gamma** survey meter reading (AN/PDR-27 or equivalent with beta shield closed) equivalent to a dose rate of approximately 0.1 mrem/hr above background close to the contaminated surface.

c.   450 µµCi per direct frisk of ^{60}Co contamination on the skin results in a dose rate to the skin of 0.1 mrem/hr.

4.   Equilibrium Formula

Airborne radioactivity levels of a radioactive daughter at equilibrium

$$A_D = \frac{G}{\lambda_P} \bullet \frac{\lambda_P}{\lambda_P + \lambda_D} \bullet \frac{\lambda_D}{\lambda_D + \lambda_V + \lambda_X}$$

with,

AD   = radioactivity level of daughter nuclides, µCi

G   = leak rate multiplied by radioactivity level of leaking fluid, µCi/sec

λ_P = decay constant of parent nuclide, sec^{-1}

λ_V = removal constant of ventilation system, equivalent to the compartment ventilation flow rate divided by the compartment volume, sec^{-1}

λ_D = decay constant of daughter nuclide, sec^{-1}

λ_X = removal constant for any other applicable equipment, equivalent to flow rate through system divided by compartment volume and multiplied by an applicable efficiency factor (e.g., spot coolers, electrostatic precipitators), sec^{-1}.

5.   Point Source

A **point source** is a source where the radiation emanates from a single point in space. If the exposure point of interest is at a distance from the source of at least 3 times the maximum dimension of the source, the source can be approximated as a point source. The dose rate (DR) is inversely proportional to the square of the distance (r^2) from the source. To determine the dose rate at a distance r_2 based on the dose rate (DR$_1$) at a distance r_1, the relationship will be:

$$DR_2 = \frac{DR_1 \bullet r_1^2}{r_2^2}$$

6.   Line/Cylinder Source

In calculating the change in dose rate with distance for a finite line source of length, L, there is assumed to be two regions of uniform changes in dose rate. Region I is from on contact with the source to a distance of L/2. In this region the source is assumed to be infinitely long and follows the dose rate relationship of:

$$\frac{DR_1}{DR_2} = \frac{r_2}{r_1}.$$

In Region II, the source can be treated as a point source (see above). To move from one region to another the dose rate is determined for the L/2 point based on the approximation for the region of the known dose rate, then the L/2 dose rate is used for distance in the other region of approximation.

This approximation is most accurate for positions equally distant from the edges of the line source.

7. Plane Source

The change in dose rate with distance for a plane source is complex. A simple approximation for a plane source consists of assuming the area of concern is circular. This is done by calculating an equivalent radius (R) of a circle based on the area of the source.

The plane source can be treated like a point source at distances r greater than 0.7 R. At smaller values of r (0.7 R $\geq$ r $\geq$ 0.1 R) the dose rate falls off so slowly that it can be assumed that the dose rate at 0.7 R is the dose rate for the entire region. The exposure rate at distances < 0.1 R are equal to 3 times the dose rate in the range of 0.1 R to 0.7 R.

8. Miscellaneous
   a. The estimate exposure rate for ^{16}N gammas is about 3 times greater than the estimated exposure rate attributed to fission produced (i.e., about 1MeV) gammas, because of the gamma energy difference.
   b. 1 µCi of ^{60}Co in the lungs or GI tract will produce a gamma dose rate of approximately 0.04 mrem/hr above background with an AN/PDR-27 or -66 on contact with the chest or abdomen, or about 100 cpm above background with a frisker.
   c. A 1 Ci gamma point source produces a dose rate of approximately 1 rem/hr at 1 meter. The radiation level depends on gamma energy such that the dose rate in rem/hr at 1 meter is equal to 0.55 CE, where C is curies of activity and E is the gamma energy of disintegration (in MeV).

7.3.4 — Shielding

Shielding calculations are commonly performed during plant operations based on a tenth-thickness approximation. The **tenth-thickness ($\times_{0.1}$)** of a given material is the thickness of a given material required to reduce the intensity of a beam of gamma radiation to one-tenth of its original value. The tenth-thickness is very dependent on the energy of the radiation of concern. In spite of this, the tenth-thickness approximation will give reasonable estimates for shielding required in the kind of situations most likely to be encountered. The following values can be used as convenient rules of thumb:
   1. The tenth-thickness of **lead** is about **2 inches** for gamma radiation.
   2. The tenth-thickness of **steel** is about **4 inches** for gamma radiation.
   3. The tenth-thickness of **water**, **polyethylene**, or **oil** is about **2 feet** for gamma radiation, and **10 inches** for neutron radiation.

The shielded dose rate (DR$_s$) can be determined from the unshielded dose rate (DR$_u$) by

$$DR_s = DR_u (0.1)^N,$$

where N is the number of tenth-thickness in the shield $(x/x_{0.1})$

(Intentionally Blank)